METHODS IN MOLECULAR BIOLOGY

Series Editor
John M. Walker
School of Life and Medical Sciences
University of Hertfordshire
Hatfield, Hertfordshire, AL10 9AB, UK

For further volumes:
http://www.springer.com/series/7651

Quantitative Proteomics by Mass Spectrometry

Second Edition

Edited by

Salvatore Sechi

NIDDK, National Institutes of Health, Bethesda, MD, USA

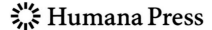 Humana Press

Editor
Salvatore Sechi
NIDDK, National Institutes of Health
Bethesda, MD, USA

ISSN 1064-3745   ISSN 1940-6029   (electronic)
Methods in Molecular Biology
ISBN 978-1-4939-3522-2   ISBN 978-1-4939-3524-6   (eBook)
DOI 10.1007/978-1-4939-3524-6

Library of Congress Control Number: 2016931330

Springer New York Heidelberg Dordrecht London

Cover image: This image is composed of 4 elements: A 3D image of a cell where the nucleus is visible, a double helix DNA that comes out of the nucleus and joins/morphs into a 3D ribbon structure of a protein. On top of these 3 elements there is a Mass Spectra.

Printed on acid-free paper

Humana Press is a brand of Springer
Springer Science+Business Media LLC New York is part of Springer Science+Business Media (www.springer.com)

Preface

Tools for Next Generation Quantitative Proteomics by Mass Spectrometry

The proteome has been defined as the entire set of proteins expressed by a genome. The genome is relatively simple, and using today's technology the human genome of an individual can be routinely sequenced. The dynamic nature of the proteome, and its heterogeneous structure with many possible post-translational modifications, make the task of fully sequencing the complete human proteome out of reach. On one hand, "gene-centric" proteomic approaches that focus on identifying the genes that code for the expressed proteins (which is distinct from characterizing the complete molecular sequences of the proteins) are becoming routine, and many laboratories have undertaken large-scale protein identification projects. On the other hand, the characterization and quantification of post-translational modifications and the accurate and reproducible quantification of proteins from tissues, biofluids, and cell lines are more challenging, and fewer specialized laboratories have the tools and expertise to perform these experiments routinely.

This volume describes prominent methodologies developed by laboratories that have been leading the field of quantitative proteomics by Mass Spectrometry (MS). The procedures for performing the experiments are described in an easy-to-understand manner, with many technical details that usually are not reported in typical research articles.

This volume is not intended to be comprehensive with respect to all currently available methodologies for performing quantitative proteomics by MS, but it should provide a broad perspective of the methodologies used for quantifying proteins and post-translational modifications in different types of biomedical specimens.

The tools for performing quantitation of proteins by MS have gone through a major revolution in the last decade, and next-generation quantitative proteomics tools are already here. I hope that this volume will facilitate the dissemination of these tools.

Bethesda, MD, USA   *Salvatore Sechi*

Contents

Preface. . . . . *v*

Contributors. . . . . *ix*

1 Increased Depth and Breadth of Plasma Protein Quantitation
   via Two-Dimensional Liquid Chromatography/Multiple Reaction
   Monitoring-Mass Spectrometry with Labeled Peptide Standards 1
   Andrew J. Percy, Juncong Yang, Andrew G. Chambers,
   and Christoph H. Borchers

2 Quantitative Analysis of the Sirt5-Regulated Lysine Succinylation
   Proteome in Mammalian Cells 23
   Yue Chen

3 Determining the Composition and Stability of Protein Complexes
   Using an Integrated Label-Free and Stable Isotope Labeling Strategy 39
   Todd M. Greco, Amanda J. Guise, and Ileana M. Cristea

4 Label-Free Quantitation for Clinical Proteomics 65
   Robert Moulder, Young Ah Goo, and David R. Goodlett

5 Proteogenomic Methods to Improve Genome Annotation 77
   Keshava K. Datta, Anil K. Madugundu, and Harsha Gowda

6 Mass Spectrometry-Based Quantitative O-GlcNAcomic Analysis 91
   Junfeng Ma and Gerald W. Hart

7 Isolating and Quantifying Plasma HDL Proteins by Sequential Density
   Gradient Ultracentrifugation and Targeted Proteomics 105
   Clark M. Henderson, Tomas Vaisar, and Andrew N. Hoofnagle

8 A Method for Label-Free, Differential Top-Down Proteomics. 121
   Ioanna Ntai, Timothy K. Toby, Richard D. LeDuc,
   and Neil L. Kelleher

9 Multiplexed Immunoaffinity Enrichment of Peptides with Anti-peptide
   Antibodies and Quantification by Stable Isotope Dilution Multiple
   Reaction Monitoring Mass Spectrometry. 135
   Eric Kuhn and Steven A. Carr

10 High-Throughput Quantitative Proteomics Enabled
   by Mass Defect-Based 12-Plex DiLeu Isobaric Tags. 169
   Dustin C. Frost and Lingjun Li

11 Isotopic *N,N*-Dimethyl Leucine (iDiLeu) for Absolute Quantification
   of Peptides Using a Standard Curve Approach 195
   Tyler Greer and Lingjun Li

12 Selecting Optimal Peptides for Targeted Proteomic Experiments
    in Human Plasma Using In Vitro Synthesized Proteins
    as Analytical Standards   207
    James G. Bollinger, Andrew B. Stergachis, Richard S. Johnson,
    Jarrett D. Egertson, and Michael J. MacCoss

13 Using the CPTAC Assay Portal to Identify and Implement Highly
    Characterized Targeted Proteomics Assays.   223
    Jeffrey R. Whiteaker, Goran N. Halusa, Andrew N. Hoofnagle,
    Vagisha Sharma, Brendan MacLean, Ping Yan, John A. Wrobel,
    Jacob Kennedy, D.R. Mani, Lisa J. Zimmerman, Matthew R. Meyer,
    Mehdi Mesri, Emily Boja, Steven A. Carr, Daniel W. Chan, Xian Chen,
    Jing Chen, Sherri R. Davies, Matthew J.C. Ellis, David Fenyö,
    Tara Hiltke, Karen A. Ketchum, Chris Kinsinger, Eric Kuhn,
    Daniel C. Liebler, Tao Liu, Michael Loss, Michael J. MacCoss,
    Wei-Jun Qian, Robert Rivers, Karin D. Rodland, Kelly V. Ruggles,
    Mitchell G. Scott, Richard D. Smith, Stefani Thomas, R. Reid Townsend,
    Gordon Whiteley, Chaochao Wu, Hui Zhang, Zhen Zhang,
    Henry Rodriguez, and Amanda G. Paulovich

14 Large-Scale and Deep Quantitative Proteome Profiling
    Using Isobaric Labeling Coupled with Two-Dimensional LC–MS/MS   237
    Marina A. Gritsenko, Zhe Xu, Tao Liu, and Richard D. Smith

15 Multiple and Selective Reaction Monitoring Using Triple
    Quadrupole Mass Spectrometer: Preclinical Large Cohort Analysis   249
    Qin Fu, Zhaohui Chen, Shenyan Zhang,
    Sarah J. Parker, Zongming Fu, Adrienne Tin, Xiaoqian Liu,
    and Jennifer E. Van Eyk

16 Methods for SWATH™: Data Independent Acquisition
    on TripleTOF Mass Spectrometers   265
    Ronald J. Holewinski, Sarah J. Parker, Andrea D. Matlock,
    Vidya Venkatraman, and Jennifer E. Van Eyk

17 Measurement of Phosphorylated Peptides with Absolute Quantification   281
    Raven J. Reddy, Timothy G. Curran, Yi Zhang,
    and Forest M. White

18 Proteomic Analysis of Protein Turnover by Metabolic
    Whole Rodent Pulse-Chase Isotopic Labeling and Shotgun
    Mass Spectrometry Analysis.   293
    Jeffrey N. Savas, Sung Kyu Park, and John R. Yates III

Index *305*

Contributors

JAMES G. BOLLINGER • *Department of Genome Sciences, University of Washington, Seattle, WA, USA*

CHRISTOPH H. BORCHERS • *University of Victoria—Genome British Columbia Proteomics Centre, Victoria, BC, Canada; Department of Biochemistry and Microbiology, University of Victoria, Victoria, BC, Canada*

EMILY BOJA • *Office of Cancer Clinical Proteomics Research, National Cancer Institute, Bethesda, MD, USA*

STEVEN A. CARR • *Broad Institute of MIT and Harvard, Cambridge, MA, USA*

ILEANA M. CRISTEA • *Department of Molecular Biology, Princeton University, Princeton, NJ, USA*

ANDREW G. CHAMBERS • *University of Victoria—Genome British Columbia Proteomics Centre, Victoria, BC, Canada*

DANIEL W. CHAN • *Clinical Chemistry Division, Department of Pathology, Johns Hopkins University School of Medicine, Baltimore, MD, USA*

JING CHEN • *Clinical Chemistry Division, Department of Pathology, Johns Hopkins University School of Medicine, Baltimore, MD, USA*

XIAN CHEN • *Department of Biochemistry and Biophysics, University of North Carolina at Chapel Hill School of Medicine, Chapel Hill, NC, USA*

YUE CHEN • *Department of Biochemistry, Molecular Biology and Biophysics, University of Minnesota, Minneapolis, MN, USA*

ZHAOHUI CHEN • *Advanced Clinical Biosystems Research Institute, The Heart Institute, Cedars-Sinai Medical Center, Los Angeles, CA, USA*

TIMOTHY G. CURRAN • *Department of Biological Engineering, Massachusetts Institute of Technology, Cambridge, MA, USA; Koch Institute for Integrative Cancer Research, Massachusetts Institute of Technology, Cambridge, MA, USA*

KESHAVA K. DATTA • *Institute of Bioinformatics, Bangalore, India; School of Biotechnology, KIIT University, Bhubaneswar, Odisha, India*

SHERRI R. DAVIES • *Department of Medicine, Washington University School of Medicine, St. Louis, MO, USA*

JARRETT D. EGERTSON • *Department of Genome Sciences, University of Washington, Seattle, WA, USA*

MATTHEW J.C. ELLIS • *Department of Medicine, Washington University School of Medicine, St. Louis, MO, USA*

DAVID FENYÖ • *Department of Biochemistry and Molecular Pharmacology, New York University School of Medicine, New York, NY, USA*

DUSTIN C. FROST • *School of Pharmacy, University of Wisconsin-Madison, Madison, WI, USA*

QIN FU • *Advanced Clinical Biosystems Research Institute, The Heart Institute, Cedars-Sinai Medical Center, Los Angeles, CA, USA*

ZONGMING FU • *Department of Medicine, Johns Hopkins University, Baltimore, MD, USA*

YOUNG AH GOO • *Mass Spectrometry Center, University of Maryland—Baltimore, Baltimore, MD, USA*

DAVID R. GOODLETT • *Mass Spectrometry Center, University of Maryland—Baltimore, Baltimore, MD, USA*

HARSHA GOWDA • *Institute of Bioinformatics, Bangalore, India; School of Biotechnology, KIIT University, Bhubaneswar, Odisha, India*

TODD M. GRECO • *Department of Molecular Biology, Princeton University, Princeton, NJ, USA*

TYLER GREER • *Department of Chemistry, University of Wisconsin-Madison, Madison, WI, USA*

MARINA A. GRITSENKO • *Biological Sciences Division and Environmental Molecular Sciences Laboratory, Pacific Northwest National Laboratory, Richland, WA, USA*

AMANDA J. GUISE • *Department of Molecular Biology, Princeton University, Princeton, NJ, USA*

GORAN N. HALUSA • *Frederick National Laboratory for Cancer Research, Leidos Biomedical Research Inc., Frederick, MD, USA*

GERALD W. HART • *Department of Biological Chemistry, The Johns Hopkins University School of Medicine, Baltimore, MD, USA*

CLARK M. HENDERSON • *Department of Laboratory Medicine, University of Washington School of Medicine, Seattle, WA, USA*

TARA HILTKE • *Office of Cancer Clinical Proteomics Research, National Cancer Institute, Bethesda, MD, USA*

RONALD J. HOLEWINSKI • *Advanced Clinical Biosystems Research Institute, The Heart Institute, Cedars-Sinai Medical Center, Los Angeles, CA, USA*

ANDREW N. HOOFNAGLE • *Department of Laboratory Medicine, University of Washington, Seattle, WA, USA*

RICHARD S. JOHNSON • *Department of Genome Sciences, University of Washington, Seattle, WA, USA*

NEIL L. KELLEHER • *Department of Chemistry, Northwestern University, Evanston, IL, USA; Department of Molecular Biosciences, Northwestern University, Evanston, IL, USA; Proteomics Center of Excellence, Evanston, IL, USA*

JACOB KENNEDY • *Clinical Research Division, Fred Hutchinson Cancer Research Center, Seattle, WA, USA*

KAREN A. KETCHUM • *Data Coordinating Center, ESAC, Inc., Rockville, MD, USA*

CHRIS KINSINGER • *Office of Cancer Clinical Proteomics Research, National Cancer Institute, Bethesda, MD, USA*

ERIC KUHN • *Broad Institute of MIT and Harvard, Cambridge, MA, USA*

RICHARD D. LEDUC • *Department of Chemistry, Northwestern University, Evanston, IL, USA; Department of Molecular Biosciences, Northwestern University, Evanston, IL, USA; Proteomics Center of Excellence, Evanston, IL, USA*

LINGJUN LI • *School of Pharmacy, University of Wisconsin-Madison, Madison, WI, USA; Department of Chemistry, University of Wisconsin-Madison, Madison, WI, USA*

DANIEL C. LIEBLER • *Department of Biochemistry, Jim Ayers Institute for Precancer Detection and Diagnosis, Vanderbilt University School of Medicine, Nashville, TN, USA*

TAO LIU • *Biological Sciences Division and Environmental Molecular Sciences Laboratory, Pacific Northwest National Laboratory, Richland, WA, USA*

XIAOQIAN LIU • *Advanced Clinical Biosystems Research Institute, The Heart Institute, Cedars-Sinai Medical Center, Los Angeles, CA, USA*

MICHAEL LOSS • *Frederick National Laboratory for Cancer Research, Leidos Biomedical Research Inc., Frederick, MD, USA*

JUNFENG MA • *Department of Biological Chemistry, The Johns Hopkins University School of Medicine, Baltimore, MD, USA*

BRENDAN MACLEAN • *Department of Genome Sciences, University of Washington, Seattle, WA, USA*

MICHAEL J. MACCOSS • *Department of Genome Sciences, University of Washington, Seattle, WA, USA*

ANIL K. MADUGUNDU • *Institute of Bioinformatics, Bangalore, India; Centre for Bioinformatics, School of Life Sciences, Pondicherry University, Puducherry, India*

D.R. MANI • *Broad Institute, Cambridge, MA, USA*

ANDREA D. MATLOCK • *Advanced Clinical Biosystems Research Institute, The Heart Institute, Cedars-Sinai Medical Center, Los Angeles, CA, USA*

MEHDI MESRI • *Office of Cancer Clinical Proteomics Research, National Cancer Institute, Bethesda, MD, USA*

MATTHEW R. MEYER • *Department of Medicine, Washington University School of Medicine, St. Louis, MO, USA*

ROBERT MOULDER • *Centre for Biotechnology, University of Turku, Turku, Finland*

IOANNA NTAI • *Department of Chemistry, Northwestern University, Evanston, IL, USA; Department of Molecular Biosciences, Northwestern University, Evanston, IL, USA; Proteomics Center of Excellence, Evanston, IL, USA*

SUNG KYU PARK • *Department of Chemical Physiology, The Scripps Research Institute, La Jolla, CA, USA*

SARAH J. PARKER • *Advanced Clinical Biosystems Research Institute, The Heart Institute, Cedars-Sinai Medical Center, Los Angeles, CA, USA*

AMANDA G. PAULOVICH • *Clinical Research Division, Fred Hutchinson Cancer Research Center, Seattle, WA, USA*

ANDREW J. PERCY • *University of Victoria—Genome British Columbia Proteomics Centre, Victoria, BC, Canada*

WEI-JUN QIAN • *Biological Sciences Division, Pacific Northwest National Laboratory, Richland, WA, USA*

RAVEN J. REDDY • *Department of Biological Engineering, Massachusetts Institute of Technology, Cambridge, MA, USA; Koch Institute for Integrative Cancer Research, Massachusetts Institute of Technology, Cambridge, MA, USA*

ROBERT RIVERS • *Office of Cancer Clinical Proteomics Research, National Cancer Institute, Bethesda, MD, USA*

KARIN D. RODLAND • *Biological Sciences Division, Pacific Northwest National Laboratory, Richland, WA, USA*

HENRY RODRIGUEZ • *Office of Cancer Clinical Proteomics Research, National Cancer Institute, Bethesda, MD, USA*

KELLY V. RUGGLES • *Department of Biochemistry and Molecular Pharmacology, New York University School of Medicine, New York, NY, USA*

JEFFREY N. SAVAS • *Department of Chemical Physiology, The Scripps Research Institute, La Jolla, CA, USA*

MITCHELL G. SCOTT • *Division of Laboratory and Genomic Medicine, Department of Pathology and Immunology, Washington University School of Medicine, St. Louis, MO, USA*

SALVATORE SECHI • *NIDDK, National Institutes of Health, Bethesda, MD, USA*

VAGISHA SHARMA • *Department of Genome Sciences, University of Washington, Seattle, WA, USA*

RICHARD D. SMITH • *Biological Sciences Division and Environmental Molecular Sciences Laboratory, Pacific Northwest National Laboratory, Richland, WA, USA*

ANDREW B. STERGACHIS • *Department of Genome Sciences, University of Washington, Seattle, WA, USA*

STEFANI THOMAS • *Clinical Chemistry Division, Department of Pathology, Johns Hopkins University School of Medicine, Baltimore, MD, USA*

ADRIENNE TIN • *Department of Epidemiology, Johns Hopkins University, Baltimore, MD, USA*

TIMOTHY K. TOBY • *Department of Chemistry, Northwestern University, Evanston, IL, USA; Department of Molecular Biosciences, Northwestern University, Evanston, IL, USA; Proteomics Center of Excellence, Evanston, IL, USA*

R. REID TOWNSEND • *Department of Medicine, Washington University School of Medicine, St. Louis, MO, USA*

TOMAS VAISAR • *Department of Medicine, University of Washington School of Medicine, Seattle, WA, USA*

JENNIFER E. VAN EYK • *Advanced Clinical Biosystems Research Institute, The Heart Institute, Cedars-Sinai Medical Center, Los Angeles, CA, USA*

VIDYA VENKATRAMAN • *Advanced Clinical Biosystems Research Institute, The Heart Institute, Cedars-Sinai Medical Center, Los Angeles, CA, USA*

FOREST M. WHITE • *Department of Biological Engineering, Massachusetts Institute of Technology, Cambridge, MA, USA; Koch Institute for Integrative Cancer Research, Massachusetts Institute of Technology, Cambridge, MA, USA*

JEFFREY R. WHITEAKER • *Clinical Research Division, Fred Hutchinson Cancer Research Center, Seattle, WA, USA*

GORDON WHITELEY • *Frederick National Laboratory for Cancer Research, Leidos Biomedical Research Inc., Frederick, MD, USA*

JOHN A. WROBEL • *Department of Biochemistry and Biophysics, University of North Carolina at Chapel Hill School of Medicine, Chapel Hill, NC, USA*

CHAOCHAO WU • *Biological Sciences Division, Pacific Northwest National Laboratory, Richland, WA, USA*

ZHE XU • *Biological Sciences Division and Environmental Molecular Sciences Laboratory, Pacific Northwest National Laboratory, Richland, WA, USA*

PING YAN • *Clinical Research Division, Fred Hutchinson Cancer Research Center, Seattle, WA, USA*

JUNCONG YANG • *University of Victoria—Genome British Columbia Proteomics Centre, Victoria, BC, Canada*

JOHN R. YATES III • *Department of Chemical Physiology, The Scripps Research Institute, La Jolla, CA, USA*

HUI ZHANG • *Clinical Chemistry Division, Department of Pathology, Johns Hopkins University School of Medicine, Baltimore, MD, USA*

SHENYAN ZHANG • *Advanced Clinical Biosystems Research Institute, The Heart Institute, Cedars-Sinai Medical Center, Los Angeles, CA, USA*

YI ZHANG • *Thermo-Fisher Scientific, San Jose, CA, USA*

ZHEN ZHANG • *Clinical Chemistry Division, Department of Pathology, Johns Hopkins University School of Medicine, Baltimore, MD, USA*

LISA J. ZIMMERMAN • *Department of Biochemistry, Jim Ayers Institute for Precancer Detection and Diagnosis, Vanderbilt University School of Medicine, Nashville, TN, USA*

Chapter 1

Increased Depth and Breadth of Plasma Protein Quantitation via Two-Dimensional Liquid Chromatography/ Multiple Reaction Monitoring-Mass Spectrometry with Labeled Peptide Standards

Andrew J. Percy, Juncong Yang, Andrew G. Chambers, and Christoph H. Borchers

Abstract

Absolute quantitative strategies are emerging as a powerful and preferable means of deriving concentrations in biological samples for systems biology applications. Method development is driven by the need to establish new—and validate current—protein biomarkers of high-to-low abundance for clinical utility. In this chapter, we describe a methodology involving two-dimensional (2D) reversed-phase liquid chromatography (RPLC), operated under alkaline and acidic pH conditions, combined with multiple reaction monitoring (MRM)-mass spectrometry (MS) (also called selected reaction monitoring (SRM)-MS) and a complex mixture of stable isotope-labeled standard (SIS) peptides, to quantify a broad and diverse panel of 253 proteins in human blood plasma. The quantitation range spans 8 orders of magnitude—from 15 mg/mL (for vitamin D-binding protein) to 450 pg/mL (for protein S100-B)—and includes 31 low-abundance proteins (defined as being <10 ng/mL) of potential disease relevance. The method is designed to assess candidates at the discovery and/or verification phases of the biomarker pipeline and can be adapted to examine smaller or alternate panels of proteins for higher sample throughput. Also detailed here is the application of our recently developed software tool—Qualis-SIS—for protein quantitation (via regression analysis of standard curves) and quality assessment of the resulting data. Overall, this chapter provides the blueprint for the replication of this quantitative proteomic method by proteomic scientists of all skill levels.

Key words Multidimensional liquid chromatography, Multiple reaction monitoring, Plasma, Protein, Proteomics, Quantitation, Sensitive, Stable isotope-labeled standard

1 Introduction

Quantitative MS is being increasingly performed in various "omics disciplines" (proteomics in this context) by academic and industrial laboratories to address biological- and medical-related queries. One of the indispensable tools in this process is "absolute" quantitation, whereby the exact concentration (typically reported in

Salvatore Sechi (ed.), *Quantitative Proteomics by Mass Spectrometry*, Methods in Molecular Biology, vol. 1410,
DOI 10.1007/978-1-4939-3524-6_1, © Springer Science+Business Media New York 2016

ng/mL) within a given patient sample (collected before or after treatment) is precisely, but not necessarily accurately, determined and then compared to the concentration of the same analyte in a sample from a healthy control (sometimes referred to as a reference sample) to obtain information on expression patterns and significance. This data can subsequently be extended to include biological pathways and protein networks, via Pathway Palette [1], for instance, for exploring disease implications. Overall, the long-standing goal of such research efforts lies in personalized medicine, wherein indicators of biological processes (i.e., biomarkers) are sought for targeted therapeutics, improved patient outcomes, and reduced healthcare expenses.

Emerging approaches for "absolute" (more appropriately referred to as precise and relative) protein quantitation center on MRM (typically performed on triple quadrupole (QqQ) mass spectrometers [2, 3]) or parallel reaction monitoring (PRM, commonly performed on hybrid quadrupole-Orbitrap mass spectrometers [4–6]) and use biologically expressed or chemically synthesized, $^{13}C/^{15}N$-labeled protein [7] or peptide [8, 9] standards for normalization [10, 11]. Comparatively, each detection and isotopic labeling technique has its merits and detractions [7], with the choice ultimately being based on the research aim, reagent cost, and supply requirements of the experiments. The "quantification concatemer" (QconCAT) technique, for example, is a recombinant genetic approach which involves incorporating nucleotides that encode for peptides from single or multiple proteins into a single synthetic gene [9, 12]. Since the QconCAT "protein" is added prior to the proteolysis step it can, in theory, measure the digestion efficiency. However, it requires that digestion be complete in order to generate an equimolar peptide mixture of intact QconCAT peptides. The digestion effectiveness is impacted by the conditions, variability (resulting from missed cleavage, or miscleavage, for instance), sequence order (which may affect trypsin specificity), and folding behavior. Alternatively, the individual labeled standards can be added post-digestion (either equimolar or balanced to the natural (NAT) levels), as is the case in our research. Although the proteolytic efficiency cannot be assessed in this approach, as long as the peptide surrogates are reproducibly generated, precise quantitative results can nonetheless be obtained.

Using bottom-up LC/MRM-MS with SIS peptides, we recently developed a method for the quantification of >140 candidate protein biomarkers across a 6 order-of-magnitude concentration range (from 31 mg/mL for serum albumin to 18 ng/mL for peroxiredoxin-2) in undepleted and non-enriched blood plasma [13]. While this concentration range is amenable to the quantification of diverse classes of high-to-moderate abundance proteins in this complex fluid, sample pre-fractionation is likely required to quantify plasma proteins of lower abundance. This has been accomplished by others with up-front, immuno-based enrichment

[14–16] or depletion [17–20]. Despite the demonstrated merits in quantifying select proteins at low ng/mL levels, enrichment is not preferred for the discovery and verification stages of the biomarker pipeline due to the prohibitive cost and long development time required to prepare well-functioning antibody-based assays for 100s to 1000s of biomarker candidates [21]. Depletion is generally disfavored due to poor reproducibility and the potential for concentration underestimation [22], among other issues related to cost and throughput. An alternative option involves employing two dimensions of RPLC, operated under alkaline and acidic pH conditions, prior to MRM-MS. This antibody-free option simplifies assay development and capitalizes on the high efficiency and orthogonality of high-pH/low-pH RPLC separation [23, 24]. Recent two-dimensional (2D) RPLC/MRM-MS quantitative proteomic methods have been developed by us [25] and others [26] to quantify proteins at or below the low ng/mL range. Key differences in our approach center on the high-pH LC eluent, the fraction pooling strategy, the LC flow rates, and the MRM data analysis strategy.

Detailed here is our targeted methodology used to quantify a sensitive and multiplexed panel of plasma proteins through the use of 2D RPLC/MRM-MS with SIS tryptic peptides as the internal standards [25]. The chromatographic dimensions utilize alkaline and acidic eluents maintained at pH 10 with constant ammonium hydroxide, instead of the conventional ammonium formate, and pH 3 with a constant concentration of formic acid, respectively. The LC systems are operated at standard-flow rates (instead of the conventional nano-flow rates) due to the analytical merits (in terms of reproducibility, multiplexing, and sensitivity) afforded by the ability to load a larger amount of plasma digest [27]. Although the current protocol requires the LC fractionation to be performed off-line, higher throughput can be obtained by reducing the low-pH LC run times and/or by reducing the number of fractions to be processed by LC/MRM-MS. The latter is applicable if a smaller panel is of interest or if the target peptides elute in a smaller number of pooled fractions. We have previously demonstrated this method to quantify 253 plasma proteins (from 625 interference-free peptide surrogates; *see* Supplemental Fig. 2 from [25] for the extracted ion chromatograms of the quantifier peptides) across 13 pooled fractions. Of note, Protein S100-B (P04271) at 450 pg/mL was the lowest amount of protein quantified, resulting in a >8 order-of-magnitude dynamic range. The endogenous (i.e., NAT) concentrations were determined largely by their relative responses (RRs), which we refer to as single point measurements (SPMs), but also recently via standard curves on a smaller panel of peptide targets. We have recently developed Qualis-SIS to automate protein quantitation and quality assessment for more rapid analysis and results interpretation [28]. The implementation of this tool in fractionated sample analyses is also described in this chapter.

2   Materials

All chemicals used in the solution/sample preparation should be of the highest analytical or LC-MS grade available. For instance, methanol, acetonitrile, and water are all LC/MS grade, while formic acid is reagent grade. These can be obtained from commercial vendors (typically Sigma-Aldrich, unless otherwise stated), with storage and handling adhering to manufacturer's instructions.

2.1 Solution Preparation

Prepare the solutions indicated in **items 1–3** then proceed to Subheading 3.1. The solution preparations outlined in **items 4–6** are to be performed immediately prior to their additions noted in Subheading 3.1, while **items 7–10** are to be conducted 30 min prior to completion of the digestion (as indicated in **step 11** of Subheading 3.1). The mobile phase preparations listed in **items 11–13** can be prepared at any time.

1. 25 mM ammonium bicarbonate: in a 50-mL Falcon tube, dissolve 43.48 mg of ammonium bicarbonate in 22 mL of LC/MS-grade water. Briefly vortex the tube to ensure complete mixing.

2. 10 % (w/v) sodium deoxycholate: in a 15-mL Falcon tube, dissolve 200 mg of sodium deoxycholate in 2 mL of 25 mM ammonium bicarbonate. Vortex until fully solubilized (*see* **Notes 1** and **2**).

3. 50 mM Tris(2-carboxyethyl)phosphine (TCEP; Thermo Scientific; Rockford, IL, USA): in a 1.5-mL microcentrifuge tube, add 100 µL of 0.5 M TCEP solution to 900 µL of 25 mM ammonium bicarbonate, then vortex briefly.

4. 25 mM Iodoacetamide: After weighing out 36.99 mg (*see* **Note 3**) of iodoacetamide in a 15-mL Falcon tube, immediately wrap in aluminum foil to prevent deactivation by light, then solubilize with 2 mL of 25 mM ammonium bicarbonate. Vortex briefly.

5. 100 mM Dithiothreitol (DTT): in a 15-mL Falcon tube, dissolve 30.85 mg of DTT in 2 mL of 25 mM ammonium bicarbonate, then vortex briefly.

6. 1 mg/mL TPCK-treated trypsin (Worthington; Lakewood, NJ, USA; *see* **Note 4**): dissolve 1 mg of trypsin in 1 mL of 25 mM ammonium bicarbonate. Vortex the microcentrifuge tube briefly before adding the trypsin to the plasma-containing protein solution.

7. 0.1 % Formic acid (FA): in a 15-mL Falcon tube, prepare 5 mL of 0.1 % FA by combining 5 µL of reagent-grade FA with 4.995 mL of LC/MS-grade water. Vortex briefly.

8. SIS peptides (UVic-Genome BC Proteomics Centre; *see* **Note 5**): After rehydrating a lyophilized SIS mix aliquot stock solution in 110 µL of 0.1 % FA, dilute the peptide mixture in a microcentrifuge tube by combining 48 µL of the stock solution with 72 µL of 0.1 % FA (*see* **Note 6**). Vortex briefly, then store on ice.

9. 3.6 % FA: In a 15-mL Falcon tube, combine 72 µL of concentrated FA with 1.928 mL of water. Vortex briefly then store on ice.

10. 55 % Acetonitrile (ACN)/0.1% FA: in a 15-mL Falcon tube, prepare 55 % ACN/0.1 % FA by combining 5 µL FA with 2.247 mL of water and 2.748 mL of ACN. Vortex briefly.

11. 100 mM ammonium hydroxide: dilute 20 µL of 5.0 M ammonium hydroxide volumetric solution (part no. 318612; Sigma-Aldrich) with 980 µL of LC-MS-grade water. This is mobile phase C for the high-pH separation.

12. 0.1 % FA in water: dilute 1 mL FA with 999 mL of LC-MS-grade water. This is mobile phase A in the low-pH system.

13. 0.1 % FA in ACN: dilute 1 mL FA with 999 mL LC-MS-grade ACN in an amber bottle. This is mobile phase B in the low-pH system.

2.2 Materials and Equipment for Sample Preparation and Processing

1. 1.5 mL MAXYMum Recovery microcentrifuge tubes (product #MCT-150-L-C; Axygen; Union City, CA, USA).

2. 15 and 50 mL polypropylene Falcon tubes (Fisher Scientific; Burlington, ON, Canada).

3. Ultra-Low Temperature Freezer (Forma 995 from Thermo Scientific or the equivalent, capable of storing samples at −80 °C).

4. Vortex mixer (catalogue no. 02-215-365 from Thermo Scientific or the equivalent).

5. Minicentrifuge (catalogue no. 05-090-100 from Thermo Scientific or the equivalent, capable of achieving a relative centrifugal force of $2200 \times g$).

6. Bench-top centrifuge (model 5415D from Eppendorf; Hamburg, Germany; or the equivalent, capable of achieving a relative centrifugal force of $12,000 \times g$).

7. Incubators (we use model 120 from Barnstead Lab-Line, Melrose Park, IL, USA, for 60 °C incubations; and model BD23 from Binder, Tuttlingen, Germany, for 37 °C incubations; *see* **Note 7**).

8. 10 mg Oasis HLB (hydrophilic-lipophilic balance) cartridges (part no. 186000383; Waters; Milford, MA, USA).

9. PrepSep 12-port vacuum manifold (catalogue no. 60104-232; Thermo Scientific).

10. Lyophilizer (catalogue no. SuperModulyo220 for the freeze dryer and catalogue no. F056 28 000 for the 16-port drum manifold; both from Thermo Scientific; *see* **Note 8**).

11. Lyophilizer accessories (wide-mouth borosilicate glass flasks, catalogue no. F056 57 000; Thermo Scientific).

12. Autosampler vials (part no. 5185-5820; Agilent Technologies; Santa Clara, CA, USA) and screw caps (part no. 5182-0715; Agilent Technologies) for sample loading in the high-pH LC system.

13. 96-Well deep-well collection plates (product #P-DW-11-C-S from Axygen or the equivalent with 1.1 mL capacity) for high-pH LC fractionation.

14. Adhesive foil (model 60941-074; VWR; Mississauga, ON, Canada) for lyophilizing the deep-well collection plates.

15. 96-Well microtiter plates (part no. 0030128.648; Eppendorf) and sealing mat (AM-96-PCR-RD; Axygen) for low-pH LC/MRM-MS sample processing.

16. 1260 Infinity Binary LC system with 80 Hz UV detector (Agilent Technologies).

17. XBridge BEH300 RPLC column (250×4.6 mm i.d., 5 μm particles; part no. 186003116; Waters).

18. 1290 Infinity UHPLC system (Agilent Technologies).

19. Zorbax Eclipse Plus C_{18} RRHD RP-UHPLC column (150×2.1 mm i.d., 1.8 μm particles; part no. 959759-902; Agilent Technologies).

20. Standard-flow ESI source (Agilent Technologies).

21. 6490 QqQ mass spectrometer (Agilent Technologies).

2.3 Data Analysis Software

1. MassHunter Quantitative Analysis (Agilent Technologies).

2. Qualis-SIS (UVic-Genome BC Proteomics Centre; *see* **Note 9**).

3. Microsoft Excel.

---

3   Methods

The general workflow used to quantify a broad and deep panel of plasma proteins is shown in Fig. 1. The method hinges on 2D LC/MRM-MS with quantitation performed via SPM or standard curves. The preparation, processing, and analysis steps required are detailed below.

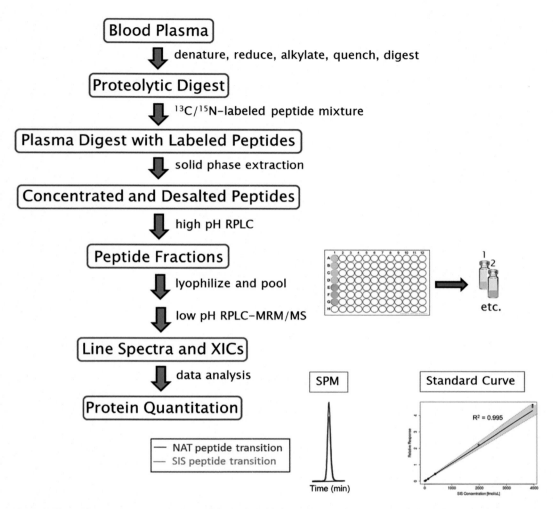

Fig. 1 Schematic of our bottom-up 2D RPLC/MRM-MS with SIS peptides workflow. The *insets* show our pooling strategy (neighboring fractions as opposed to regular intervals) and our quantitative analysis strategy (SPM and standard curves, aided by our Qualis-SIS software tool)

3.1 Sample Preparation: Plasma Proteolysis, Acidification, and Concentration

IMPORTANT: The solution preparations described in **items 4–6** of Subheading 2.1 are to be performed immediately prior to their independent addition to the sample, as noted below. **Items 7–10** of Subheading 2.1 are to be performed 30 min prior to completion of the digestion (i.e., **step 11** of Subheading 3.1).

1. In a 1.5-mL microcentrifuge tube, add 20 μL of raw blood plasma (catalogue no. HMPLEDTA2; Bioreclamation; Westbury, NY, USA) to 590 μL of 25 mM ammonium bicarbonate and 50 μL of 10 % (w/v) sodium deoxycholate. Vortex for 5 s.

2. Add 80 μL of the 50 mM TCEP solution (**item 3** of Subheading 2.1) to the tube then vortex for 5 s.

3. Incubate this microcentrifuge tube in a dry air incubator at 60 °C for 30 min to accelerate chemical (unfolding via the deoxycholate surfactant and disulfide bond reduction via TCEP) and physical (via thermal) denaturation.

4. Add 80 µL of 100 mM iodoacetamide to the protein solution to alkylate the reduced cysteine residues.

5. Vortex for 5 s before incubating for 30 min at 37 °C in the dark in a dry air incubator.

6. Quench the remaining iodoacetamide in the sample by adding 80 µL of 100 mM DTT.

7. Vortex for 5 s then incubate at 37 °C in a dry air incubator for 30 min.

8. Add a 100 µL aliquot of the trypsin solution to the plasma-containing solution to give a 10:1 substrate:enzyme ratio.

9. After vortexing for 5 s, allow digestion to proceed overnight for 16 h at 37 °C. The final concentrations of deoxycholate, TCEP, iodoacetamide, and DTT during proteolysis are 0.5 %, 4 mM, 8 mM, and 8 mM, respectively.

10. Prior to digestion termination, return to Subheading 2.1 to prepare the solutions indicated in **items 7–10**.

11. Upon digestion completion (*see* **Note 10**), immediately place the tube on ice then spike in 100 µL of the diluted and chilled SIS mix (Subheading 2.1, **item 8**).

12. Vortex briefly, then add 180 µL of 3.6 % chilled FA.

13. Pellet the acid insoluble surfactant by centrifugation at $12,000 \times g$ for 10 min.

14. Desalt and concentrate 1196 µL of the peptide supernatant by solid phase extraction (SPE) using traditional vacuum manifold processing on a 10 mg Oasis HLB cartridge. The extractions are performed with the vacuum bleed valve set to −25 kPa and the screw valve in the SPE port rotated to provide no higher than a 1 mL/min flow rate. The extraction steps are as follows:

    (a) wash with 1 mL LC/MS-grade methanol,

    (b) condition with 1 mL LC/MS-grade water,

    (c) load with 2× of 400 µL of 0.1 % FA and 598 µL of digest supernatant (*see* **Note 11**),

    (d) wash with 1 mL water, and

    (e) elute with 600 µL of 55 % ACN/0.1 % FA into a microcentrifuge tube.

15. Place parafilm over the tube then freeze the digest and the lyophilizer container at −80 °C for a minimum of 4 h.

16. Once frozen, puncture two small holes in the parafilm with a small needle and lyophilize to dryness overnight.

17. Rehydrate and transfer two rounds of 10 mM ammonium hydroxide (each at 800 µL) to an autosampler vial for high-pH LC fractionation.

3.2 High-pH LC Fractionation

1. Turn on the UV lamp and select the 214 nm wavelength. Allow at least an hour for the lamp to warm up.

2. Set the RPLC column temperature to 40 °C.

3. Use mobile phase compositions of 100 % water for A, 100 % ACN for B, and 100 mM ammonium hydroxide for C (*see* **item 11** in Subheading 2.1).

4. Set the volume flow rate to 1 mL/min.

5. After purging the system and running a matrix-free blank, inject 1.5 mL of the peptide digest (equates to approximately 0.8 mg NAT) onto the HPLC column.

6. Separate the peptides with a 31-min ACN gradient and a constant concentration of 10 mM ammonium hydroxide. The specific gradient is as follows (time in min, B composition in %): 0, 3; 3, 3; 3.05, 6.5; 4, 6.5; 4.05, 10; 5, 10; 5.05, 13.5; 6, 13.5; 6.05, 17; 7, 17; 7.05, 20.5; 8, 20.5; 8.05, 24; 9, 24; 9.05, 27.5; 10, 27.5; 10.5, 31; 11, 31; 11.05, 36; 12, 36; 12.05, 41; 13, 41; 13.05, 46; 14, 46; 14.05, 51; 20, 51; 21, 80; 30, 80; and 31, 3.

7. Collect fractions into a 96-well plate every 0.5 min from 3 to 26.5 min.

8. Place adhesive foil on the plate then freeze and lyophilize as described above (*see* **steps 15** and **16** of Subheading 3.2).

9. After each run, re-equilibrate the HPLC column with 3 % mobile phase B for 14 min.

10. Once all samples have been fractionated, store the column in 80 % B and turn off the UV lamp to help prolong its lifetime.

3.3 Low-pH LC/ MRM-MS Sample Processing

The low-pH LC system is interfaced to the 6490 QqQ mass spectrometer via a standard-flow ESI source. Both systems are controlled by Agilent's MassHunter Workstation software (version B.07.00).

1. Thermostat the column and autosampler compartments at 50 and 4 °C, respectively.

2. Set up the autosampler section of the LC-MRM acquisition method with a 20 µL/min draw speed, 40 µL/min eject speed, and vial/well bottom sensing activated.

3. Use mobile phase compositions of 0.1 % FA in water for A and 0.1 % FA in ACN for B (*see* **items 12** and **13** in Subheading 2.1).

4. Set up the binary pump section of the LC-MRM acquisition method with the following ACN gradient (time in min, B composition in %): 0, 2.7; 1.5, 6.3; 16, 13.5; 18, 13.8; 33, 22.5; 38, 40.5; 39, 81; 42.9, 81; and 43, 2.7.

5. Use a 0.4 mL/min flow rate.

6. Operate the ESI source in the positive ion mode with a capillary voltage of 3.5 kV.

7. Set up the MS portion of the LC-MRM acquisition method with the following general parameters: 300 V nozzle voltage, 11 L/min sheath gas at 250 °C, 15 L/min drying gas at 150 °C, and 30 psi nebulizer gas flow. Ultrahigh purity nitrogen serves as the carrier gas in all settings.

8. Use the dynamic MRM mode for improved sensitivity with the diverter valve directed to the mass spectrometer for the first 40 min of the LC gradient and to waste in the last 3 min.

9. Enter the scheduled transition information in the compound portion of the acquisition method. This includes specific (i.e., precursor/product ion m/z values, retention times, and collision energies; indicated in Supplemental Table 2 of [25]) and general (i.e., unit mass resolution in the first quadrupole (Q1) and the third quadrupole (Q3) mass analyzers, 1 min retention time window, 380 V fragmentor voltage, and 5 V cell accelerator potential) MRM parameters.

10. Adjust the cycle time in each fractions acquisition method to yield approximately 10 ms dwells (see **Note 12**).

11. Set up the acquisition worklist with 20 µL injection volumes for each reconstituted sample.

12. Prepare for LC/MRM-MS by first purging the binary pumps with 50 % mobile phase B for 5 min at 10 mL/min.

13. Once the LC system re-equilibrates at the starting mobile phase conditions (2.7 % B at 0.4 mL/min), run a blank to flush the sample loop, connecting tubing, and LC-MS system then perform a quality control (QC) check (see **Note 13**).

14. Additionally prepare for fraction analysis by confirming peptide retention times with the SIS mix in buffer (100 fmol column load; see **Note 14**).

15. Once the peptide elution times have been re-scheduled, rehydrate each well with 100 µL of 0.1 % FA then pool the neighboring fractions into a separate 96-well plate as follows (fraction number, wells): 1, A2 to A6; 2, A7 to B8; 3, B9; 4, B10; 5, B11; 6, B12; 7, C1; 8, C2; 9, C3; 10, C4; 11, C5 and C6; 12, C7 to C9; and 13, C10 to D12. Independently and sequentially load 20 µL of the desired fraction(s) onto the UHPLC column.

16. After each LC/MRM analysis, re-equilibrate the UHPLC column with the starting conditions for 4 min.

17. Upon completion of fraction processing, reduce the eluent consumption and preserve the column stationary phase by adjusting the mobile phase composition to 50 % B for continuous isocratic flow at 0.02 mL/min. It is additionally beneficial at this point to divert the eluate to waste and to set the nebulizer and capillary voltage to 0 until the next analysis.

3.4 Plasma Protein Quantitation via SPM

1. Manually inspect the peaks in the MassHunter Quantitative Analysis software to ensure correct peak detection and accurate integration (see **Note 15**). In this process, use a peak area threshold of 300 counts per second (in the "Integration Parameters Setup"), Gaussian smoothing (in the "Smoothing Setup"), and NAT-SIS labels as Target-ISTD (in the "ISTD Setup").

2. Export the RRs (as NAT/SIS peak area ratios) then manually calculate the endogenous protein concentration(s) in the control and patient samples by taking into account each peptides RR, its synthetic peptide concentration (corrected with characterization values from amino acid analysis and capillary zone electrophoresis; in fmol/µL), its protein molecular weight, and a conversion factor (1000).

Using the SPM quantitation strategy, the 2D LC/MRM-MS method with SIS peptides was demonstrated to quantify 253 plasma proteins over an 8 order-of-magnitude concentration range [25]. This represents an improvement of at least an order of magnitude over unfractionated analysis conducted by us [25, 29] and others [18]. In fact, 41 proteins were able to be quantified that were below the lowest quantifiable protein in the unfractionated LC/MRM-MS analysis of a matched plasma digest (myeloblastin at 19 ng/mL). Figure 2 shows the extracted ion chromatograms (XICs) of quantifier peptides from two mid-abundance and two low-abundance proteins whose NAT could not be measured by 1D LC/MRM-MS. Considering the low-abundance range (defined by us as being <10 ng/mL), 31 plasma proteins could be quantified with 19 being present at or below 5 ng/mL. Our antibody-free 2D LC method can also improve the breadth of plasma protein quantitation, with 31 high-to-moderate abundance proteins being additionally quantified with fractionation. Figure 3 illustrates the concentration range expected from this analysis and highlights those proteins quantified in the absence and presence of LC fractionation, for comparison.

3.5 Plasma Protein Quantitation via Standard Curves

To obtain more comprehensive information about the assay (in terms of the limits of quantitation, the dynamic range, and the regression equation), standard curves are required. This is conducted in our laboratory by preparing a series of peptide solutions with constant NAT and variable SIS concentrations across a 3–4

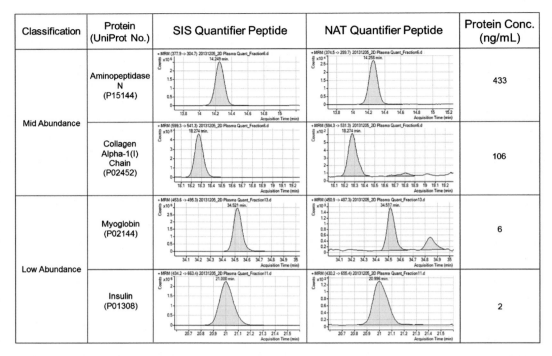

Fig. 2 Representative XICs for moderate- and low-abundance proteins quantified *only* by 2D LC/MRM-MS in human plasma. Shown are quantifier peptides that could not be quantified by unfractionated analysis of a matched plasma digest due to undetectable NAT. In all parts, the SIS peptides are displayed in the *left panel* and the NAT on the *right*

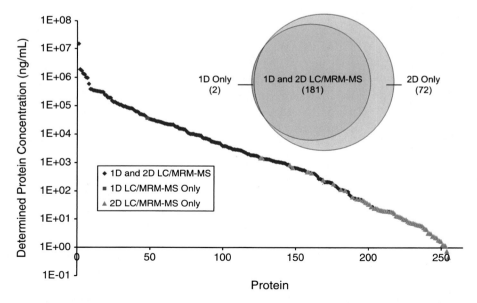

Fig. 3 Protein quantitation from a matched plasma control processed by 1D and 2D LC/MRM-MS. Plotted are the concentrations of each protein's quantifier peptide, with the differences in quantitation further illustrated by the area-proportional Venn diagram shown in the *inset*. Reprinted from [25], with permission

order-of-magnitude range (*see* **Note 9**). To increase the NAT levels for multiple high-pH LC injections, repeat the method described above with additional control plasma aliquots then pool the peptide mixture following digestion before aliquoting into microcentrifuge tubes and spiking in the SIS peptide dilution series. The high-pH LC fractionation and low-pH LC/MRM-MS sample processing steps proceed as described above.

1. Manually inspect the peaks in the MassHunter Quantitative Analysis software as indicated above (*see* **step 1** of Subheading 3.4 and **Note 15**).

2. Export the response data from the MassHunter Quantitative Analysis software then re-orient as shown in Table 1 for input into the Qualis-SIS software [28]. A protein input file (*see* the example in Table 2 and **Note 16**) is additionally required for the tool to generate standard curves and extract assay-related metrics (e.g., lower/upper limits of quantitation, dynamic range, coefficient of determination, endogenous protein concentration).

3. The following user-defined global parameters are typically employed for curve generation in our laboratory: $1/x^2$ (x = concentration) regression weighting, low-to-high concentration level removal strategy [22], <20 % deviation in a given levels precision and accuracy, and a minimum of three consecutive qualified levels.

4. If patient samples were processed, input their corresponding peptide/protein files into Qualis-SIS to calculate their endogenous protein concentrations by regression analysis of the control curve (*see* **Note 17** for a description on how these input files change with patient sample data).

5. Assess the quality of the patient sample measurement in the "Quality Assessment" page of Qualis-SIS. Acceptable quantitation values are indicated in green, while ones that require caution are denoted in yellow and those that should be rejected are in red (*see* **Note 18** for explanations).

Using bottom-up 2D LC/MRM-MS and the standard curve strategy, a condensed panel of 27 plasma proteins from the first four pooled fractions was quantified. With the exception of apolipoprotein C-II (determined concentrations of 10.8 and 9 μg/mL for the quantifier ESLSSYWESAK and qualifier TAAQNLYEK peptide surrogates, respectively), the proteins were targeted by only their quantifier peptide. Overall, their endogenous concentrations were found to spread between 427 μg/mL (for complement C3) and 6.8 ng/mL (for osteopontin), and demonstrated strong correlation to the SPMs, as evidenced by a 0.998 coefficient of determination. Figure 4 illustrates two representative curves generated

Table 1
Basic "peptide input file" format for Qualis-SIS used in the generation of peptide standard curves

Conc. Levels	Matrix metalloproteinase-9. AVIDDAFAR.2.y7		Osteopontin.GDSVVYGLR.2.y5	
	SIS response	NAT response	SIS response	NAT response
L1	0	3715	0	2754
L1	0	4226	0	2943
L1	0	3703	0	2605
L2	0	3814	0	2843
L2	0	4288	0	2949
L2	0	3974	0	2845
L3	728	3079	1019	2348
L3	639	3161	846	2102
L3	820	3176	863	2114
L4	1705	3611	1659	2577
L4	2052	4294	1702	2775
L4	1855	3905	1717	2756
L5	5017	4075	3865	2373
L5	5078	3877	3878	2690
L5	4889	4037	3923	2758
L6	24823	3896	16,845	2789
L6	23222	3600	17,081	2684
L6	24083	4135	17,205	2713
L7	37984	3031	30,936	2325
L7	40413	2911	29,698	2477
L7	39525	3107	30,365	2197

Shown are the SIS and NAT responses for two quantifier peptides (precursor/product ion type indicated in the compound name headers) of two lower abundance proteins obtained from the 2D LC/MRM-MS analysis of control plasma. Additional columns of metric parameters (e.g., peak width and retention time) can be added at the user's discretion.

for two lower abundance proteins—matrix metalloproteinase-9 (at 16 ng/mL) and osteopontin (at 6.8 ng/mL). As is evident for the two cases illustrated, all determined protein concentrations lay within the assay's dynamic range (10^2–10^3 on average) and were an average of 61-fold higher than the lower limit of quantitation and 10-fold lower than the upper limit.

Table 2
"Protein input file" for Qualis-SIS used in the generation of peptide standard curves

Compound name	UniProt KB Acc. No.	Protein MW (Da)	SIS peptide concentration at most concentrated level (fmol/uL)	No. of conc. levels	Dilution series from most concentrated
Matrix metalloproteinase-9. AVIDDAFAR.2.y7	P14780	76,370.56	3.032	7	1:2:5:3:3:4:5
Osteopontin.GDSVVYGLR.2.y5	P10451	33,713.53	3.346	7	1:2:5:3:3:4:5

The file requires information related to the target peptides and the proteins it corresponds to, along with curve details. Shown are the particulars for the quantifier peptides listed in **Table 1**. The curves comprised a maximum of 7 concentration levels with the SIS concentrations varying by 1800-fold from either 3.0 fmol/μL (for matrix metalloproteinase-9) or 3.3 fmol/μL (for osteopontin).

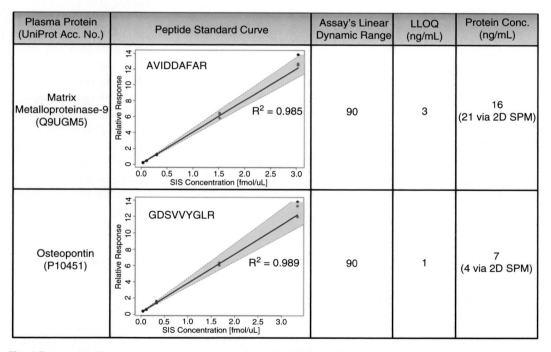

Fig. 4 Representative standard curves processed by 2D LC/MRM-MS and analyzed by Qualis-SIS. The curves and extracted information correspond to the quantifier peptides of two lower abundance proteins—matrix metalloproteinase-9 (determined concentration: 16 ng/mL) and osteopontin (determined concentration: 7 ng/mL). The peptide and protein input file format for Qualis-SIS are indicated in Tables 1 and 2, respectively, and are to be uploaded as CSV files

3.6 Discussion

Strategies for enhanced depth and breadth of plasma protein quantitation are being increasingly developed for evaluation and stratification of candidate biomarkers. Quantitation methods for lower abundance proteins commonly utilize up-front immunoenrichment together with targeted MRM technology and labeled standards implemented within a bottom-up proteomic workflow. A potential alternative that requires less development time and is less resource-intensive, involves inserting high-pH RPLC fractionation between the tryptic proteolysis and low-pH RPLC/MRM-MS sample processing steps. We have recently demonstrated the utility of this approach in quantifying 253 plasma proteins over 13 pooled fractions. Of note, the concentration distribution comprises 31 plasma proteins below 10 ng/mL, which is our internal threshold for low abundance. The benefit of our developed method lies in screening putative biomarkers at the discovery and verification stages, through the use of the entire panel or a subset therein. In the latter application, the low-pH LC gradient should be re-optimized for shortened run times, while the number of pooled fractions to be processed can also be reduced if the elution distribution of the targets is accommodating. It is important here to bear in mind that if a peptides signal is spread across two or more pooled fractions, it should be monitored by dynamic MRM only in the pooled fraction that provides the greatest response. As for analysis, the endogenous protein concentrations can be determined by SPM or standard curves (with regression analysis conducted for the patient samples). We recommend using SPM in discovery studies and standard curves in verification. Regardless of the strategy, quantitation should be conducted on each peptides quantifier transition, with the concentration obtained from the quantifier peptide serving as the concentration of the protein. Although the described methodology relates to the quantitation of 253 specific plasma proteins, it can be readily adapted to other targets, and potentially other biofluids, provided that their elution distribution for pooling is determined, the low-pH LC gradient optimized, and the retention times scheduled.

4   Notes

1. Deoxycholate is used for protein denaturation over common alternatives, such as urea or sodium dodecyl sulfate, due to its facile removal post-digestion through acidification. This helps preserve the LC system and prevent MS-related issues concerning decreased ionization efficiency and ion suppression from the surfactant itself. Side-by-side comparison studies conducted previously demonstrate its better proteolytic efficiency than other chaotropes and surfactants in digested plasma [30] and cerebrospinal fluid [22].

2. Because vortexing the deoxycholate solution produces foam, we transfer a portion (e.g., 500 µL) to a microcentrifuge tube prior to aliquoting to the plasma-containing solution.

3. We use an analytical balance for all weighings, and a micropipetter for accurate measurements of small volumes of liquid.

We use graduated cylinders for larger volumes, such as those used in the mobile phase preparations.

4. The TPCK (L-1-tosylamide-2-phenylethyl chloromethyl ketone)-treated trypsin is used in contrast to the traditional sequencing-grade trypsin due to the significant reduction in cost it affords in the preparation of larger digest batches for 2D LC. To overcome its diminished enzyme activity, a 10:1 substrate:enzyme ratio is employed during proteolysis.

5. In our laboratory, the SIS peptides are synthesized (via Fmoc chemistry on an Overture peptide synthesizer from Protein Technologies) and purified (via RPLC with fractions of interest confirmed by MALDI-TOF-MS on an Ultraflex III TOF/TOF mass spectrometer from Bruker Daltonik) in-house, with characterization (via amino acid analysis for composition and capillary zone electrophoresis for purity) being presently conducted externally. The $^{13}C/^{15}N$ labels are incorporated on the C-terminal residue of tryptic peptides, which are proteotypic and adhere to the set of selection rules documented previously [31].

6. The described preparation for the SIS mix refers to quantitation by SPM. In quantitation via standard curves, a SIS peptide dilution series is prepared for addition to a series of tubes containing a constant concentration of the NAT peptides. In this application (described further in Subheading 3.5), 7-point curves were prepared here with the spiked-in SIS concentrations varying between 250 fmol/μL (for standard G) and 0.139 fmol/μL (for standard A). The specific ratios for the different steps of the serial dilution were 1:2:5:3:3:4:5, starting from the highest concentration.

7. Dry block incubators can alternatively be used provided that it can be set to the required temperatures. If utilized, preference selection toward ones with heated lids for greater precision.

8. If a SpeedVac system is employed, ensure that it is operated at room temperature and is not interrupted during the solvent evaporation process. Also, the peptide mixtures need not be frozen as they are concentrated from solution in an open plate. Note that we prefer lyophilization due to the improved efficiency when drying large volumes.

9. The Qualis-SIS software is available at bioinformatics.proteincentre.com:3838/qualis-sis/ under the Creative Commons Attribution-ShareAlike 4.0 Unported License (CC BY-SA).

10. The peptide digest solution should be clear at this point, with the deoxycholate surfactant only becoming insoluble upon the sequential additions of the SIS mix (prepared in 0.1 % FA) and 3.6 % FA. Do not proceed if the solution appears cloudy at any

step pre- and immediately post-digestion. If precipitation is observed, check the chemical purities then repeat with fresh solutions.

11. Carefully remove the supernatant by withdrawing the solution along one side of the tube to prevent disrupting the deoxycholate precipitate.

12. Ensure that the cycle/dwell times selected enable 10–15 points to be collected across each chromatographic peak for optimal ion statistics. Although this depends on the number of concurrent MRM transitions in each cycle and the total number of transitions monitored, we typically employ <900 ms cycle times with <10 ms dwell times.

13. MS performance checks are first conducted with the check/ auto tune feature in Agilent's MassHunter Workstation software for m/z value verification/re-alignment at different mass resolution settings (unit is of interest here). LC-MS platform performance is then assessed with our QC kits (commercialized through MRM Proteomics), which have previously demonstrated their utility in several intra-/inter-laboratory studies [32, 33].

14. Peptide retention time verification is first conducted in buffer due to simplicity then transferred to plasma where a maximum shift of ca. 0.5 min is expected. SIS and NAT transitions have identical MRM acquisition parameters and retention times since they behave identically in the LC column and mass spectrometer. The only difference lies in their precursor/product ion m/z values.

15. Skyline (available at: https://skyline.gs.washington.edu/labkey/project/home/software/Skyline/begin.view; [34]) can alternatively be used to obtain the peak-related information for input into Qualis-SIS.

16. The compound names listed in the peptide (Table 1) and protein (Table 2) input file must be identical for Qualis-SIS to function. Also, there must not be an apostrophe or comma in the protein name.

17. In the case of peptide quant data from patient samples, the peptide input file should contain responses oriented from left to right as before with the control (*see* Table 1), but with the sample IDs indicated instead in the concentration level column. The protein input file should contain information related to the protein (i.e., the accession number and molecular weight) and SIS peptide concentration (in fmol/μL), on a per-compound basis. For additional assistance, *see* input file examples provided in the "File" page of Qualis-SIS.

18. The color-coded matrix shown in the "Quality Assessment" page provides clues as to whether the quantitative data of the patient samples should be trusted. Green is acceptable since the derived concentration lies within the linear dynamic range of the assay, while red is unacceptable since it falls outside a user-specified deviation from the lower or upper limits of quantitation. If a value is outside the assay range but within the threshold deviation, it is denoted in yellow. It is then left to the user's discretion whether to trust it or to take the SPM value instead.

Acknowledgments

We wish to thank Genome Canada and Genome BC for STIC (Science and Technology Innovation Centre) funding and support. Carol Parker (UVic-Genome BC Proteomics Centre) is acknowledged for assisting in the manuscript editing process.

Competing Interests: Christoph Borchers is the director of the Centre and the Chief Scientific Officer of MRM Proteomics, which has commercialized the performance kits noted above for system/platform assessment.

References

1. Askenazi M, Li S, Singh S, Marto JA (2010) Pathway Palette: a rich internet application for peptide-, protein- and network-oriented analysis of MS data. Proteomics 10:1880–1885. doi:10.1002/pmic.200900723

2. Villanueva J, Carrascal M, Abian J (2014) Isotope dilution mass spectrometry for absolute quantification in proteomics: concepts and strategies. J Proteomics 96:184–199. doi:10.1016/j.jprot.2013.11.004

3. Gillette MA, Carr SA (2013) Quantitative analysis of peptides and proteins in biomedicine by targeted mass spectrometry. Nat Methods 10:28–34

4. Gallien S, Duriez E, Crone C, Kellmann M, Moehring T et al (2012) Targeted proteomic quantification on quadrupole-orbitrap mass spectrometer. Mol Cell Proteomics 11:1709–1723. doi:10.1074/mcp.O112.019802

5. Gallien S, Bourmaud A, Kim SY, Domon B (2014) Technical considerations for large-scale parallel reaction monitoring analysis. J Proteomics 100:147–159. doi:10.1016/j.jprot.2013.10.029

6. Peterson AC, Russell JD, Bailey DJ, Westphall MS, Coon JJ (2012) Parallel reaction monitoring for high resolution and high mass accuracy quantitative, targeted proteomics. Mol Cell Proteomics 11:1475–1488. doi:10.1074/mcp.O112.020131

7. Brun V, Dupuis A, Adrait A, Marcellin M, Thomas D et al (2007) Isotope-labeled protein standards: toward absolute quantitative proteomics. Mol Cell Proteomics 6:2139–2149

8. Gerber SA, Rush J, Stemman O, Kirschner MW, Gygi SP (2003) Absolute quantification of proteins and phosphoproteins from cell lysates by tandem MS. Proc Natl Acad Sci U S A 100:6940–6945

9. Pratt JM, Simpson DM, Doherty MK, Rivers J, Gaskell SJ et al (2006) Multiplexed absolute quantification for proteomics using concatenated signature peptides encoded by QconCAT genes. Nat Protoc 1:1029–1043

10. Boja ES, Rodriguez H (2012) Mass spectrometry-based targeted quantitative proteomics: achieving sensitive and reproducible detection of proteins. Proteomics 12:1093–1110. doi:10.1002/pmic.2011003871093

11. Picotti P, Aebersold R (2012) Selected reaction monitoring-based proteomics: workflows, potential, pitfalls and future directions. Nat Methods 9:555–566. doi:10.1038/nmeth.2015

12. Simpson DM, Beynon RJ (2012) QconCATs: design and expression of concatenated protein standards for multiplexed protein quantification. Anal Bioanal Chem 404:977–989. doi:10.1007/s00216-012-6230-1

13. Percy AJ, Chambers AG, Yang J, Hardie DB, Borchers CH (2014) Advances in multiplexed MRM-based protein biomarker quantitation toward clinical utility. Biochim Biophys Acta 1844:917–926. doi:10.1016/j.bbapap.2013.06.008

14. Berna M, Ott L, Engle S, Watson D, Solter P et al (2008) Quantification of NTproBNP in rat serum using immunoprecipitation and LC/MS/MS: a biomarker of drug-induced cardiac hypertrophy. Anal Chem 80:561–566. doi:10.1021/ac702311m

15. Whiteaker JR, Zhao L, Lin C, Yan P, Wang P et al (2012) Sequential multiplexed analyte quantification using peptide immunoaffinity enrichment coupled to mass spectrometry. Mol Cell Proteomics 11:M111.015347. doi:10.1074/mcp.M111

16. Whiteaker JR, Zhao L, Frisch C, Ylera F, Harth S et al (2014) High-affinity recombinant antibody fragments (Fabs) can be applied in peptide enrichment immuno-MRM assays. J Proteome Res 13:2187–2196. doi:10.1021/pr4009404

17. Keshishian H, Addona T, Burgess M, Kuhn E, Carr SA (2007) Quantitative, multiplexed assays for low abundance proteins in plasma by targeted mass spectrometry and stable isotope dilution. Mol Cell Proteomics 6:2212–2229

18. Huttenhain R, Soste M, Selevsek N, Rost H, Sethi A et al (2012) Reproducible quantification of cancer-associated proteins in body fluids using targeted proteomics. Sci Transl Med 4:142ra94. doi:10.1126/scitranslmed.3003989

19. Liu T, Hossain M, Schepmoes AA, Fillmore TL, Sokoll LJ et al (2012) Analysis of serum total and free PSA using immunoaffinity depletion coupled to SRM: correlation with clinical immunoassay tests. J Proteomics 75:4747–4757. doi:10.1016/j.jprot.2012.01.035

20. Rezeli M, Végvári A, Ottervald J, Olsson T, Laurell T et al (2011) MRM assay for quantitation of complement components in human blood plasma—a feasibility study on multiple sclerosis. J Proteomics 75:211–220. doi:10.1016/j.jprot.2011.05.042

21. Paulovich AG, Whiteaker JR, Hoofnagle AN, Wang P (2008) The interface between biomarker discovery and clinical validation: the tar pit of the protein biomarker pipeline. Proteomics Clin Appl 2:1386–1402

22. Percy AJ, Yang J, Chambers AG, Simon R, Hardie DB et al (2014) Multiplexed MRM with internal standards for cerebrospinal fluid candidate protein biomarker quantitation. J Proteome Res 13:3733–3747. doi:10.1021/pr500317d

23. Gilar M, Olivova P, Daly AE, Gebler JC (2005) Two-dimensional separation of peptides using RP-RP-HPLC system with different pH in first and second separation dimensions. J Sep Sci 28:1694–1703

24. Song C, Ye M, Han G, Jiang X, Wang F et al (2010) Reversed-phase-reversed-phase liquid chromatography approach with high orthogonality for multidimensional separation of phosphopeptides. Anal Chem 82:53–56. doi:10.1021/ac9023044

25. Percy AJ, Simon R, Chambers AG, Borchers CH (2014) Enhanced sensitivity and multiplexing with 2D LC/MRM-MS and labeled standards for deeper and more comprehensive protein quantitation. J Proteomics 106:113–124. doi:10.1016/j.jprot.2014.04.024

26. Shi T, Sun X, Gao Y, Fillmore TL, Schepmoes AA et al (2013) Targeted quantification of low ng/mL level proteins in human serum without immunoaffinity depletion. J Proteome Res 12:3353–3361. doi:10.1021/pr400178v

27. Percy AJ, Chambers AG, Yang J, Domanski D, Borchers CH (2012) Comparison of standard- and nano-flow liquid chromatography platforms for MRM-based quantitation of putative plasma biomarker proteins. Anal Bioanal Chem 404:1089–1101. doi:10.1007/s00216-012-6010-y

28. Mohammed Y, Percy AJ, Chambers AG, Borchers CH (2015) Qualis-SIS: automated standard curve generation and quality assessment for multiplexed targeted quantitative proteomic experiments with labeled standards. J Proteome Res 14:1137–1146. doi:10.1021/pr5010955

29. Domanski D, Percy AJ, Yang J, Chambers AG, Hill JS et al (2012) MRM-based multiplexed quantitation of 67 putative cardiovascular disease biomarkers in human plasma. Proteomics 12:1222–1243. doi:10.1002/pmic.201100568

30. Proc JL, Kuzyk MA, Hardie DB, Yang J, Smith DS et al (2010) A quantitative study of the effects of chaotropic agents, surfactants, and solvents on the digestion efficiency of human plasma proteins by trypsin. J Proteome Res 9:5422–5437

31. Kuzyk MA, Parker CE, Domanski D, Borchers CH (2013) Development of MRM-based assays for the absolute quantitation of plasma proteins. Methods Mol Biol 1023:53–82. doi:10.1007/978-1-4614-7209-4_4

32. Percy AJ, Chambers AG, Yang J, Borchers CH (2013) Multiplexed MRM-based quantitation of candidate cancer biomarker proteins in undepleted and non-enriched human plasma. Proteomics 13:2202–2215. doi:10.1002/pmic.201200316

33. Percy AJ, Chambers AG, Smith DS, Borchers CH (2013) Standardized protocols for quality control of MRM-based plasma proteomic workflow. J Proteome Res 12:222–233. doi:10.1021/pr300893w

34. MacLean B, Tomazela DM, Shulman N, Chambers M, Finney GL et al (2010) Skyline: an open source document editor for creating and analyzing targeted proteomics experiments. Bioinformatics 26:966–968. doi:10.1093/bioinformatics/btq054

Chapter 2

Quantitative Analysis of the Sirt5-Regulated Lysine Succinylation Proteome in Mammalian Cells

Yue Chen

Abstract

Lysine (Lys) succinylation is a recently discovered protein posttranslational modification pathway that is evolutionarily conserved from bacteria to mammals. It is regulated by Sirt5, a member of the class III histone deacetylases (HDACs) or the Sirtuins. Recent studies demonstrated that Lys succinylation and Sirt5 are involved in diverse cellular metabolic processes including urea cycle, ammonia transfer, and glucose metabolism. In this chapter, we describe the general protocol to identify Sirt5-regulated Lys succinylation substrates and a computational method to calculate the absolute modification stoichiometries of Lys succinylation sites. The strategy employs Stable Isotope Labeling of Amino acid in Cell culture (SILAC) and the immunoaffinity enrichment of Lys succinylated peptides to identify the Lys succinylation sites that are significantly upregulated in Sirt5 knockout mouse embryonic fibroblast cells.

Key words SILAC, Quantification, Sirt5, Immunoaffinity purification, Lysine succinylation, Nano-HPLC-mass spectrometry

1 Introduction

Protein posttranslational modifications (PTMs) provide a critical mechanism for cells to regulate key processes and fine-tune protein functions [1]. Lysine, with a nucleophilic side chain, has been identified as the targets of numerous protein modifications. In addition to the well-known Lys acetylation, methylation, ubiquitination, biotinylation, and SUMOylation, we and others have recently identified Lys propionylation, butyrylation, crotonylation, succinylation, malonylation, glutarylation, and 2-hydroxyisobutyrylation [2–9]. The diversity of Lys modifications presents a complex regulatory network that links cellular metabolism, enzymatic activities with protein functions, and epigenetic regulations [4, 9–11]. System-wide characterization of these PTM pathways is needed to comprehensively understand their functional significance in cellular processes and disease progression.

Salvatore Sechi (ed.), *Quantitative Proteomics by Mass Spectrometry*, Methods in Molecular Biology, vol. 1410, DOI 10.1007/978-1-4939-3524-6_2, © Springer Science+Business Media New York 2016

Sirtuins are a class of enzymes that are traditionally considered as histone deacetylases (HDACs). They use nicotinamide adenosine dinucleotide (NAD+) as the cofactor and the family is comprised of seven members in mammalian cells, Sirt1 to Sirt7. However, other than Sirt1-3, Sirt4-7 exhibits little or no deacetylase activity in vitro. Recent studies discovered that Sirt5 catalyzes an efficient Lys desuccinylation reaction both in vitro and in vivo, which established Sirt5 as the first regulatory enzyme of the Lys succinylation pathway [7, 12]. Sirt5 is known to be involved in the urea cycle through regulating the activity of CPS1, the enzyme that catalyzes the rate-limiting step by converting ammonia to carbamoyl phosphate in mitochondria [8, 13]. Loss of Sirt5 leads to the increase of Lys acylation abundance on CPS1 and decreased enzymatic activity, which results in the accumulation of ammonia in blood [14]. Sirt5 also regulates glutamine metabolism and ammonia production in nonliver cells through regulating glutaminase activity [15]. A deep understanding of the crucial role of Sirt5 in cellular functions requires a system-wide effort to identify the Sirt5-regulated Lys succinylation substrates, best achieved using quantitative proteomics approach.

Global identification and quantification of posttranslational modification pathways are technically challenging, largely due to the low abundance and transient nature of modified proteins in the cell. Therefore, a highly efficient enrichment strategy is required for sensitive and comprehensive analysis of protein modifications. We initially developed an efficient strategy for global analysis of Lys acetylation substrates that employs immunoaffinity enrichment of Lys acetylated peptides with pan anti-acetyl Lys polyclonal antibody [16]. Comparing to monoclonal antibody, the polyclonal antibody has broad substrate specificity capable of targeting peptides with various flanking sequences. Our strategy led to the first system-wide characterization of the Lys acetylation proteome in mammalian cells and in bacteria [16, 17]. Given the structural similarities between Lys acetylation and succinylation, we applied a similar strategy for global analysis of Lys succinylation proteome. A pan anti-succinyl Lys polyclonal antibody was successfully developed for immunoaffinity enrichment and liquid chromatography mass spectrometry (LCMS) to identify Lys succinylated peptides [5]. We combined this strategy with SILAC-based quantitative proteomics analysis for system-wide identification of Sirt5-responsive Lys succinylation substrates in mammalian cells [18]. Our studies identified more than 2500 Lys succinylation sites in MEF cells and mouse liver tissue. Over 90 % of sites quantified with the SILAC approach showed increased abundance, which strongly suggests that Sirt5 plays a central role in regulating Lys succinylation abundance in cells. We further calculated the absolute stoichiometries of Lys succinylation sites using a modified method of previously published approach [19]. The average stoichiometries of Lys succinylation is less than 10 %, but upon Sirt5 knockout, over half of the Lys succinylation sites have more than 10 % stoichiometries, which further suggests the critical role of Sirt5 in regulating global Lys succinylation abundance.

In this chapter, we will present our experimental protocols in detail, and list important steps during sample analysis. Briefly, our strategy involves the following four steps: (1) Culture Sirt5 wild type (Sirt5$^{+/+}$) and Sirt5 knock out (Sirt5$^{-/-}$) mouse embryonic fibroblast (MEF) cells in SILAC heavy and light media; (2) Perform cell lysate, tryptic digestion, and then high pH reversed phase chromatography based peptide fractionation; (3) Immunoaffinity precipitation of Lys succinylated peptides; (4) NanoLC-mass spectrometry analysis; (5) Quantitative data analysis (Fig. 1).

Two steps during the procedure are critically important for the successful analysis. First, immunoaffinity enrichment efficiency of peptides bearing Lys succinylation is crucial for the experiments. As pan PTM antibodies typically have much lower affinity toward modified peptides, it is important to choose a high-quality antibody for the enrichment experiment. In addition, we found that longer incubation time and stringent washing condition may also help improve the efficiency of the experiment. Second, extensive peptide fractionation is crucial to improve the sensitivity of the analysis. The modified peptides from highly abundant proteins may interfere with the affinity enrichment and identification of the modification from low abundant proteins. To overcome this challenge, it is important to apply an efficient approach to fractionate the sample prior to the immunoaffinity enrichment and reduce the sample complexity. Recently, high pH reversed phase chromatography has been demonstrated with superior separation and fractionation efficiency for global proteomics analysis [20]. We describe in this chapter our method to apply high pH reversed phase chromatography to allow deep Lys succinylation proteome coverage and quantitative analysis.

2 Materials

2.1 SILAC Cell Culture, Labeling Efficiency Check, Cell Lysis, and Tryptic Digestion

1. SILAC cell culture media—Dulbecco's Modified Eagle Medium (D-MEM) without l-Lysine or l-Arginine (pH 7.0–7.4) with glucose (4.5 g/L), l-glutamine (2 mM), and dialyzed fetal bovine albumin (FBS) (10 %) (Life Technology SILAC metabolic labeling kit).

2. Stable isotope encoded amino acids—$^{13}C_6$-Lysine and $^{13}C_6{}^{15}N_2$-Arginine with 99 % isotope purity.

3. Cell lines—Sirt5$^{+/+}$ and Sirt5$^{-/-}$ mouse embryonic fibroblast (MEF) cells.

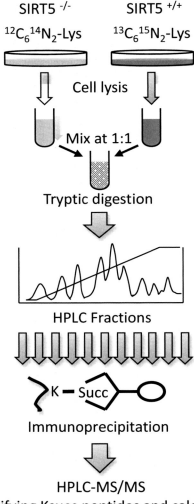

Fig. 1 Schematic workflow for SILAC-based quantification of the Lys succinylation proteome in mammalian cells in response to Sirt5 knockout. The strategy employs high pH reverse phase peptide fractionation prior to immunoaffinity enrichment of Lys succinylated peptides

4. NETN cell lysis buffer—NaCl (200 mM), EDTA (2 mM), Tris–HCl (40 mM, pH 8.0), NP-40 (1 %).

5. 60 mm and 150 mm cell culture plates.

6. Bradford assay kit.

7. Humidified 37 °C cell culture incubator maintaining 5 % carbon dioxide level.

8. Laminar biosafety cabinet class II A/B3.

9. Sterile falcon tubes and pipettes.

10. Cell scraper.

11. 37 °C water bath.

12. UV spectrometer.

13. 70 % Ethanol (v/v).

14. Orbital shaker.

15. 4 °C Refrigerator.

16. Phosphate-buffered saline (PBS) (pH 7.4).

17. Benchtop centrifuge at room temperature.

18. Ultrasonicator.

19. 4 °C Microcentrifuge.

20. Microfuge tubes (1.5 mL).

21. Litmus paper (pH 1–14) or pH meter.

22. Trichloroacetic acid (TCA).

23. Acetone (store at –20 °C).

24. Digestion buffer: 50 mM ammonium bicarbonate (pH 8.0).

25. Iodoacetamide (IA) stock solution—300 mM in digestion buffer. Dissolve 55.5 mg of IA in 1 mL digestion buffer.

26. Dithiothreitol (DTT) stock solution—100 mM in digestion buffer. Dissolve 15.4 mg of DTT in 1 mL digestion buffer.

27. l-Cysteine stock solution—300 mM in digestion buffer. Dissolve 36.3 mg of Cys in 1 mL digestion buffer.

28. Trypsin (sequencing grade).

2.2 Peptide Fractionation with High pH Reversed phase chromatography

1. HPLC buffer A: 10 mM ammonium formate in water, pH 7.8 in water.

2. HPLC buffer B: 10 mM ammonium formate in 90 % acetonitrile and 10 % water, pH 7.8.

3. Vacuum-assisted filtration device with 0.22 μm membrane.

4. Water-bath sonicator.

5. Preparative HPLC system with UV detector and fraction collector.

6. C18 semi-preparative HPLC column (Luna C18 10 mm×250 mm, 5 μm particle, 100 Å pore size).

7. Syringe filter system (0.22 μm).

8. Buchi rotary evaporator.

9. Water aspirator vacuum pump.

2.3 Immunoaffinity Purification (IP)

1. Agarose beads conjugated with pan anti-succinyllysine antibody (PTM Biolabs).

2. Benchtop centrifuge.

3. NETN IP buffer: NaCl (100 mM), EDTA (1 mM), Tris–HCl (50 mM, pH 8.0), NP-40 (0.5 %, w/v).

4. ETN buffer: EDTA (1 mM), Tris–HCl (50 mM, pH 8.0), NaCl (100 mM).

5. Elution buffer: TFA in water (0.1 %, v/v).

2.4 Reverse-Phase NanoLC (RPLC)-Mass Spectrometry

1. Nano-flow HPLC such as Eksigent 1D-plus NanoLC or Proxeon Easy 1000 nLC system.

2. RPLC solvent A: Formic acid in water (0.1 %, v/v).

3. RPLC solvent B: Formic acid in acetonitrile (0.1 %, v/v).

4. Commercially available fused silica capillary C18 HPLC column (75 μm ID, 12 cm length) with nanospray needle tip or in-house packed capillary C18 HPLC column (Luna C18, 5 μm, 100 Å pore size, 75 μm ID, 12 cm length) with laser-pulled tip.

5. Nanospray source.

6. High resolution mass spectrometer. For example, Orbitrap or Q-TOF mass spectrometer.

2.5 Quantitative Data Analysis

1. PC with at least 2 GB RAM, multi-core processor, installed with 32-bit or 64-bit Windows operating system and .NET framework 4.5.

2. MSFileReader.exe software (Thermofisher, compatible with instrument data).

3. Uniprot mouse protein sequence database (FASTA file).

4. Maxquant software (http://maxquant.org, latest version).

3 Methods

3.1 SILAC Cell Culture

1. Prepare D-MEM SILAC media following manufacturer's instructions (see **Notes 1–4**).

2. Pass the media through 0.22 μm filter, and warm up in 37 °C water bath.

3. Take the cryovials of Sirt5$^{-/-}$ and Sirt5$^{+/+}$ mouse embryonic fibroblast (MEF) cells from storage tank, and put in 37 °C water bath to thaw for 1 min.

4. Wash the cells with 15-mL sterile conical centrifuge tubes with SILAC media. Use light SILAC medium to Sirt5$^{-/-}$ MEF cells and heavy SILAC media to Sirt5$^{+/+}$ MEF cells. Discard the supernatant after washing. Repeat this step once more time to ensure complete removal of DMSO.

5. Culture the cells in 60 mm cell culture plate with 4 mL SILAC media in 37 °C humidified incubator (5 % carbon dioxide).

6. Culture cells for half a day before changing medium once to further remove DMSO.

7. Split cells when the confluence reaches 90 %, and allow the cells to grow for at least six doublings in heavy medium.

8. Check labeling efficiency (see Subheading 3.2) and if the cells reach over 97 % labeling efficiency, start culturing the cells in 150 mm large dishes.

9. Grow at least five 150 mm dishes of cells with light amino acid labeling and five 150 mm dishes of cells in heavy amino acid labeling with less than 90 % confluence.

3.2 SILAC Labeling Efficiency Test, Cell Lysis, and Tryptic Digestion

1. Discard the media and rinse the cells in ice-cold PBS buffer to remove culture medium. To check the labeling efficiency in SILAC media, only a fraction of heavy labeled cells need to be collected.

2. Scrape the cells off the plate with cold PBS and collect in 15-mL conical tubes.

3. Wash the cells with ice-cold PBS two times, and resuspend the cell pellet in two volumes of NETN cell lysis buffer. Transfer the cells in lysis buffer to a 1.5-mL microfuge tube.

4. Keep the cells in lysis buffer for 30 min on ice.

5. Centrifuge the lysate at $16,000 \times g$ for 10 min at 4 °C.

6. Collect the supernatant into a new tube and add additional one volume of NETN buffer to the pellet.

7. Sonicate the pellet for a 5 s pulse followed by a 15 s pause at 10 % power in cold room. Complete 5–10 cycles till the lysate no longer viscous.

8. Centrifuge at $16,000 \times g$, 10 min, 4 °C.

9. Pool the supernatants and measure protein concentration with Bradford assay.

10. To check labeling efficiency, follow **step 11** for protein precipitation. Otherwise, take 5–8 mg of proteins from heavy and light labeled cell lysate and mix thoroughly by equal amount together prior to protein precipitation.

11. Perform TCA protein precipitation by adding TCA dropwise to a final concentration of 20 % (v/v).

12. Precipitate proteins for 4 h at 4 °C.

13. Centrifuge at $16,000 \times g$ for 10 min at 4 °C and discard the supernatant.

14. Wash the protein pellets in ice-cold acetone and centrifuge at $16,000 \times g$ for 5 min at 4 °C. Discard the supernatant.

15. Repeat **step 14** two more times.

16. Add sufficient digestion buffer to the protein pellet to have a final concentration less than 5 mg/mL and measure the pH with litmus paper to make sure the pH is around 8.

17. Resuspend trypsin in digestion buffer and add to the protein buffer at enzyme-to-substrate ratio of 1:50 (w/w).

18. Incubate overnight in 37 °C.

19. Perform disulfide bond reduction by adding DTT stock solution to final 5 mM and incubate at 55 °C for 30 min.

20. Perform Cys alkylation by adding iodoacetamide stock solution to a final concentration of 15 mM. Incubate for 30 min in the dark at room temperature.

21. Add Cysteine stock solution to block the excess iodoacetamide with a final concentration of 15 mM and incubate for 30 min at room temperature.

22. To ensure complete digestion, add additional trypsin at enzyme-to-substrate ratio of 1:100 (w/w) and incubate for 3 h at 37 °C.

23. Stop the digestion by adding TFA stock solution to the digestion buffer till a final concentration of 0.1 % (v/v).

24. Desalt the peptides with C18 tip or cartridge following manufacturer's protocol.

25. Dry the eluted peptides in vacuum centrifuge system.

26. To check labeling efficiency, perform LC-MS/MS analysis (*see* Subheading 3.6). Perform protein quantification with Maxquant. To determine labeling efficiency, manually inspect heavy to light peak ratios of the most intense heavy isotope labeled peptides (*see* **Note 5**). Ideally the labeling efficiency should reach 97–98 %. In case that the labeling efficiency is still low, culture cells for one or two more cycles and check again to ensure complete labeling.

27. To perform peptide fractionation before immunoprecipitation (Fig. 1), follow Subheading 3.3. Samples can be stored at −80 °C for later use.

3.3 Peptide Fractionation with High pH Reversed Phase Chromatography

1. Prepare 1 L HPLC buffer A with 10 mM ammonia formate in water. Adjust pH to 7.8 (*see* **Note 6**).

2. Prepare 1 L HPLC buffer B with 10 mM ammonia formate in 90 % acetonitrile and 10 % water. Adjust pH to 7.8.

3. Filter HPLC buffer A and B through 0.22 μm filter before use.

4. Degas the solvents by sonicating the buffers in water-bath sonicator for 20 min. The step can be skipped if the HPLC is equipped with an online degasser.

5. Dissolve the tryptic peptides in 2–3 mL HPLC buffer A.

6. Centrifuge at $16,000 \times g$.

7. Carefully extract the supernatant and pass it through 0.22 µm filter. Discard the precipitates.

8. Prepare glass vials for fractionation collector and set up the following HPLC application: Flow rate 4 mL/min; gradient: 0 min—2 % B, 40 min—30 % B, 55 min—90 % B, 60 min—90 % B (*see* **Note 7**).

9. Wash the HPLC column with 100 % HPLC buffer B for at least three bed volumes.

10. Wash the column with 100 % buffer A for at least five bed volumes till the UV absorption at 215 nm stabilizes.

11. Load samples through manual injector or autosampler. When using manual injector, if the sample volume is larger than half of the loop size, divide the sample into several fractions and load each fraction sequentially. Following each loading, wait until the UV absorption stabilizes again before next sample loading. Make sure to collect all the flow-through solvents in a 250 mL glass container as one fraction in case that the loading amount exceeds the column binding capacity. When using autosampler for sample loading, the sample volume needs to be less than half of the loop size for complete loading.

12. Start the HPLC and collect a fraction roughly every minute.

13. When HPLC fractionation is finished, wash the column with 10–90 % B for at least ten bed volumes (*see* **Note 8**).

14. Pool fractions together through concatenation scheme [20], for example, by combining fraction 1, 11, 21, … to new fraction 1 and so on, to obtain a total of ten final fractions.

15. Evaporate the solvents with Buchi rotavapor till about 1 mL.

16. Transfer the fraction to 1.5-mL Eppendorf tube and dry in Speed Vac.

17. Perform immunoaffinity purification of Lys succinylated peptides from each fraction according to Subheading 3.4.

3.4 Immunoaffinity Purification (IP)

1. Dissolve the peptides from each fraction in 1 mL NETN buffer.

2. Centrifuge the peptide solution at $16,000 \times g$ for 10 min and carefully take the supernatant.

3. Keep 50 µl of the supernatant prior to IP for LCMS and protein quantification later on.

4. Add 50 µl anti-succinyllysine antibody-conjugated agarose beads to the peptide solution and incubate at 4 °C overnight on an orbital shaker.

5. Centrifuge at $1000 \times g$ and remove the supernatant as the IP soup fraction.

6. Wash the beads with 1 mL ice-cold NETN buffer three times.

7. Wash the beads with 1 mL ice-cold ETN buffer once.

8. Wash the beads with 1 mL ice-cold water twice.

9. Add 50 µl Elution buffer for elution.

10. Centrifuge at $1000 \times g$ 1 min, and collect the eluted peptides into a new tube.

11. Repeat elution two more times and pool eluted peptide solution together.

12. Dry the eluted peptides in SpeedVac.

13. Desalt the peptides with C18 cartridge following manufacturer's protocol.

14. Perform Nano-HPLC-MS/MS analysis following Subheading 3.6.

3.5 Nano-HPLC-Mass Spectrometry Analysis

Our nano-HPLC/MS/MS analysis is performed on a Proxeon Easy nLC 1000 (Thermofisher) coupled online through a nanospray source to Q-Exactive Orbitrap mass spectrometer (ThermoFisher). Capillary C18 fused silica column with nanospray tip can be purchased commercially or packed in-house.

1. Set up nano-HPLC method in Xcalibur software. Both column equilibration volume and sample loading volume are set to 6 µl with constant pressure loading. Elute the peptides with a linear gradient of 5–30 % solvent B for 60 min and then 30–80 % solvent B for 15 min at a constant flow rate of 200 nL/min.

2. Set up mass spectrometer instrument method (*see* **step 5** for parameters).

3. Dissolve peptides in 3 µl solvent A and place the vial in the autosampler.

4. Start the instrument acquisition method. Capillary column is equilibrated first with 100 % solvent A, followed by sample loading and gradient elution.

5. Tryptic peptides are analyzed by a Q-Exactive Orbitrap mass spectrometer operated in data-dependent acquisition mode acquiring top 12 most intense ions. Other instrument parameters are: spray voltage 2.0 kV, heating capillary temperature 200 °C, full MS resolution 70,000 at 200 m/z with a mass range of 300–1800 m/z, MS/MS resolution 17,500 at 200 m/z using high energy collusion dissociation (HCD) at the 30 % normalized collision energy. An ion from ambient air (m/z 445.120024) is used as lock mass ion for internal calibration. Dynamic exclusion is set with a repeat count of 2, repeat duration of 8 s, exclusion duration of 40 s, maximum exclusion size of 500, and exclusion mass width ±10 ppm.

3.6 Quantitative Data Analysis

Protein identification and quantification is analyzed by MaxQuant [21] software with the Andromeda search engine.

1. Install MSFileReader.exe software (Thermofisher) on a Windows-based PC.

2. Install .Net framework 4.5 (Microsoft).

3. Install the latest version of Maxquant following the instructions online (http://www.maxquant.org).

4. Download and install the mouse fasta database (http://www.Uniprot.org) in Maxquant following the online instructions.

5. Configure modifications in the Maxquant software for database search (Lys succinylation, $K + C_4H_4O_3$, 100.01604 Da).

6. Copy all raw data into a folder on the same computer installed with Maxquant and Xcalibur. For Lys succinylation quantification, include all raw data for the LCMS analysis of IP-enriched Lys succinylated peptides and the peptide fractions collected prior to IP for protein quantification.

7. Start the MaxQuant software and load all raw files.

8. For quantification parameters, select "Doublets" as SILAC type and check "Lys6" for $U-^{13}C_6$ Lysine labeling (check both "Lys6" and "Arg10" if double labeling with $U-^{13}C_6$ Lysine and $U-^{13}C_6^{15}N_2$ Arginine was performed). For protein quantification, use only unmodified peptides as well as peptides with N-terminal acetylation.

9. Select methionine oxidation, acetylation on protein N-terminus, and Lys succinylation as variable modifications and Cys carbamidomethylation as fixed modification. Select the mouse database concatenated with reverse decoy database and common contaminant proteins. Select trypsin as enzyme, maximum number of missing cleavages as 2, MS/MS tolerance as 0.5 Da, and top six peaks per 100 Da window.

10. Set 1 % FDR threshold at proteins, peptides, and sites levels. Minimum peptide length is 7 and minimum modified peptide score is 40.

11. Select the number of cores on the computer as the number of threads to maximize search speed and then start the analysis.

12. The program will perform peak generation, recalibrate precursor mass, extract SILAC pairs, identify peptides, group protein identifications, and calculate SILAC ratios.

13. Upon the completion of Maxquant analysis, a "combined" folder will be generated under the Raw files folder. The txt folder in the combined folder contains the identification and quantification data in different tab-delimited txt files.

14. The file "Succinyl (K) Sites.txt" (actual file name depending on the configuration of the modification in the Maxquant software) contains all the identification and quantification for Lys

succinylation sites passing the cutoff criterion (*see* **Note 9**). The file "evidence.txt" contains all the peptide-to-spectra identification and quantification, assigned protein groups, and modification state. The file "proteinGroups.txt" contains all the protein identifications, quantifications, and available functional annotations. Maxquant is also capable of analyzing data generated with other fractionation and quantification strategies (*see* **Note 10**). For additional information on result interpretation, please check online tutorial of the Maxquant software (http://www.maxquant.org).

15. Calculation of site-specific stoichiometries of Lys succinylation should follow the algorithm described previously that calculates site-specific stoichiometries of phosphorylation sites based on the SILAC ratios of modified peptides, unmodified peptides, and corresponding proteins (*see* **Note 11**) [19]. For phosphorylated peptides, modified and unmodified peptides share the same tryptic peptide sequence. However, for Lys succinylated peptides, the unmodified peptides would be cleaved by trypsin. Therefore, we select the longest fully cleaved peptides of the Lys succinylated peptides as the unmodified counterparts and use their ratios for the calculation of Lys succinylation site stoichiometry.

4   Notes

1. Cell lines such as Hela and HEK293 cells may convert arginine to proline when arginine is present in excess in the cell, which will alter the quantification ratios of heavy isotope labeled peptides. Therefore, if double SILAC labeling with both heavy Lys and Arg is used, it will be necessary to perform a titration experiment with different concentration of Arg to minimize the conversion.

2. The purity of amino acids is critical for cell growth and quantification accuracy. Always use amino acids with high purity for SILAC cell culture.

3. Regular FBS contains amino acids that may reduce the labeling efficiency. Always use high-quality dialyzed FBS from reputable vendors. If the labeling efficiency is unsatisfactory, test dialyzed FBS from other vendors.

4. Penicillin-streptomycin can be added to the culture media if they are compatible with the cell growth. It is important to prevent bacterial or fungal contamination due to the cost of SILAC experiments.

5. To check the labeling efficiency manually, search the data with Maxquant software specifying SILAC quantification. In the

evidence.txt, sort the data with modification state column. The modification state of 1 in doublet SILAC experiment indicates the identification of heavy isotope labeled peptides. Sort all the heavy isotope labeled peptides based on Intensity column and select 10–20 most intense peptides. Manually check the intensities of the heavy ions and the light ions. In Xcalibur software, be sure to find the scan with the strongest precursor intensities for calculation. If a light peptide ion intensity is 5 % relative to the heavy peptide ion, the labeling efficiency would be 95.2 % $(100/(100+5)\times100$ %). For double labeling experiments, select the proline-containing peptides to check $[U\text{-}^{13}C_6{}^{15}N_2]$ arginine-to-$[U\text{-}^{13}C_5{}^{15}N_1]$ proline conversion.

6. Adjust the buffer pH in water. For HPLC buffer B, mix with organic solvents after pH adjustment.

7. HPLC gradient and flow rate should vary based on the complexity of the protein lysates and selected column configuration in order to achieve satisfactory performance.

8. Wash the pumps and HPLC system with different percentages of acetonitrile sufficiently after each analysis to remove residual peptide sample and other hydrophobic contaminants.

9. It is important to remove all IDs indicated as identifications from decoy or contaminant proteins ("+" in the "Reverse" column and "+" signs in the "Contaminant" column). Since succinylated Lys cannot be recognized by trypsin, the C-terminal Lys succinylation identifications must be false positive unless peptides are located on the protein C-terminal and should be filtered.

10. Other strategies for peptide fractionation include strong-cation-exchange (SCX) and isoelectric focusing (IEF) when high pH reverse phase fractionation is not feasible. Other quantification strategies include label-free quantification, reductive isotopic dimethylation labeling, or isotope-tagging quantification like iTRAQ and TMT labeling methods when SILAC labeling is not feasible.

11. Calculation of absolute PTM site stoichiometries is technically challenging. The previous study achieves this goal using an elegant mathematical model. The method is widely applicable to quantitative proteomics studies of diverse types of PTMs, but it is limited by two basic assumptions of the mathematical model. First, the method requires the accurate quantification of both modified peptide and its unmodified counterpart. Quantification of the unmodified peptide is particularly challenging due to large protein dynamic range in the whole cell lysate. Therefore, extensive fractionation or long-HPLC gradient analysis is typically needed to allow the identification and quantification of the corresponding unmodified peptide.

Second, the method assumes that each peptide only exists in two sequence forms—completely modified and completely unmodified. Therefore, the method cannot be applied to calculate the stoichiometries of multiply-modified peptides if the stoichiometries of each site are different.

Acknowledgements

The author would like to thank Drs. Yingming Zhao, Timothy Griffin, and Minjia Tan for their critical reading of the manuscript. The work is supported by Minnesota Medical Foundation Grant and Startup funding from the University of Minnesota at Twin Cities.

References

1. Walsh CT, Garneau-Tsodikova S, Gatto GJ Jr (2005) Protein posttranslational modifications: the chemistry of proteome diversifications. Angew Chem Int Ed Engl 44(45):7342–7372. doi:10.1002/anie.200501023

2. Chen Y, Sprung R, Tang Y, Ball H, Sangras B, Kim SC, Falck JR, Peng J, Gu W, Zhao Y (2007) Lysine propionylation and butyrylation are novel post-translational modifications in histones. Mol Cell Proteomics 6(5):812–819. doi:10.1074/mcp.M700021-MCP200, M700021-MCP200 [pii]

3. Garrity J, Gardner JG, Hawse W, Wolberger C, Escalante-Semerena JC (2007) N-lysine propionylation controls the activity of propionyl-CoA synthetase. J Biol Chem 282(41):30239–30245. doi:10.1074/jbc.M704409200, M704409200 [pii]

4. Tan M, Luo H, Lee S, Jin F, Yang JS, Montellier E, Buchou T, Cheng Z, Rousseaux S, Rajagopal N, Lu Z, Ye Z, Zhu Q, Wysocka J, Ye Y, Khochbin S, Ren B, Zhao Y (2011) Identification of 67 histone marks and histone lysine crotonylation as a new type of histone modification. Cell 146(6):1016–1028. doi:10.1016/j.cell.2011.08.008, S0092-8674(11)00891-9 [pii]

5. Zhang Z, Tan M, Xie Z, Dai L, Chen Y, Zhao Y (2011) Identification of lysine succinylation as a new post-translational modification. Nat Chem Biol 7(1):58–63. doi:10.1038/nchembio.495, nchembio.495 [pii]

6. Colak G, Xie Z, Zhu AY, Dai L, Lu Z, Zhang Y, Wan X, Chen Y, Cha YH, Lin H, Zhao Y, Tan M (2013) Identification of lysine succinylation substrates and the succinylation regulatory enzyme CobB in Escherichia coli.

Mol Cell Proteomics 12(12):3509–3520. doi:10.1074/mcp.M113.031567, M113.031567 [pii]

7. Peng C, Lu Z, Xie Z, Cheng Z, Chen Y, Tan M, Luo H, Zhang Y, He W, Yang K, Zwaans BM, Tishkoff D, Ho L, Lombard D, He TC, Dai J, Verdin E, Ye Y, Zhao Y (2011) The first identification of lysine malonylation substrates and its regulatory enzyme. Mol Cell Proteomics 10(12):M111.012658. doi:10.1074/mcp.M111.012658, M111.012658 [pii]

8. Tan M, Peng C, Anderson KA, Chhoy P, Xie Z, Dai L, Park J, Chen Y, Huang H, Zhang Y, Ro J, Wagner GR, Green MF, Madsen AS, Schmiesing J, Peterson BS, Xu G, Ilkayeva OR, Muehlbauer MJ, Braulke T, Muhlhausen C, Backos DS, Olsen CA, McGuire PJ, Pletcher SD, Lombard DB, Hirschey MD, Zhao Y (2014) Lysine glutarylation is a protein posttranslational modification regulated by SIRT5. Cell Metab 19(4):605–617. doi:10.1016/j.cmet.2014.03.014, S1550-4131(14)00118-1 [pii]

9. Dai L, Peng C, Montellier E, Lu Z, Chen Y, Ishii H, Debernardi A, Buchou T, Rousseaux S, Jin F, Sabari BR, Deng Z, Allis CD, Ren B, Khochbin S, Zhao Y (2014) Lysine 2-hydroxyisobutyrylation is a widely distributed active histone mark. Nat Chem Biol 10(5):365–370. doi:10.1038/nchembio.1497, nchembio.1497 [pii]

10. Xie Z, Dai J, Dai L, Tan M, Cheng Z, Wu Y, Boeke JD, Zhao Y (2012) Lysine succinylation and lysine malonylation in histones. Mol Cell Proteomics 11(5):100–107. doi:10.1074/mcp.M111.015875, M111.015875 [pii]

11. Hirschey MD, Zhao Y (2015) Metabolic regulation by lysine malonylation, succinylation and glutarylation. Mol Cell Proteomics 14(9):2308–

2315. doi:10.1074/mcp.R114.046664, mcp. R114.046664 [pii]

12. Du J, Zhou Y, Su X, Yu JJ, Khan S, Jiang H, Kim J, Woo J, Kim JH, Choi BH, He B, Chen W, Zhang S, Cerione RA, Auwerx J, Hao Q, Lin H (2011) Sirt5 is a NAD-dependent protein lysine demalonylase and desuccinylase. Science 334(6057):806–809. doi:10.1126/science.1207861

13. Meijer AJ, Lamers WH, Chamuleau RA (1990) Nitrogen metabolism and ornithine cycle function. Physiol Rev 70(3):701–748

14. Yefimenko I, Fresquet V, Marco-Marin C, Rubio V, Cervera J (2005) Understanding carbamoyl phosphate synthetase deficiency: impact of clinical mutations on enzyme functionality. J Mol Biol 349(1):127–141. doi:10.1016/j.jmb.2005.03.078, S0022-2836(05)00380-3 [pii]

15. Polletta L, Vernucci E, Carnevale I, Arcangeli T, Rotili D, Palmerio S, Steegborn C, Nowak T, Schutkowski M, Pellegrini L, Sansone L, Villanova L, Runci A, Pucci B, Morgante E, Fini M, Mai A, Russo MA, Tafani M (2015) SIRT5 regulation of ammonia-induced autophagy and mitophagy. Autophagy 11(2):253–270. doi:10.1080/15548627.2015.1009778

16. Kim SC, Sprung R, Chen Y, Xu Y, Ball H, Pei J, Cheng T, Kho Y, Xiao H, Xiao L, Grishin NV, White M, Yang XJ, Zhao Y (2006) Substrate and functional diversity of lysine acetylation revealed by a proteomics survey.

Mol Cell 23(4):607–618. doi:10.1016/j.molcel.2006.06.026, S1097-2765(06)00454-0 [pii]

17. Zhang K, Chen Y, Zhang Z, Zhao Y (2009) Identification and verification of lysine propionylation and butyrylation in yeast core histones using PTMap software. J Proteome Res 8(2):900–906. doi:10.1021/pr8005155

18. Park J, Chen Y, Tishkoff DX, Peng C, Tan M, Dai L, Xie Z, Zhang Y, Zwaans BM, Skinner ME, Lombard DB, Zhao Y (2013) SIRT5-mediated lysine desuccinylation impacts diverse metabolic pathways. Mol Cell 50(6):919–930. doi:10.1016/j.molcel.2013.06.001, S1097-2765(13)00438-3 [pii]

19. Olsen JV, Vermeulen M, Santamaria A, Kumar C, Miller ML, Jensen LJ, Gnad F, Cox J, Jensen TS, Nigg EA, Brunak S, Mann M (2010) Quantitative phosphoproteomics reveals widespread full phosphorylation site occupancy during mitosis. Sci Signal 3(104):ra3. doi:10.1126/scisignal.2000475

20. Yang F, Shen Y, Camp DG 2nd, Smith RD (2012) High-pH reversed-phase chromatography with fraction concatenation for 2D proteomic analysis. Expert Rev Proteomics 9(2):129–134. doi:10.1586/epr.12.15

21. Cox J, Mann M (2008) MaxQuant enables high peptide identification rates, individualized p.p.b.-range mass accuracies and proteome-wide protein quantification. Nat Biotechnol 26(12):1367–1372. doi:10.1038/nbt.1511, nbt.1511 [pii]

Chapter 3

Determining the Composition and Stability of Protein Complexes Using an Integrated Label-Free and Stable Isotope Labeling Strategy

Todd M. Greco, Amanda J. Guise, and Ileana M. Cristea

Abstract

In biological systems, proteins catalyze the fundamental reactions that underlie all cellular functions, including metabolic processes and cell survival and death pathways. These biochemical reactions are rarely accomplished alone. Rather, they involve a concerted effect from many proteins that may operate in a directed signaling pathway and/or may physically associate in a complex to achieve a specific enzymatic activity. Therefore, defining the composition and regulation of protein complexes is critical for understanding cellular functions. In this chapter, we describe an approach that uses quantitative mass spectrometry (MS) to assess the specificity and the relative stability of protein interactions. Isolation of protein complexes from mammalian cells is performed by rapid immunoaffinity purification, and followed by in-solution digestion and high-resolution mass spectrometry analysis. We employ complementary quantitative MS workflows to assess the specificity of protein interactions using label-free MS and statistical analysis, and the relative stability of the interactions using a metabolic labeling technique. For each candidate protein interaction, scores from the two workflows can be correlated to minimize nonspecific background and profile protein complex composition and relative stability.

Key words Affinity isolation, Immunoprecipitation, Protein complexes, Protein interactions, Stable isotope labeling quantification, Label-free quantification, SAINT, I-DIRT

1 Introduction

Significant progress in the functional understanding of protein complexes has been made, due in large part to improvements in rapid biochemical affinity isolations using high-affinity epitope tags [1–5], such as FLAG, GFP and biotin, and to the increased sensitivity of detection by mass spectrometry (MS). Detection of low-abundance and/or transient interactions, along with posttranslational modifications within complexes, is now feasible [6–12]. Yet, increased sensitivity of detection also provides a greater number of identified proteins that co-isolate nonspecifically during the affinity capture of protein complexes. Therefore, it is

Salvatore Sechi (ed.), *Quantitative Proteomics by Mass Spectrometry*, Methods in Molecular Biology, vol. 1410,
DOI 10.1007/978-1-4939-3524-6_3, © Springer Science+Business Media New York 2016

critical that experimental designs employ appropriate negative controls. Such controls usually include isolation of the epitope tag alone from cell/tissue lysates and/or quantitative MS strategies to measure the specificity of interactions, i.e. enrichment of isolated proteins relative to contaminant datasets or to the background proteome [13, 14]. Since protein complexes are not static structures, MS-based proteomics has also been useful for studying the dynamics of protein complex regulation, which involve time-dependent changes in complex composition and subcellular localization in response to external and internal stimuli [8, 15–17].

A key technical advancement in the study of protein complex composition and dynamics was the development of quantitative MS for studying biological systems. Quantitative MS provides two technical approaches that are useful for studying proteins complexes. In one approach, quantification is achieved through the use of stable isotopes. The most prevalent strategies use ^{13}C- and/or ^{15}N-containing reagents to label whole-cell proteomes prior to sample processing [18], for example, using stable isotope labeling by amino acids in cell culture (SILAC) [19]. Alternatively, isotope-coded labeling reagents can be used, which are integrated after isolation of the protein complexes [20]. In a second approach, quantification is achieved using tandem MS fragmentation events (spectral counts) or high-resolution MS^1 signals of intact peptide ions (peak area). Spectral counting- and peak area-based approaches are collectively termed "label-free" approaches [21, 22]. Both isotope labeling and label-free approaches can measure differences in relative protein abundance between different samples. These quantitative comparisons can inform on the changes in protein interactions under different biological conditions, such as different cell cycle stages, disease stages, or environmental stimuli. They also are invaluable for defining the composition of a protein complex by assessing the likely specificity of protein interactions. In this case, the comparison is performed between one sample representing the affinity isolated protein complexes and one representing a nonspecific control. Currently, label-free approaches are the most widely employed for determining interaction specificity as these methods (1) can be integrated into existing qualitative proteomic workflows with minimal modification, (2) have a broad range of applications, not being restricted to certain biological model systems, (3) are cost-effective when larger sample amounts are required, and (4) are well-suited for detection of large abundance differences, which are expected for the majority of specific interactions. Moreover, several computational tools for measuring and evaluating interaction specificities have been optimized for label-free metrics and are applicable to affinity isolation studies ranging from small-scale (a few target proteins) to large-scale (hundreds of targets) studies [21, 23, 24]. One such algorithm is SAINT (*Significance Analysis of INTeractome*) [25], which we have employed in our studies [8, 26] and is described in this chapter. Following immunoaffinity purification and proteomic analysis, SAINT employs a probabilistic approach using mixture models to analyze the spectral count or peak area distribution of a background (negative control) dataset compared to the experimental samples (Fig. 1a). For each identified protein, SAINT assigns a score representing its likelihood of being a specific interaction.

Whole-cell metabolic stable isotope labeling paired with affinity purification-MS has several key advantages relative to label-free

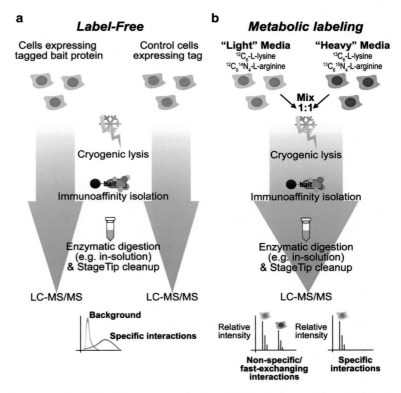

Fig. 1 Integrated label-free and metabolic label-based approach for profiling protein interactions. (**a**) *Label-free workflow*. Cells expressing the tagged protein of interest (bait protein) are cultured in parallel to cells expressing the tag alone. Following cryogenic lysis, immunoaffinity isolation of the tagged protein (with its interactions) is performed using antibody-conjugated magnetic beads. Captured proteins are subjected to enzymatic digestion, sample clean-up and mass spectrometry analysis. Mass spectrometry signals (e.g., spectrum counts or peak intensities) from the bait protein isolation versus control isolations are analyzed to set cut-offs for high-confidence specific interactions. (**b**) *Metabolic labeling workflow*. Cells expressing tagged bait protein are cultured in media containing "light" amino acids, while control wild type cells (not expressing a tag) are cultured in "heavy" amino acid media. Cells are mixed in a 1:1 ratio and subjected to cryogenic lysis, immunoaffinity purification, and proteomic analysis as in (**a**), except the MS data analysis calculates the relative ion intensities of the "heavy" and "light" peptide signals to determine the specificity of interactions

approaches. First, the quantitative precision and accuracy of labeled approaches are often superior, allowing for detection of lower abundance interactions. Second, metabolic labeling can control for variability introduced during protein extraction and affinity isolation since the differentially labeled (light and heavy) samples are mixed prior to subsequent sample processing. This feature of metabolic labeling was leveraged in the I-DIRT (*I*sotopic *D*ifferentiation of *I*nteractions as *R*andom or *T*argeted) technique [27], which has been used to determine interaction specificity for affinity isolated epitope-tagged protein complexes in diverse biological contexts [7, 28–32]. In this approach, an unlabeled (light) whole-cell proteome expressing epitope-tagged protein complexes is mixed with an isotope-labeled (heavy) whole-cell proteome devoid of the epitope tag (wild-type control). After mixing, cell lysis and affinity purification and enzymatic digestion of epitope-tagged protein complexes are performed followed by nLC-MS/MS analysis (Fig. 1b). I-DIRT can also be used for analysis of nontagged protein complexes, but requires that the isotope-labeling be performed in a background control cell line with targeted knockdown of the protein of interest. By using I-DIRT affinity isolation, identified proteins that are nonspecific would derive equally from the light and heavy samples, while specific interactions would be detected only with light isotope signals. Overall, label-free and isotope-labeled approaches each have specific advantages in characterizing protein complexes.

In this chapter, we describe an approach that integrates label-free analysis using the SAINT algorithm and metabolic labeling using I-DIRT to provide increased confidence in the specificity of interactions as well as to provide a profile of the relative stability of interactions within isolated protein complexes. Although the I-DIRT approach provides a powerful tool for defining a core set of stable interactions, one caveat is that *bona fide* interactions that exchange on-and-off the complex during cell lysis and affinity isolation are excluded as nonspecific associations. In contrast, label-free affinity isolation approaches do not preclude fast-exchanging proteins from being detected as specific interactions. Therefore, when performed in parallel, these approaches can identify candidate interactions that are specific but may be less stable. Together, with functional studies or with prior knowledge about the function of the complex of interest, this complementary method can inform on the potential impact that an interaction's relative stability has on its functional roles within the complex. Here, we illustrate this for the case of chromatin remodeling complexes containing human histone deacetylases in T cells, as we have reported in [8]. However, this integrated label-free and metabolic labeling approach is broadly applicable to studies of diverse protein complexes in a variety of cell types.

2 Materials and Equipment

2.1 Metabolic Labeling of CEM T Cells for I-DIRT Analysis

1. Custom "Heavy" isotope culture medium: l-arginine/l-lysine deficient RPMI-1640 media (Life Technologies) supplemented 10 % with *dialyzed* fetal bovine serum (Gibco, Life Technologies), 100 mg/L $^{13}C_6$-l-lysine (Cambridge Isotopes), 100 mg/L $^{13}C_6^{15}N_4$-l-arginine (Cambridge Isotopes), and 1 % penicillin-streptomycin (Life Technologies).

2. Custom "Light" isotope culture medium: l-arginine/l-lysine deficient RPMI-1640 media (Life Technologies) supplemented 10 % with *dialyzed* fetal bovine serum (Life Technologies), 80 mg/L $^{12}C_6$-l-lysine (Sigma), 80 mg/L $^{12}C_6^{14}N_4$-l-arginine (Sigma), and 1 % penicillin-streptomycin (Life Technologies).

3. Cell line: Human peripheral blood derived T lymphoblasts (CCRF-CEM, ATCC).

4. T75 flasks.

5. T300 flasks.

6. 50 mL conical tubes.

7. Swinging bucket rotor (prechilled).

8. Dulbecco's Phosphate Buffered Saline (D-PBS) (ice cold).

9. Protease inhibitor cocktail, 100× (Sigma).

10. Cell freezing buffer: 10 mM HEPES-NaOH, pH 7.4, containing 1.2 % polyvinylpyrrolidine. Supplement with protease inhibitor cocktail to 10× immediately before use.

11. Liquid nitrogen.

12. Styrofoam container with 50 mL conical tube rack insert.

2.2 CEM T Cell Culture for Label-Free Proteomic Analysis

1. Same reagents as above, *except* cells are passaged in the standard culture medium: RPMI-1640 media (Life Technologies) supplemented with 10 % fetal bovine serum (Life Technologies) and 1 % penicillin-streptomycin (Life Technologies).

2.3 Cell Lysis

1. Retsch MM 301 Mixer Mill with 2 × 10 mL jars and 2 × 20 mm (tungsten carbide or stainless steel) grinding balls (Retsch, Newtown, PA).

2. Liquid nitrogen.

3. Foam ice bucket.

4. Long forceps.

5. Windex.

6. Methanol.

7. 10 % bleach solution

8. Ultrapure water.

9. Spatula (chilled by liquid nitrogen).

10. Dry ice.

11. 50 mL conical tubes.

2.4 Affinity Isolation of Protein Complexes

2.4.1 Conjugation of Magnetic Beads

1. Dynabeads M-270 Epoxy (Invitrogen). Store at 4 °C.

2. Affinity purified antibodies against an epitope tag or protein of interest (e.g., anti-GFP antibodies described below for the isolation of GFP-tagged proteins) or Immunoglobulin G (for isolation of Protein A-tagged proteins). Store at −80 °C.

3. 0.1 M Sodium Phosphate buffer, pH 7.4 (4 °C, filter sterilized). Prepare as 19 mM NaH_2PO_4, 81 mM Na_2HPO_4. Adjust pH to 7.4, if necessary.

4. 3 M Ammonium Sulfate (filter sterilized). Prepare in 0.1 M Sodium Phosphate buffer, pH 7.4.

5. 100 mM Glycine–HCl, pH 2.5 (4 °C, filter sterilized). Prepare in water and adjust to pH 2.5 with HCl.

6. 10 mM Tris, pH 8.8 (4 °C, filter sterilized). Prepare in water and adjust to pH 8.8 with HCl.

7. 100 mM Triethylamine: Prepare fresh in water. *CAUTION*: Triethylamine is toxic and extremely flammable, and must be handled in a chemical hood and disposed appropriately.

8. DPBS, pH 7.4 (Dulbecco's Phosphate-Buffered Saline (1×), liquid) (Invitrogen).

9. DPBS containing 0.5 % Triton X-100. Prepare fresh in DPBS.

10. DPBS containing 0.02 % sodium azide. Prepare fresh in DPBS. *CAUTION*: Sodium azide is a toxic solid compound and must be handled in a chemical hood and disposed appropriately.

11. Rotator (at 30 °C).

12. Magnetic separation tube rack (Invitrogen).

13. Tube shaker e.g., TOMY micro tube mixer.

14. Safe-Lock tubes, 2 mL round bottom (Eppendorf).

15. Ultrapure water (e.g., from a Milli-Q Integral Water Purification System).

2.4.2 Immunoaffinity Isolation

1. Frozen cell powder (*see* Subheading 3.3.1). Store at −80 °C.

2. Optimized lysis buffer (*see* Subheading 3.3.2) prepared fresh prior to each experiment. Store on ice.

3. Magnetic beads conjugated with antibodies (*see* Subheading 3.4.1). Store at 4 °C.

4. 50 mL conical tubes.

5. Polytron for tissue homogenization (e.g., PT 10–35 Polytron from Kinematica).

6. Centrifuge and rotor, compatible with 50 mL conical tubes and capable of $8000 \times g$ at 4 °C.

7. Tube rotator at 4 °C.

8. Ultrapure H_2O.

9. Safe-Lock tubes, 2 mL round bottom (Eppendorf).

10. Safe-Lock tubes, 1.5 mL (Eppendorf).

11. Bar magnets (for conical tubes) and magnetic separation rack (for micro tubes) (Invitrogen).

12. 4× LDS elution buffer: Dissolve 0.666 g of Tris–HCl, 0.682 g of Tris–Base, 0.8 g of LDS, and 0.006 g of EDTA (free acid) in ultrapure H_2O to a final volume of 10 mL. Aliquot and store at –20 °C.

13. 10× reducing agent: 0.5 M TCEP, pH neutral (Pierce).

14. 10× alkylating agent: 0.5 M chloroacetamide in water. Aliquot and store at –20 °C.

15. Heat block at 70 °C.

2.5 In-Solution Digestion of Immunoisolated Proteins

Store stock solutions in glass containers that have been thoroughly rinsed with ultrapure water. Avoid using glassware that has been washed with detergents.

1. Primary eluate from label-free affinity isolation (*see* Subheading 3.4.2).

2. Primary eluate from I-DIRT affinity isolation (*see* Subheading 3.4.2).

3. Refrigerated microcentrifuge capable of $14,000 \times g$ (maintain at 20 °C).

4. LoBind pipet tips, 200 µL (Eppendorf).

5. Amicon Ultra-0.5 centrifugal filters, 30 kDa NMWL (Millipore).

6. MS grade water (Fisher).

7. 10 % sodium deoxycholate (DOC): Prepare in MS grade water and protect from light.

8. Tris–HCl buffer: 0.2 M Tris–HCl, pH 8.0, in MS grade water. Store at 4 °C.

9. TUD wash buffer: Prepare buffer fresh before use by mixing 1.5 mL of 0.2 M Tris–HCl, 0.6 mL of 10 % DOC, and 1.44 g urea. Yields ~3 mL of buffer, sufficient for two samples.

10. ABC-DOC wash buffer: 0.05 M ABC, 2 % DOC.

11. Trypsin, lyophilized MS-grade (Pierce). After suspension, store at –80 °C.

12. Digestion buffer: Prepare 100 µL per sample by mixing 1 µL of 0.5 µg/µL trypsin stock and 99 µL of 0.05 M ABC. Prepare fresh immediately before use.

13. 10 % trifluoroacetic acid (TFA). Prepare in MS grade water and store at 4 °C.

14. Ethyl acetate. *CAUTION*: Ethyl acetate is flammable and toxic. Handle in a chemical hood and dispose appropriately.

15. SpeedVac Concentrator.

2.6 Peptide Clean-Up and Fractionation Using SDB-RPS StageTips

1. Microcentrifuge.

2. LoBind pipet tips, 200 μL (Eppendorf).

3. 14 gauge needle (Hamilton #90514).

4. Syringe plunger, 100 μL (Hamilton #1162-02).

5. Empore SDB-RPS disks (3 M #2241).

6. 50 % ethyl acetate/0.5 % TFA in MS grade water.

7. 0.5 % TFA in MS grade water.

8. Buffer 1: 0.10 M ammonium formate, 0.5 % formic acid, 40 % acetonitrile in water.

9. Buffer 2: 0.15 M ammonium formate, 0.5 % formic acid, 60 % acetonitrile in water.

10. Buffer 3: 5 % ammonium hydroxide and 80 % acetonitrile in water.

11. FA solution: 1 % formic acid and 4 % acetonitrile in water.

12. Autosampler vials.

2.7 Nanoliquid Chromatography Tandem Mass Spectrometry Analysis

1. Nanoflow HPLC system, e.g., Dionex Ultimate, Waters nano Acuity, or Agilent 1200 series.

2. Mobile phase A (MPA): 0.1 % FA/99.9 % water. Store in amber bottle for up to 6 months.

3. Mobile phase B (MPB): 0.1 % FA/97 % ACN/2.9 % water. Store in amber bottle for up to 6 months.

4. Analytical column, e.g., Acclaim PepMap RSLC 75 μm ID × 25 cm (Dionex).

5. LTQ-Orbitrap Velos hybrid mass spectrometer (Thermo Fisher Scientific).

6. Nanospray ESI source (Thermo Fisher Scientific).

7. SilicaTip Emitter, Tubing (OD × ID) 360 μm × 20 μm; Tip (ID) 10 μm (New Objective).

2.8 Data Analysis

1. Multi-core/multi-CPU 64-bit PC workstation with at least 12 GB of RAM and 2 TB of storage.

2. Software for generating peaklists and scoring PSMs, with support for precursor ion quantification e.g., Proteome Discoverer 1.4 (Thermo Fisher Scientific), Mascot 2.3 (Matrix Science), Scaffold 4.0 (Proteome Software).

3. SAINT (http://www.crapome.org/).

4. Spreadsheet software (e.g., Microsoft Excel).

3 Methods

This protocol involves two different quantitative MS workflows (*see* Fig. 1). The most significant differences are in the steps for cell culture; so, the respective steps for label-free and isotope-labeling workflows are described separately (Subheadings 3.1 and 3.2). The cell lysis, immunoaffinity isolation, and in-solution digestion are performed identically, independent of the quantitative workflow.

3.1 Metabolic Labeling of CEM T Cells for I-DIRT Analysis

1. Aliquot 1.2×10^7 wild-type CEM T cells into conical tube and pellet at $200 \times g$ (*see* **Note 1**).

2. Resuspend the pellet in 10 mL of heavy isotope media and aliquot equally into $2 \times$ T75 flasks.

3. Add 25 mL of heavy media to each flask and culture in incubator ($37 °C/5 \% CO_2$) until cell concentration is $2 \times 10^6/mL$ (~4 days).

4. Transfer each cell suspension (~30 mL) into a T300 flask and add 150 mL of heavy isotope media. Culture in incubator as above.

5. Divide total cell suspension (360 mL) equally into 50 mL conical tubes (8×45 mL).

6. Pellet cells at $200 \times g$ at 4 °C.

7. Aspirate media and resuspend cell pellets in ice-cold D-PBS (10 mL). Pool cell suspensions in 2×50 mL conical tubes.

8. Pellet cells as above and aspirate media.

9. Wash cell pellets with D-PBS (20 mL), pool into a single pre-weighed 50 mL conical tube.

10. Pellet cells as above and aspirate media.

11. Repeat **steps 9** and **10**.

12. Weigh conical tube to determine wet cell pellet weight.

13. Keep cells on ice while preparing liquid nitrogen freezing bath.

14. Place a fresh 50 mL conical tube in Styrofoam container/rack. Fill conical tube with liquid nitrogen about halfway and leave uncovered. Fill bottom of rack with liquid nitrogen to slow evaporation in tube.

15. Add 100 μL of cell freezing buffer per gram of cells and pipet drop wise into the conical tube containing liquid nitrogen (*see* **Note 2**).

16. CRITICAL: Use a needle to create holes in the conical tube cap before re-capping the tube containing the frozen cell material.

Secure the cap and gently agitate in a fume hood to allow liquid nitrogen evaporation. Heavy-labeled frozen cell pellets can be stored at –80 °C for up to several years.

17. Cell material for the complementary "light" I-DIRT sample (*see* Fig. 1) is generated as above, *except* the CEM T cells that stably express the affinity-tagged protein of interest are cultured in the custom "light" isotope medium (*see* **Note 3**).

3.2 CEM T Cell Culture for Label-Free Proteomic Analysis

1. Cell material for the label-free affinity isolation experiment is generated as described above in Subheading 3.1, *except* two separate CEM T cell lines, both grown in standard culture media, are required:

    (a) One cell line should stably express the affinity-tagged protein of interest.

    (b) The other cell line is the control, which stably expresses the affinity tag alone or an empty vector.

3.3 Cell Lysis

The procedures described below for immunoaffinity purification of protein complexes utilize mammalian cells as the starting material, which is cryogenically disrupted using a Mixer Mill. However, cell lysis can also be carried out using several alternative approaches, including direct homogenization in a detergent-containing lysis buffer or passage of the lysate through a needle. We prefer the method of cryogenic disruption described below, as we have observed that it leads to an increased efficiency of extraction and decreased level of nonspecific associations. This method has provided us with a reliable and effective means of cell lysis for isolating varied protein complexes [2, 5, 6, 8, 12, 15, 17, 26, 33–39] and has been described in detail elsewhere [40].

3.3.1 Cryogenic Cell Disruption

1. Clean one spatula, the Retsch Mixer Mill jars, and the grinding balls sequentially with Windex, ultrapure H_2O, 10 % bleach solution, ultrapure dH_2O, and 100 % methanol. Allow all parts to dry completely.

2. Cool the jars and balls in liquid nitrogen (e.g., using a foam ice bucket filled with liquid nitrogen). Once the liquid nitrogen no longer appears to be bubbling, the jars are sufficiently cool and can be removed from the liquid nitrogen using a pair of long forceps.

3. Quickly place the frozen cell pellets into the jar. For the label-free affinity isolation workflow, use cell pellets collected from a standard CEM T cell culture expressing the desired affinity-tagged protein (*see* Subheading 3.2). For the I-DIRT affinity isolation workflow, combine equal amounts of cell pellets collected from "light" and "heavy" CEM T cell cultures (*see* Subheading 3.1). Cell pellets can fill up to a maximum of one-third of the total volume of the jar for optimal cryogenic

grinding (e.g., ~2.5 g frozen tissue pellets per 10 mL jar). Place a single chilled ball on top of the cell pellets, close the jar, and cool in the liquid nitrogen container.

4. Place the jars in the Retsch Mixer Mill holders. If only processing one sample, use an empty jar (without a ball) as a balance. Cryogenically lyse cells using ten cycles of 2 min 30 s each at a frequency of 30 Hz. In between each cycle, re-cool jars in liquid nitrogen and check that the jars remain securely closed.

5. Open the jar and use a chilled spatula to transfer the frozen cell powder to a 50 mL conical tube chilled on dry ice. Proceed rapidly to avoid thawing of the ground sample. Store the powder at –80 °C until the affinity isolation is to be performed.

3.3.2 Optimization of Lysis Buffer and Isolation Conditions

During cell lysis and protein isolation, the efficient extraction of the targeted protein in a soluble fraction, while maintaining its interactions, is the primary goal. As a result, the lysis buffer conditions should be optimized for each protein of interest before performing larger scale immunoaffinity isolations for proteomics studies. It is therefore recommended that small-scale experiments be performed to assess the efficiency of protein solubilization and efficiency of isolation by western blotting using at least three lysis buffer conditions with varied levels of stringency, as described in detail previously [13, 40] .

3.4 Affinity Isolation of Protein Complexes

3.4.1 Conjugation of Magnetic Beads

This protocol has been optimized for the conjugation of M-270 Epoxy Dynabeads, but can also be applied for conjugation of additional types of magnetic beads with larger or smaller diameters (e.g., M-450 or MyOne Dynabeads). In such cases, the amount of antibody used for conjugation should be adjusted based on the binding capacity of the bead. This protocol can be used for conjugating beads with either high-affinity purified antibodies or commercially available antibodies. It is important to note that the storage of antibodies in buffers containing free amines (e.g., Tris) will limit the amount of antibody that will be covalently conjugated to the surface epoxy groups; so it is best to avoid such buffers.

It is optimal to begin this protocol in the afternoon and perform all washing steps (**step 9**) in the morning of the following day. All steps should be performed at room temperature, unless otherwise indicated. During the washing steps, the beads must not be allowed to dry out (i.e., proceed immediately from one wash step to the next and do not allow the beads to sit without a washing solution between individual steps).

1. Weigh out the necessary amount of magnetic Dynabeads in a round-bottom tube (*see* **Note 4**).

2. Add 1 mL Sodium Phosphate buffer (pH 7.4) over the top of the beads. Mix by vortexing for 30 s, followed by 15 min on a tube shaker (vigorous setting).

3. Place the tube on a magnetic rack or against a magnet. Remove and discard the buffer once the beads have settled towards the magnetic side.

4. Remove the tube from the rack. Add 1 mL Sodium Phosphate buffer (pH 7.4) and mix by vortexing for 30 s and remove the buffer as above.

5. Remove the tube from the rack. In the following order, add the necessary amount of antibodies, Sodium Phosphate buffer (pH 7.4), and Ammonium Sulfate solution (3 M).

   (a) The optimal total volume of the bead conjugation solution, which includes the antibodies, Sodium Phosphate buffer, and Ammonium Sulfate solution, is ~20 μL/mg beads.

   (b) The amount of antibody conjugated is 3–5 μg Ab/mg M-270 epoxy beads. If another type (or size) of bead is used, the amount of Ab should be optimized, as the binding capacity may be different.

   (c) The 3 M Ammonium Sulfate solution is added last to give a final concentration of 1 M (i.e., added at one-third of total final volume).

   (d) For example, a total volume of 360 μL is used to conjugate 18 mg beads. For an antibody:bead ratio of 3:1000, first add 54 μg antibody to the beads. Second, add 0.1 M Sodium Phosphate Buffer (such that the volume of 0.1 M Sodium Phosphate Buffer is equal to 360 μL minus the volumes of antibody and 3 M Ammonium Sulfate used). Finally, add 120 μL of 3 M Ammonium Sulfate.

6. Secure the tube with parafilm and incubate the bead slurry overnight on a rotator at 30 °C.

7. The next morning, place the tube with bead slurry against the magnetic rack.

8. OPTIONAL: Retain the supernatant to assess the efficiency of bead conjugation by SDS-PAGE.

9. Wash the beads sequentially with the following buffers:

   (a) 1 mL of Sodium Phosphate buffer.

   (b) 1 mL 100 mM Glycine–HCl, pH 2.5 (FAST).

   (c) 1 mL 10 mM Tris–HCl pH 8.8.

   (d) 1 mL 100 mM Triethylamine solution (FAST).

   (e) 4 × 1 mL DPBS.

   (f) 1 mL DPBS containing 0.5 % Triton X-100. Mix on a Tomy shaker with gentle agitation for 15 min.

   (g) 1 mL DPBS.

10. Resuspend washed beads in 12.5 μL DPBS containing 0.02 % NaN$_3$ per mg of beads. Measure the final volume of the bead slurry to determine the bead concentration (mg of beads/μL DPBS).

11. Beads can be used immediately or stored for up to 2 weeks at 4 °C. After 1 month of storage, their efficiency for isolation decreases by approximately 40 %.

3.4.2 Immunoaffinity Isolation

1. Prepare an appropriate volume of optimized lysis buffer as determined in Subheading 3.3.2. Pre-cool the buffer to 4 °C. Add protease inhibitors immediately prior to use. Prepare 10 mL of wash buffer per sample (used in **steps 8** and **14–17**), which is typically identical in composition to the optimized lysis buffer but lacks protease and phosphatase inhibitor cocktails.

2. Incubate the frozen cell/tissue powder on ice for 1 min. Proceed immediately to **step 3**.

3. Resuspend the frozen cell/tissue powder in the lysis buffer by first adding a small amount of lysis buffer and swirling the homogenate to solubilize pellet. Continue to add lysis buffer and gently mix by swirling or inversion until the powder is completely solubilized (*see* **Note 5**).

4. Run the Polytron homogenizer for 10 s in ultrapure dH$_2$O to wash.

5. To avoid the spilling of the sample, ensure that the homogenate occupies a maximum of a 1/3 of the conical tube volume. Subject lysates to Polytron homogenization for 2 × 15 s (speed = 22.5 k), briefly incubating the sample on ice between homogenizations.

6. If processing additional samples, rinse the homogenizer with ultrapure dH$_2$O and run the Polytron in ultrapure dH$_2$O to wash out any excess lysate residue. When finished with homogenization steps, perform a final methanol rinse.

7. Centrifuge the lysate at 8000 × g at 4 °C for 10 min.

8. While the lysates are centrifuging, place the tube containing antibody-conjugated magnetic beads against a magnetic rack for 30–60 s. Discard the storage buffer and wash with 3 × 1 mL wash buffer by gently pipetting up and down to resuspend the beads. *Do not vortex the beads.* Resuspend the beads in 100–200 μL of wash buffer.

9. Carefully pour the clarified lysates (supernatant) into new 50 mL conical tubes (*see* **Note 6**). RETAIN (1) the insoluble cell/tissue pellet and (2) 40 μL of the supernatant to serve as the input fraction for further analysis.

10. Mix the beads in solution by gently flicking the tube of antibody-conjugated beads. Pipette the appropriate amount of beads into the tube containing the clarified lysates.

11. Rotate the lysate–bead solution on a rotator at 4 °C for 1 h (*see* **Note 7**).

12. During the incubation step, prepare 1× LDS elution buffer.

13. Use a rubber band to attach a bar magnet to the tube holding the lysate-bead suspension. Incubate on ice for 5 min. RETAIN the flow-through (unbound) fraction by pouring the supernatant into a clean conical tube for further analysis.

14. Resuspend the beads in 1 mL of wash buffer and transfer the bead slurry to a clean round-bottom tube.

15. Place the tube against the magnetic rack to separate the beads from the buffer. Discard wash buffer. *Perform this step between all subsequent wash steps.*

16. Wash the beads 3 × 1 mL wash buffer. On the third wash, transfer the bead slurry to a clean round-bottom tube.

17. Wash the beads 2 × 1 mL with wash buffer.

18. Add 1 mL DPBS to beads and transfer slurry to a third clean round-bottom tube.

19. Wash once more with 1 mL of DPBS to remove residual detergent. Completely remove DPBS wash.

20. Add 40 μL of 1× LDS elution buffer to beads.

21. Incubate for 10 min at 70 °C, then 10 min at RT with agitation (*see* **Note 8**).

22. Isolate beads on the magnetic rack and transfer the primary eluate to a microcentrifuge tube. RETAIN the bead fraction.

23. Resuspend the bead fraction in 40 μL sample buffer and repeat step 21, except incubating the beads at 95 °C for 5 min.

24. Add 5 μL of 10× TCEP and 5 μL of 10× CAM to primary and secondary eluates. Heat at 95 °C for 5 min. RETAIN 10 % of primary and secondary eluates for analysis of isolation efficiency.

25. If immediately performing an in-solution digestion, proceed directly to Subheading 3.5 with the remaining 90 % of the *primary* eluate. Otherwise, samples can be stored at ≤−20 °C.

26. To assess the efficiency of immunoisolation, analyze equal percentages of the following fractions by Western blotting.

    (a) Cell pellet (**step 9**).

    (b) Input supernatant (**step 9**).

    (c) Flow-through (**step 13**).

    (d) Primary eluate (**step 22**).

    (e) Secondary eluate (**step 23**).

3.5 In-Solution Digestion of Immunoisolated Proteins

The in-solution digestion protocol described below uses a filter-aided sample preparation method [41] incorporating urea [42] and sodium deoxycholate [43] wash buffers to remove the LDS detergent and limit protein/peptide losses, respectively. Sodium deoxycholate is removed by organic phase extraction post-digestion [44]. Other digestion protocols may be used, but they must be capable of removing the LDS detergent prior to MS analysis.

Day 1

1. Set temperature of microcentrifuge to 20 °C (all subsequent spins are performed at this temperature).

2. Add 400 µL of TUD buffer to each unlabeled and isotope-labeled primary eluate (from Subheading 3.4.2).

3. Transfer each sample to a separate Amicon-0.5 filter and centrifuge at $14,000 \times g$ for 10 min, or until volume is reduced to the minimum (~25 µL).

4. Discard flow-through and add 400 µL of TUD buffer to filter. Centrifuge as above.

5. Discard flow-through and add 300 µL of TUD buffer to filter. Centrifuge as above.

6. Discard flow-through and add 300 µL of TUD buffer to filter. Centrifuge as above.

7. Add 200 µL of ABC-DOC buffer to filter. Centrifuge at $14,000 \times g$ for 10 min. Ensure that the retained volume is at the minimum.

8. Transfer filter units to fresh collection tubes and add 100 µL of Digestion buffer to the filter. Mix on TOMY shaker for 1 min.

9. Wrap the top of each tube in parafilm. Incubate in water bath overnight at 37 °C.

Day 2

1. Centrifuge filters at $14,000 \times g$ for 5 min to recover digested peptides.

2. Add 25 µL of MS-grade water and centrifuge as above.

3. Add 25 µL of MS-grade water and centrifuge as above.

4. Discard filter unit and retain flow-through containing peptides.

5. Add an equal volume of ethyl acetate to the sample.

6. Adjust each sample to 0.5 % TFA.

7. Vortex, then mix on TOMY shaker for 2 min.

8. Centrifuge at $14,000 \times g$ for 5 min.

9. Recover the denser aqueous phase, while avoiding the top organic (ethyl acetate) phase and interphase.

10. Proceed to "Sample Clean-up and Peptide Fractionation using StageTips" (Subheading 3.6) with the recovered aqueous phase.

3.6 Sample Clean-Up and Fractionation Using StageTips

1. For each sample, prepare one StageTip by depositing a single Empore SDB-RPS disk (cut using a 14 gauge needle) into the bottom of a 200 μL pipette tip using a syringe plunger (*see* **Note 9**).

2. Add half of the sample to the StageTip and centrifuge at $2000 \times g$ until all solution has passed through the Empore disk (*see* **Note 10**).

3. Add the remaining sample to the StageTip and repeat centrifugation.

4. Wash disk with 100 μL of 50 % ethyl acetate/0.5 % TFA.

5. Wash disk with 100 μL of 0.5 % TFA.

6. Pass 50 μL of Elution buffer 1 over the disk and collect the eluate in an autosampler vial.

7. Repeat **step 6** using Elution buffer 2 and again using Elution buffer 3. Collect each eluate in a separate autosampler vial.

8. Concentrate samples by vacuum centrifugation to near-dryness.

9. Add FA solution to achieve a final volume of 9 μL. Vortex briefly to mix.

10. Proceed immediately to nLC-MS/MS analysis (Subheading 3.7) or store at −80 °C for future analysis.

3.7 Nanoliquid Chromatography Tandem Mass Spectrometry Analysis

Many HPLC and MS system configurations are suitable for analyzing label-free and isotope-labeled peptides. However, to achieve optimal depth of analysis and quantitative precision, an LC system capable of low flow rates (<0.5 μL/min) and high pressure support (>400 bar) is highly preferable, as these capabilities allow the highest sensitivity of peptide detection using analytical columns with inner diameters ≤75 μm and lengths ≥25 cm. Additionally, a high-resolution and high-mass accuracy MS system with tandem MS fragmentation capability is required.

1. Ensure that the system is properly calibrated according to the manufacturer's specifications.

2. Using MS instrument software, create an appropriate data-dependent acquisition method (*see* **Note 11**). For isotope-labeled samples it is critical that the precursor (MS^1) scan be performed with high resolution (e.g., 60,000 at m/z 400).

3. Using the LC instrument software, create a reverse-phase method for label-free analysis. Program the method to separate peptides over 3 h using a linear gradient of 4–40 % mobile phase B.

4. For the analysis of isotope-labeled samples, the method should separate peptides over 6 h using the same mobile phase gradient parameters listed in **step 3**. The increase in LC run time is intended to compensate for increased spectral complexity due to isotopic labeling.

5. Create a shorter length (e.g., 60 min) gradient method to use for analysis of standard/quality control samples.

6. Perform duplicate (at a minimum) injections of a peptide standard to ensure that the system is performing at an acceptable level prior to injecting experimental samples.

7. For experimental samples, inject 4 μL of each fraction using the appropriate LC-MS/MS method designed above. The injection order should be selected to analyze the label-free samples first, followed by the isotope-labeled samples (*see* **Note 12**).

8. After experimental sample injections are complete, inject the standard peptide mixture to confirm that instrument performance has been maintained throughout the analysis.

3.8   Data Analysis

3.8.1   Peptide Identification and Protein Assignment for Label-Free Datasets

1. Extract all MS/MS spectra from raw mass spectrometry data, removing MS/MS spectra that do not contain at least ten peaks.

2. Generate instrument and experiment-specific database search parameters.

   (a) Define static peptide modification for cysteine carbamidomethylation.

   (b) Define variable modification for methionine oxidation (*see* **Note 13**).

3. Submit spectra to an appropriate workflow to obtain pep-tide spectrum matches and protein group assignments (*see* **Note 14**).

4. Select peptide and protein scoring filters to achieve a desired false discovery rate (e.g., ≤1 %).

5. Export data tables containing, at minimum, protein group descriptions with respective accession numbers and total spectrum counts. This output will be used interaction specificity analysis using the SAINT algorithm.

3.8.2   I-DIRT Isotope Labeling Quantification

1. To analyze I-DIRT datasets, which include both light- and heavy-labeled lysine and arginine containing peptides, create a duplicate analysis workflow from the workflow created above, and modify as follows:

   (a) Define additional variable modifications for heavy $^{13}C_6$ lysine and heavy $^{13}C_6$-$^{15}N_4$ arginine.

   (b) Include additional necessary modules and associated parameters for the extraction of light and heavy peptide signals and their integration over the LC elution peak.

2. Calculate I-DIRT peptide ratios as (Heavy Peptide Signal)/(Light Peptide + Heavy Peptide Signal).

3. Calculate protein group level ratios as the median of peptide ratios.

4. If available in the software, protein ratios can be median normalized to correct for non-equal mixing of light and heavy-labeled cells. This option should only be used if the majority of identified proteins are nonspecific interactions.

5. Export data tables containing, at minimum, protein group descriptions, accession numbers, I-DIRT protein ratios, ratio variances, and number of quantified peptides.

3.8.3   SAINT Interaction Specificity Analysis Using Label-Free Spectral Counts

1. Access the website www.crapome.org and register for a free user account to enable the full SAINT analysis functionality (*see* **Note 15**).

2. Select "Workflow 3: Analyze Your Data".

3. OPTIONAL: If desired (e.g., if control isolations have not been performed), select negative controls from the CRAPOME database (*see* **Note 16**).

4. Using the label-free analysis data tables exported above, generate a compatible SAINT matrix input file, as specified in the workflow **step 2** (Upload Data).

5. Upload SAINT matrix file and proceed to **step 3**, Data Analysis.

6. Under the "Analysis Options", enable "Probability Score", choose the "SAINT" model, and increase the "n-iter" option to 10,000 (*see* **Note 17**). Run Analysis.

7. After the analysis has completed, save and open output file, which reports the individual and average SAINT scores (AvgP) for each identified protein. Scores range from 0 to 1 (least to most specific).

8. Evaluate the performance of SAINT in distinguishing between specific interactions and nonspecific background. If many interactions are already known for a particular protein of interest, the sensitivity and specificity of the analysis can be estimated by constructing ROC plots. If no prior interaction knowledge is available, then construct a histogram for the distribution of SAINT scores (*see* **Note 18** and Data Interpretation section below). Use these analyses to select a SAINT score cut-off that eliminates the majority of nonspecific interactions (false positives), while retaining the highest scoring interactions.

3.8.4   Interpretation of Label-Free and Labeling Results

For each candidate protein interaction, the output of SAINT and I-DIRT provides a score ranging from 0-to-1 and 0.5-to-1, respectively. As illustrated in the graph in Fig. 2, higher SAINT scores represent increased probability of a specific interaction, while

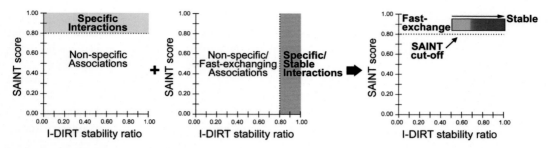

Fig. 2 Integration of SAINT and I-DIRT methods allows the simultaneous investigation of specificity and relative stability of protein interactions. Specificity determination using the SAINT algorithm assigns interaction specificity scores to individual proteins (SAINT score), while I-DIRT metabolic labeling distinguishes between stable/specific interactions and nonspecific/background contaminants by calculation of isotope ratios (I-DIRT stability ratio). Integrating these two methods (*right panel*) provides insight into the relative stability and/or fast-exchanging nature of specific protein interactions of a given bait protein

higher I-DIRT scores represent greater interaction stability and specificity. An important step that requires careful consideration is the selection of appropriate cutoff scores to both maximize true positive interactions and minimize false positives. Also, it should be noted that this workflow does measure exchange rates or binding affinities directly. If more rigorous determination of interaction stability is required, one could measure the isotope exchange in a time series after mixing a light-labeled immunoisolated complex with a heavy-labeled whole-cell lysates. Alternatively, a more targeted analysis of relative protein abundance under different isolation stringencies can be performed by Western blotting.

For SAINT, the selected threshold defines nonspecific versus specific interactions. One of the most effective approaches for selecting SAINT scoring thresholds is to generate a Receiver Operating Characteristic (ROC) curve for each bait protein using previously known interactions (e.g. from the BioGRID repository [45]). This approach allows empirical selection of a SAINT score threshold to balance true versus false positives, as in [8]. However, this is not always feasible, especially in the case of proteins that lack known protein interactions. Alternatively, the distribution of SAINT scores can be examined to select an appropriate threshold [26]. Overall, for the majority of datasets, we have found as a general guide that a SAINT score threshold between 0.8 and 0.95 is appropriate.

For I-DIRT datasets, protein scores represent the fraction of the protein abundance from the tagged bait condition versus the total abundance (tagged bait + wild-type background). Therefore, values closer to 1.0 represent specific interactions that are very stable. Values less than 1.0 and decreasing progressively down to 0.5 reflect increasingly nonspecific interactions. However, a subset of these proteins may exhibit fast-exchange within their respective complexes and represent false negatives. Therefore, an important aspect of interpreting results from I-DIRT experiments is that they alone cannot distinguish nonspecific versus fast-exchanging interactions.

However, since the label-free SAINT workflow is performed separately for control and experimental samples, exchange does not influence SAINT scoring. Therefore, by cross-comparing proteins with low I-DIRT values but high SAINT scores, candidate interactions that are specific but fast-exchanging can be classified with higher confidence. A proof-of-concept example is illustrated for well-established proteins interactions of histone deacetylase 5 (HDAC5), such as 14-3-3 and the nuclear receptor corepressor 1 protein, which have low I-DIRT scores but high SAINT scores (Fig. 3a). These results are consistent with the known regulation and cellular roles of HDAC5, which involve shuttling of HDAC5 between the nucleus and the cytoplasm concomitant with dynamic changes in its interactions (Fig. 3b) [6, 17]. In addition, using the well-studied histone deacetylase 1 protein, we illustrate that, if prior functional knowledge of known and unknown protein interactions is available (Fig. 3d), then the specificity and the stability data can be used together to form hypotheses about the roles of these proteins within particular complexes (Fig. 3c). Finally, if these data are acquired for interactions shared between different affinity enriched proteins, one could formulate hypotheses whether a given protein is similarly or differentially regulated within distinct complexes.

Another advantage of this complementary interaction workflow is the identification of a subclass of candidate interactions with low specificity scores by SAINT but with high I-DIRT values (Fig. 2, lower-right quadrant). When using spectrum counts to assess the specificity of interaction, small proteins or low-abundance interactions may be detected as false negatives (low SAINT scores). Yet, these candidates are "rescued" by the I-DIRT analysis, which can provide reliable quantification with only a few sequenced peptides. Alternatively, this subset of candidates can also be environmental sample contaminants, as they would be present only with a light isotope signal. These environmental contaminants can be excluded by performing an additional I-DIRT experiment in which the isotope labels are swapped.

Overall, this integrative label-free and isotope-labeling approach generates interaction datasets that inform on both the specificity and the relative stability of protein interactions. Additionally, isotope-labeling experiments are now more cost-effective, making integrating both complementary approaches into a single experimental design feasible. This hybrid approach is broadly applicable to investigating the dynamic interactions within protein complexes in different cell types and across different biological conditions. We expect that the future development of improved quantitative mass spectrometry techniques will continue to shed light on the intricacies of protein complex regulation in space and time.

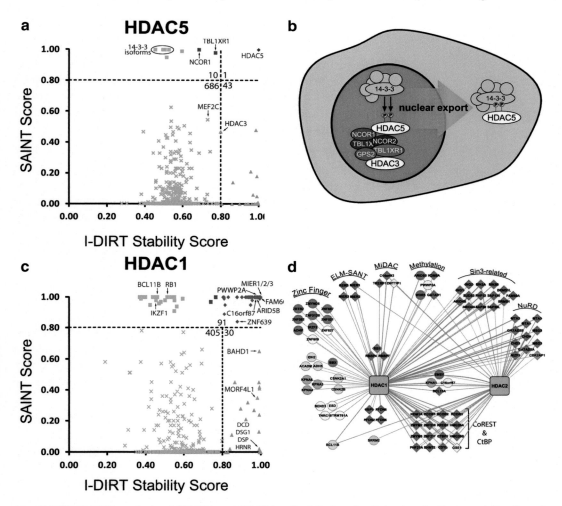

Fig. 3 SAINT/I-DIRT analysis of HDAC5 and HDAC1 reflect their subcellular localizations and functions in transcriptional regulation and chromatin remodeling. (**a**) SAINT/I-DIRT plot for immunoisolated HDAC5 highlights fast-exchanging interactions with 14-3-3 chaperone proteins and components of the nuclear co-repressor complex. (**b**) Transient interactions predicted by SAINT/I-DIRT scoring are consistent with the nucleo-cytoplasmic shuttling of HDAC5. In the nucleus, HDAC5 associates with the nuclear NCoR proteins (*purple*). During nuclear export, HDAC5 can dissociate from the NCoR complex and increase its interaction with 14-3-3 chaperone proteins. (**c**) SAINT/I-DIRT plot for immunoisolated HDAC1 allows classification of known and novel protein associations of HDAC1, highlighting transient association with proteins associated with transcription and stable association with numerous chromatin remodeling complexes. (**d**) HDAC1 associates specifically with chromatin remodeling complexes (e.g. NuRD, Sin3a/b, CoREST), the transcriptional regulatory complex CtBP, and the mitotic deacetylase complex MiDAC. The integrated SAINT/I-DIRT method provides functional insight into the relative stabilities of individual proteins within known complexes reported to associate with HDAC1

4 Notes

1. This protocol describes the characterization of protein complexes from a specific cell type: mammalian CEM T cells, which are grown in suspension. However, this method can be applied in any cellular/tissue model system that achieves sufficient incorporation of metabolic labels.

2. For cell amounts less than 1 g, add 100 μL of cell freezing buffer.

3. To generate the "light" cell material in the I-DIRT workflow, standard culture medium with nondialyzed serum can be used as an alternative method. However, the growth rates and signaling pathways may vary significantly when compared to methods using dialyzed serum, depending on the cell type used.

4. Round-bottom tubes are the preferred tube shape, which minimizes bead trapping during the conjugation. The required amount of beads is dependent on both the experimental objective and the abundance of the protein to be immunoaffinity purified. As an approximate guide, 1–2 mg beads are appropriate for small-scale optimization experiments (as described in Subheading 3.2), 5–7 mg beads are usually sufficient for single immunoaffinity purifications, and 10–20 mg beads may be suitable for proteins of high abundance.

5. After suspending the cell/tissue powder in lysis buffer, the lysate solution may be slightly turbid; however, the solution should be devoid of cell/tissue clumps or aggregates. Do not proceed to Polytron (**step 4**) homogenization until a homogenous suspension is observed. If necessary, additional rotation for 10–20 min at 4 °C can be performed to promote solubilization.

6. If insoluble particles are present in supernatant after centrifugation, a pipette can be used to selectively transfer supernatant to a clean 50 mL conical tube.

7. Longer incubation times tend to promote the accumulation of nonspecific binders and the loss of weak interacting partners [2].

8. We have found that when using high-affinity antibodies (e.g. anti-GFP) stringent heat and detergent denaturing conditions are required for efficient recovery of the target proteins from the beads.

9. Ensure that the disk makes a seal with the walls of the pipette tip and is located a few mm above the tapered end of the tip. Each disk can bind ~25 μg of peptides. If greater capacity is needed, additional disks can be layered in the same StageTip and the number of washes increased to be equal to the number of total disks used.

10. Sample loading and washing of StageTips can be performed manually by applying pressure with a small plastic syringe or by centrifugation of the StageTip in a collection tube with an adapter.

11. Many considerations are required when designing an LC-MS/MS method, many of which are instrument-specific. However, in general the MS acquisition cycle should be designed based on the performance characteristics of the LC system. It is critical that the MS cycle time, determined largely by the number of full and tandem MS scans, permits acquisition of multiple full scans over the average LC elution peak. For example, given LC peak widths of 15–30 s, an optimal time for a single acquisition cycle would be in the range of 2–3 s for data-dependent methods.

12. It is recommended to perform several "blank" or "wash" runs between unlabeled and isotope-labeled samples and/or samples from different target protein isolations to minimize run-to-run carry-over.

13. Other variable modifications may be included in the primary database search, such as phosphorylation, acetylation, or deamidation. However, as addition of modifications increases both search time and space, it is recommended to retain only those modifications that are present in the sample.

14. When selecting an analysis workflow, ensure that it incorporates the ability to control for false positive sequence matches, e.g., by performing database searching against reversed protein sequences to estimate false discovery rates. If available, it is highly recommended to use a software platform that also controls false identification rates at the protein level.

15. An alternative strategy to using the online SAINT algorithm is to download the latest version of the SAINT source files (www.sourceforge.com) and compile it for your appropriate operating system. This strategy allows the SAINT algorithm to be run locally in the command-line, but requires additional computational knowledge. For a more detailed description of the underlying SAINT algorithm and its associated parameters *see* [25].

16. To compute meaningful SAINT specificity scores for the unlabeled/label-free datasets, at least two biological replicates of the experimental and control isolations are required. Ideally, control isolations are "user" controls performed in parallel to the experimental samples; however, user controls can be replaced and/or supplemented with negative control data from the CRAPOME database [24] to provide additional stringency. These datasets are easily added when using the online SAINT workflow #3.

17. Several user-defined options are available when running SAINT. A thorough discussion of their recommended usage can be found in [25].

18. For illustrative examples using ROC curves and histogram distributions for evaluating SAINT scoring *see* [8] and HIN200 [26].

Acknowledgements

We are grateful for funding from NIH grants R01GM114141, R21AI102187, and R21 HD073044, an NJCCR postdoctoral fellowship to TMG, and a NSF graduate research fellowship to AJG.

References

1. Rigaut G et al (1999) A generic protein purification method for protein complex characterization and proteome exploration. Nat Biotechnol 17(10):1030–1032

2. Cristea IM et al (2005) Fluorescent proteins as proteomic probes. Mol Cell Proteomics 4(12):1933–1941

3. Sueda S, Tanaka H, Yamagishi M (2009) A biotin-based protein tagging system. Anal Biochem 393(2):189–195

4. Lambert JP et al (2015) Proximity biotinylation and affinity purification are complementary approaches for the interactome mapping of chromatin-associated protein complexes. J Proteomics 118:81–94

5. Kaltenbrun E et al (2013) A Gro/TLE-NuRD corepressor complex facilitates Tbx20-dependent transcriptional repression. J Proteome Res 12(12):5395–5409

6. Greco TM et al (2011) Nuclear import of histone deacetylase 5 by requisite nuclear localization signal phosphorylation. Mol Cell Proteomics 10(2):M110.004317

7. Domanski M et al (2012) Improved methodology for the affinity isolation of human protein complexes expressed at near endogenous levels. Biotechniques 0(0):1–6

8. Joshi P et al (2013) The functional interactome landscape of the human histone deacetylase family. Mol Syst Biol 9:672

9. Pflieger D et al (2008) Quantitative proteomic analysis of protein complexes: concurrent identification of interactors and their state of phosphorylation. Mol Cell Proteomics 7(2):326–346

10. Hubner NC et al (2010) Quantitative proteomics combined with BAC TransgeneOmics reveals in vivo protein interactions. J Cell Biol 189(4):739–754

11. Blagoev B et al (2003) A proteomics strategy to elucidate functional protein-protein interactions applied to EGF signaling. Nat Biotechnol 21(3):315–318

12. Mathias RA et al (2014) Sirtuin 4 is a lipoamidase regulating pyruvate dehydrogenase complex activity. Cell 159(7):1615–1625

13. Miteva YV, Budayeva HG, Cristea IM (2013) Proteomics-based methods for discovery, quantification, and validation of protein-protein interactions. Anal Chem 85(2):749–768

14. Nesvizhskii AI (2012) Computational and informatics strategies for identification of specific protein interaction partners in affinity purification mass spectrometry experiments. Proteomics 12(10):1639–1655

15. Cristea IM et al (2006) Tracking and elucidating alphavirus-host protein interactions. J Biol Chem 281(40):30269–30278

16. Collins BC et al (2013) Quantifying protein interaction dynamics by SWATH mass spectrometry: application to the 14-3-3 system. Nat Methods 10(12):1246–1253

17. Guise AJ et al (2012) Aurora B-dependent regulation of class IIa histone deacetylases by mitotic nuclear localization signal phosphorylation. Mol Cell Proteomics 11(11):1220–1229

18. Oda Y et al (1999) Accurate quantitation of protein expression and site-specific phosphorylation. Proc Natl Acad Sci U S A 96(12):6591–6596

19. Ong SE, Kratchmarova I, Mann M (2003) Properties of 13C-substituted arginine in sta-

ble isotope labeling by amino acids in cell culture (SILAC). J Proteome Res 2(2):173–181

20. Ranish JA et al (2003) The study of macromolecular complexes by quantitative proteomics. Nat Genet 33(3):349–355

21. Sardiu ME et al (2008) Probabilistic assembly of human protein interaction networks from label-free quantitative proteomics. Proc Natl Acad Sci U S A 105(5):1454–1459

22. Tate S et al (2013) Label-free quantitative proteomics trends for protein-protein interactions. J Proteomics 81:91–101

23. Choi H et al (2011) SAINT: probabilistic scoring of affinity purification-mass spectrometry data. Nat Methods 8(1):70–73

24. Mellacheruvu D et al (2013) The CRAPome: a contaminant repository for affinity purification-mass spectrometry data. Nat Methods 10(8):730–736

25. Choi H et al (2012) Analyzing protein-protein interactions from affinity purification-mass spectrometry data with SAINT. Curr Protoc Bioinformatics. Chapter 8, Unit8.15

26. Diner BA et al (2015) The functional interactome of PYHIN immune regulators reveals IFIX is a sensor of viral DNA. Mol Syst Biol 11(2):787

27. Tackett AJ et al (2005) I-DIRT, a general method for distinguishing between specific and nonspecific protein interactions. J Proteome Res 4(5):1752–1756

28. Byrum SD et al (2012) ChAP-MS: a method for identification of proteins and histone post-translational modifications at a single genomic locus. Cell Rep 2(1):198–205

29. Bottermann K et al (2013) Systematic analysis reveals elongation factor 2 and alpha-enolase as novel interaction partners of AKT2. PLoS One 8(6), e66045

30. Di Virgilio M et al (2013) Rif1 prevents resection of DNA breaks and promotes immunoglobulin class switching. Science 339(6120):711–715

31. Ramanagoudr-Bhojappa R et al (2013) Physical and functional interaction between yeast Pif1 helicase and Rim1 single-stranded DNA binding protein. Nucleic Acids Res 41(2):1029–1046

32. Tsai YC et al (2012) Functional proteomics establishes the interaction of SIRT7 with chromatin remodeling complexes and expands its role in regulation of RNA polymerase I transcription. Mol Cell Proteomics 11(5):60–76

33. Moorman NJ et al (2010) A targeted spatial-temporal proteomics approach implicates multiple cellular trafficking pathways in human cytomegalovirus virion maturation. Mol Cell Proteomics 9(5):851–860

34. Goldberg AD et al (2010) Distinct factors control histone variant H3.3 localization at specific genomic regions. Cell 140(5):678–691

35. Carabetta VJ, Silhavy TJ, Cristea IM (2010) The response regulator SprE (RssB) is required for maintaining poly(A) polymerase I-degradosome association during stationary phase. J Bacteriol 192(14):3713–3721

36. Niepel M et al (2013) The nuclear basket proteins Mlp1p and Mlp2p are part of a dynamic interactome including Esc1p and the proteasome. Mol Biol Cell 24(24):3920–3938

37. Mann JM et al (2013) Complex formation and processing of the minor transformation pilins of Bacillus subtilis. Mol Microbiol 90(6):1201–1215

38. Castellana M et al (2014) Enzyme clustering accelerates processing of intermediates through metabolic channeling. Nat Biotechnol 32(10):1011–1018

39. Miteva YV, Cristea IM (2014) A proteomic perspective of Sirtuin 6 (SIRT6) phosphorylation and interactions and their dependence on its catalytic activity. Mol Cell Proteomics 13(1):168–183

40. Conlon FL et al (2012) Immunoisolation of protein complexes from Xenopus. Methods Mol Biol 917:369–390

41. Manza LL et al (2005) Sample preparation and digestion for proteomic analyses using spin filters. Proteomics 5(7):1742–1745

42. Wisniewski JR et al (2009) Universal sample preparation method for proteome analysis. Nat Methods 6(5):359–362

43. Erde J, Loo RR, Loo JA (2014) Enhanced FASP (eFASP) to increase proteome coverage and sample recovery for quantitative proteomic experiments. J Proteome Res 13(4):1885–1895

44. Chen EI et al (2007) Optimization of mass spectrometry-compatible surfactants for shotgun proteomics. J Proteome Res 6(7):2529–2538

45. Chatr-Aryamontri A et al (2015) The BioGRID interaction database: 2015 update. Nucleic Acids Res 43(database issue):D470–D478

Chapter 4

Label-Free Quantitation for Clinical Proteomics

Robert Moulder, Young Ah Goo, and David R. Goodlett

Abstract

Label-free quantification (LFQ) has emerged as a viable option for quantitative LC-MS/MS-based proteomic analyses for use on the scale of hundreds of samples such as are encountered in clinical analysis. Notably, sample preparation, sample loading, HPLC separations, and mass spectrometric performance must be highly reproducible for this approach to be effective. The following protocols describe the key steps in the methods related to sample preparation and analysis for LC-MS/MS-based label-free quantitation using standard data-dependent acquisition.

Key words Label-free quantification, Proteomics, Mass spectrometry, Area under the curve (AUC)

1 Introduction

In combination with chromatographic separation, electrospray ionization mass spectrometry (ESI-MS) provides a means of characterization and comparison of highly complex mixtures in a concentration-dependent manner. For ionizable analytes in general and peptides in particular, the intensity of the signal produced by ESI-MS will be proportional to its concentration in the eluting chromatographic peak [1] and thus depends on the peak volume. In circumstances where the quantity of sample is limited, analyte detectability can be enhanced by using separation columns with a smaller radial dimension. In such instances the benefit stems from the reduced flow rates and physical volume of the eluting peaks. In this manner the gain in sensitivity/peak intensity can be estimated from the ratio of the square of the column diameter [2]; e.g., from a 2 mm to a 75 µm i.d. column the estimated gain in signal intensity is 700-fold. The use of so-called nano-flow systems has been particularly efficacious in applications with limited sample amounts, where flow rates of 50–300 nl/min have been typically used with columns of 50–75 µm i.d. In the past 20 years such nano-flow separations have grown as the mainstay of qualitative and quantitative proteomics experiments. Moreover, with the maturation of liquid

Salvatore Sechi (ed.), *Quantitative Proteomics by Mass Spectrometry*, Methods in Molecular Biology, vol. 1410,
DOI 10.1007/978-1-4939-3524-6_4, © Springer Science+Business Media New York 2016

chromatographic systems capable of providing reproducible separation gradients at flow rates in the order of hundreds of nanoliters per minute, proteomics profiles can, with appropriate attention to detail, be produced in a reproducible manner suitable for quantitative comparisons [3].

From LC-MS/MS analysis, the chromatographic profiles of the precursor/isotope envelope of identified peptides can be determined from precursor ion scans and integrated as an area under the curve (AUC) measurement proportional to the abundance of the analyte. Software has been developed for the alignment, integration, and normalization of LC-MS profiles, which can be subsequently compared at the protein or peptide level. Amongst the common platforms that have been used, Progenesis (NonLinear Dynamics) and MaxQuant currently remain popular [4].

At this stage the key procedures involve defining the peak profile for integration whilst not including nearby interferences from neighboring isotope clusters. With the Progenesis software the integrated peaks can be previewed and modified to remove errors/overlap in the integration. The collected peptide features for each protein are summed. Whilst the software may provide an option to use all peptides associated with each protein, it is important to base the integration on peptides that are unique to each protein, which means culling homologous peptide matches to more than one parent protein. For normalization of the data several options are possible, depending on whether you have used a spiked standard or prefer some *housekeeping* or reference protein. For a well-characterized sample, where the injected amount has been matched, the use of total ion intensity of the identified proteins is a suitable choice.

For the analysis of plasma or serum samples a depletion step can be beneficial for extending the detectable dynamic range for the purpose of discovery workflows, as shown in Fig. 1 from collection to depletion and data analysis. The workflow is thus broken down into sections: depletion (Subheading 3.1); concentration of the depleted proteins (Subheading 3.2); digestion (Subheading 3.3); desalting (Subheading 3.4); sample concentration adjustment (Subheading 3.5); LC-MS-MS and data analysis (Subheading 3.6). Additional suggestions and alternatives are included under Notes in Subheading 4.

2   Materials

All reagents should be HPLC grade purity or at least 99 %. This is specifically required for the following: MilliQ water, ammonium bicarbonate (NH_4HCO_3), 1,4-dithiothreitol (DTT), Iodoacetamide (IAA), Trypsin (Modified Sequencing Grade, e.g., from Promega V5111), and Urea.

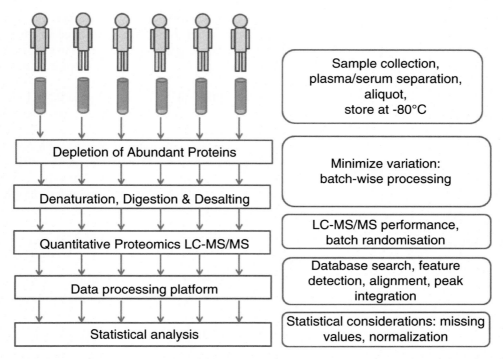

Fig. 1 A schematic of the sample preparation and analysis pipeline for label-free quantification of depleted serum

2.1 Denaturing and Digestion Reagents

1. 50 mM NH_4HCO_3 buffer: Dissolve 350 mg NH_4HCO_3 in 100 ml water.

2. 8 M Urea: Dissolve 24 g of urea in 50.0 ml of 50 mM NH_4HCO_3 solution.

3. Reducing reagent: 200 mM DTT in NH_4HCO_3 as above.

4. Alkylating reagent 200 mM IAA in NH_4HCO_3 as above.

5. Trypsin solution: Select a sufficient quantity of trypsin to digest at a ratio of 1:30. With Promega sequence grade trypsin each vial contains 20 μg of trypsin, mix multiple vials such that they are digested with an equivalent batch. Typically for 30 μg of protein add 10 μl of a 0.1 μg/μl trypsin solution.

2.2 HPLC Buffers

1. Mobile phase buffer A: 0.2 % formic acid in 98 % water 2 % acetonitrile.

2. Mobile phase buffer B: 0.2 % formic acid in 5 % water 95 % acetonitrile.

3 Methods

3.1 Depletion

Depletion of the top 12 most abundant serum proteins with Pierce spin columns (Prod # 85164, 85165). These are single-use depletion cartridges that can be used to achieve parallel processing of multiple samples in a reproducible cost-effective fashion. Make sure that you have sufficient columns that have been produced in the same batch/lot number when planning your experiments in case there is significant lot-to-lot variations that will affect quantitative results.

3.1.1 Proteins Targeted for Depletion

α1-Acid glycoprotein (P02763), Fibrinogen (P02761), α1-antitrypsin (P01009), Haptoglobin (P00738), α2-macroglobulin (P01023), IgA (P01876), Albumin (P02768), IgG (P01857, P01859-61; major subclasses of gamma globulin), Apolipoprotein A-I (P02647), IgM (P01871), Apolipoprotein A-II (P02652), Transferrin (P02787). Although the column capacity is 10 μl (~600 μg of protein), the use of 8 μl is recommended. Columns should be stored at 4 °C.

3.1.2 Depletion Protocol

1. Equilibrate depletion spin columns (Pierce) to room temperature.

2. Add 8 μl of serum or plasma into each column.

3. Close the caps and manually invert the columns until the resin is completely suspended in solution.

4. Place the columns to an Eppendorf tube rotator and rotate for 1 h at room temperature.

5. Once their end closures have been twisted off, place the columns into 2 ml Eppendorf tubes. Loosen the caps of the columns.

6. Centrifuge the columns at $1000 \times g$ for 2 min (at room temperature). The collected filtrates are the depleted fraction ($V \sim 500$ μl).

7. The cartridges contain the bound fraction, which can, if required, be saved for further analysis.

3.2 Concentration of the Sera After Depletion

To facilitate the handling and implementation of the denaturing and digestion protocols for the depleted serum fraction, it is necessary to change the buffer composition and volume (*see* **Note 1**). As an alternative to the preferred ultracentrifugation-buffer exchange method please *see* **Note 2**.

3.2.1 Ultracentrifugation-Buffer Exchange

1. Rinse a sufficient number (include extra) of Ultrafiltration spin columns for the samples (Sartorius-Stedim, Vivaspin, 4 ml, 5 kDa cut-off) with 1 ml of buffer equivalent to the Pierce slurry solution for the depletion cartridge (10 mM PBS, 0.15 M NaCl, 0.02 % sodium azide, pH 7.4): +4 °C, $3000 \times g$, ~20 min. Check the remaining volume and remove any columns that have performed at a slower rate. It is recommended that sufficient cartridges be purchased at the start of the project to conduct all the intended work, as there can be batch-to-batch variations which can cause anomalies in results.

2. Concentrate depleted serum samples to a volume of ~100 µl using the washed Ultrafiltration spin columns (+4 °C, 3000 × g, ~20 min).

3. Perform buffer exchange with 8 M Urea in 50 mM NH_4HCO_3.

   (a) 1200 µl of 8 M Urea: +4 °C, 3000 × g, 35 min.

   (b) 500 µl of 8 M Urea: +4 °C, 3000 × g, 35 min.

   (c) 500 µl of 8 M Urea: +4 °C, 3000 × g, 30 min.

   Final volume of sample should be ~100 µl.

4. To ensure that the concentrated proteins are in solution, ultrasonicate the spin columns for 5 min on ice and withdraw the liquid with a pipette.

5. To reduce the losses from transfer of the concentrate, wash the spin columns with 50 µl of 8 M Urea by sonicating 5 min on ice. Combine this with the rest of the concentrated sample.

3.3 Digestion (See Note 1)

1. Add 5 µl of reducing solution to reach the final concentration of ~5 mM DTT. Incubate 1 h at +37 °C.

2. Add 10 µl of alkylating solution to reach the final concentration of ~13 mM IAA. Vortex, incubate for 30 min in the dark at room temperature without an added IAA quenching step.

3. Add 850 µl of 50 mM NH_4HCO_3 to the samples to dilute the urea concentration below 1.5 M before trypsin digestion.

4. Add 10 µl of trypsin (~1:30) to each sample. Incubate at +37 °C overnight (16 h).

3.4 Desalting

Desalting with Sep Pak 50 mg cartridge (part number: WAT054955). Dried tryptic digested peptides are reconstituted with 1 ml of 1 % trifluoroacetic acid (TFA). Check if the pH is acidic. *See* **Note 3** concerning potential sources of contamination and interference.

1. Wet the column with 1 ml of 100 % methanol.

2. Equilibrate with 1 ml of 80 % Acetonitrile + 0.1 % TFA.

3. Equilibration with 2 × 1 ml of 0.1 % TFA.

4. Apply sample. Repeat with flow through.

5. Wash the cartridge with 3 × 1 ml of 0.1 % formic acid (FA) in 2 % acetonitrile.

6. Elute the peptides with 1 ml of 80 % acetonitrile + 0.1 % FA.

3.5 Sample Concentration Adjustment

1. Speed vac to dryness.

2. Reconstitute with 20 µl of 2 % formic acid + 2 % acetonitrile.

3. Use a Nanodrop detector (Thermo Scientific) to determine the UV absorbance spectrum (200–350 nm).

(a) Observe the estimated concentration based on the generalization that 1 absorbance unit is equivalent to 1 µg/µl of protein.

(b) Note the 260/280 nm ratio, this should be in the order of 0.7 for proteins/peptides not contaminated with nucleotides.

4. Spiking iRT peptides (Biognosis) 30:1 in the LC-MS/MS analyzed solution.

5. Using a 20 µl sample loop for 5 µl injections for a total of 200 ng, the target concentration should be 40 ng/µl. As the Nanodrop detector only gives reliable measurements down to 200 ng/µl, the dilutions should be made on the basis of solution at least at this concentration.

3.6 LC-MS/MS (See Note 3)

1. With an EasyNano-LC: A 20 × 0.1 mm i.d. pre-column packed with 5 µm Magic C18 (Michrom) silica connected by a New Objective two-way union together with a 75 µm × 150 mm analytical column packed with 5 µm Magic C18 (Michrom).

2. A separation gradient from 5 % B (95 % A) to 35 % B (65 % A) in 65 min at a flow rate of 300 µl/min.

3. Autosampler loop size and injection size: 20 µl sample loop for 5 µl injections.

4. To remove the influence of injection order, the samples are randomized for batches of single injections (each sample separated by a 15 min blank), with three/four replicate injections in total (three batches) with the system performance monitored between batches using a lab standard sample. A pool of the samples in the batch is analyzed at the start of each batch. The maintenance of constant/accountable instrument performance is essential for a successful LFQ experiment.

5. Using an Orbitrap-Velos Pro, to perform data-dependent MS/MS data acquisition, the following are typical for a proteomics analysis: Ionization in positive ion mode with CID of the 15 most intense ions (m/z 300–2000, charge states > 1+). Dynamic exclusion *30 s. Orbitrap precursor ion scan resolution 60,000 (at m/z 400), with a target value of 1,000,000 ions and a maximum injection time of 100 ms. For the ion trap the target values and maximum injection time values are set to 500,000 and 50 ms [5, 6]. When making the selection of the top "*n*" most intense ions, it is important to consider the associated duty cycle and the width of the *chromatographic peak, i.e., ensuring that there are sufficient MS1 data points to describe the peak elution profile.

3.7 Data Analysis

The following describes a standard workflow built around the Proteome Discoverer and Progenesis software. *See* **Note 4** addressing an alternative approach using the open-source platform MaxQuant.

1. *General data evaluation*: Even though proprietary data analysis software is in constant evolution, in addition to Xcalibur and Proteome Discoverer (Thermo Scientific data acquisition and analysis software), the free software *RawMeat* (Vast Scientific) provides a quick and useful tool to gain an overview of the data attributes and the suitability of the sample and applied method (Fig. 2). For example, for a "top *n*" method it provides an indication of the suitability of the selected method in terms of sample complexity, although note that it is important for quantification to aim to have sufficient MS1 scans to define the elution profile of the chromatographic peaks (vide supra). In terms of the efficiency of the trypsin digestion, the charge distribution provides a good indication of the general success: The doubly charged precursors should be the most frequent

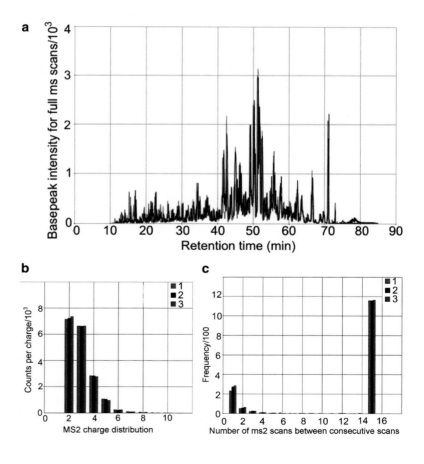

Fig. 2 Data evaluation using the RawMeat software. (**a**) Base peak chromatograms for three replicate LC-MS/MS analyses. (**b**) Charge distribution, the horizontal axis is precursor charge and the vertical axis is counts per charge. (**c**) Top *n* usage: number of ms2 scans in between consecutive scans, shown for a top 15 method with CID fragmentation

for a tryptic digest. The current version of Progenesis, Progenesis QI, includes similar QC metrics for sample preparation and instrument performance (*see* **step 3j**).

2. *Using ProteomeDiscoverer (V. 1.4) with Mascot 2.1.*

   (a) *Database selection*: Select the most recent release of the human Swissprot database. If the work is part of an extended study, then be sure to use the same version of the database for the duration of the study (i.e., avoid automatic updates which will change search results). Whilst isoform databases can be very informative, the overlap due to differences in protein inference can be problematic when combining data from many searches. Include with the database a contaminant list (e.g., http://www.crapome.org/), then create a concatenated forward and reverse database so that the estimates of the false discovery rate can be accountable after the search has been conducted.

   (b) *Search parameters*: Carboamidyl methylation fixed, methionine oxidation variable, suitable mass tolerances: e.g., 6 ppm precursor tolerance, 0.6 Da fragment tolerance, one missed cleavage, fragmentation type = ESI trap. With a concatenated database and Proteome Discover, use the *Fixed value PSM validator* for false positive estimation.

   (c) *Multi-consensus report*: When the searches are complete, use Proteome Discoverer to create a multi-consensus report of the collected search results, and export this to Excel format including protein group and PSM results (layer 1 and 2).

3. *Alignment, normalization, and integration using Progenesis*:

   (a) Create an experiment and load the RAW data files to the program as described in the Progenesis operation instructions.

   (b) Observe the ion maps of the loaded files and pay attention to any irregularities.

   (c) Select the alignment file or allow Progenesis to do this automatically. You might choose from a pooled standard sample that you have used, which gives a good representation of all the detected peaks.

   (d) When the alignment is complete, confirm that the selected vectors are appropriate. Figure 3a displays a representation of the detected features with the vectors used for alignment. Progenesis provides a color-coded quality assessment of the alignment results, indicating regions that are in need of attention.

   (e) Filtering: choose which region of the chromatogram and mass range to use in the quantitative comparisons and select the precursor charge states to consider, e.g., 1–5.

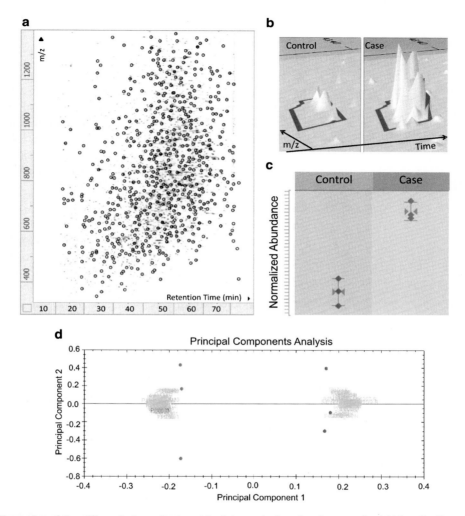

Fig. 3 Examples of the different steps displayed in data analysis using Progenesis. (**a**) Visualization of an ion map. (**b**) Display of the precursor intensity and chromatographic profiles of a pair of differential features. (**c**) Difference detected between two proteins. (**d**) PCA analysis (illustrated for two samples)

(f) Check the normalization results, noting any irregularities in the normalization factors; a lower intensity chromatogram will produce larger values. Be wary of files with large differences.

(g) Design the experiment: select which analyses are replicates and define groups for comparison.

(h) Check the integration of significant peaks, this is particularly critical when dealing with large fold differences and limited quantitative evidence. Figure 3b is taken from a 3D montage view of a differential peak that was detected from triplicate analyses of serum sample from two individuals.

(i) To assign identifications to the features, upload the multi-consensus identification report created with Proteome discoverer.

(j) At this stage in the workflow the quality control (QC) metrics are available. These can be used to confirm the quality of analysis in terms of differences that might occur during sample preparation and in instrument performance. For example, the charge distribution, number of missed cleavages, and modification frequency between samples are reported, as are the chromatographic peak widths, mass accuracy, and the scan rates.

(k) The Progenesis statistics include PCA analysis and hierarchical clustering to identify differentially abundant features/proteins. Figure 3d shows the PCA separation displayed by Progenesis for data from the triplicate analysis of two distinct serum samples. Examples of a differentially abundant feature and a protein are shown in Fig. 3b and c, respectively. Lists describing the protein and peptide intensities (normalized and un-normalized) may be exported in csv format so that more advanced statistical analyses and comparisons can be made.

3.8 Time Line for a LFQ Serum Proteomics Experiment

Here follows a time line for a typical experiment that could be based on the preparation of a sub-batch of 20 samples.

Day 1:

1. Deplete.
2. Precipitate/concentrate.

Day 2:

3. Reconstitute sample in approximately 150 μl of 8.0 M urea in a 1.5 ml polypropylene centrifuge tube.
4. Add 5 μl of Reducing Reagent and mix the sample by gentle vortex.
5. Reduce the mixture for 1 h at room temperature or in an oven at 37 °C.
6. Add 10 μl of Alkylating Reagent and alkylate for 0.5 h at room temperature in the dark (use aluminum foil to cover the sample).
7. Add 850 μl of NH_4HCO_3 solution to dilute the urea before digesting it with trypsin.
8. Add trypsin in appropriate ratio (1:30) to approximate amount of protein by weight. Digest overnight at 37 °C.

Day 3:

9. Adjust pH for desalting.

10. Desalt.

11. Dry, reconstitute, assay and dilute the solution for analysis.

12. Design randomization experiment and start batch analyses.

4 Notes

1. *Volume reduction.* Starting with the processing of 8 μl of serum, the flow through fraction of interest is isolated at the expense of about 60-fold volume expansion, i.e., to 500 μl in PBS. To accommodate the digestion protocol, changes of the buffer composition and volume are needed. Precipitation can be performed in a "hands off" fashion, but may suffer from differences in performance and requires some degree of visual judgement. We have found that the buffer exchange approach is more reproducible. However, we have at times encountered problems with membranes that limit the passage of the liquid such that the procedure is slowed considerably.

2. *Alternative to the ultracentrifugation-buffer exchange.* As an alternative to the method described under Subheading 3.2, one may use the following protein precipitation method on the depleted serum flow through:

   (a) Add 2 ml of cold acetone (−20 °C) to the sample (4 volumes).

   (b) Invert several times and keep at −20 °C for at least 4 h (overnight).

   (c) Invert several times and keep at −20 °C for at least 4 h (overnight). Centrifuge for 15 min at 4 °C at $1300 \times g$.

   (d Remove the supernatant and dry the sample at room temperature

   (e) Add 150 μl of 8 M urea to the sample, vortex to dissolve.

3. *Instrument performance Ion suppression.* One of the largest threats to a successful LFQ experiment is the occurrence of ion suppression. This may arise from the contamination of the sample, buffers, or originate from the LC system. Importantly, one should avoid detergents and sources of plasticisers. If you use facility labware that is routinely washed, in-house detergents may occur as harmful residues in these. Be sure to rinse with appropriate solvents and make sure that all components of the container are compatible with the solvents and acids used. Note that concentrated acetonitrile and/or formic acid can release residues from low quality/inappropriate labware. Internet discussion groups, such as the ABRF, frequently discuss and advise on such issues (http://www.abrf.org/).

It is essential to monitor instrument operations (both chromatographic and mass spectrometric) during the data acquisition. For a commercial or in-house standard digest, the ion intensities, peak areas, sequence coverage, and proteins/peptides identified can be used to gauge success or failure. Similarly, the retention times of a simple mixture are easily spotted and compared.

Trifluoroacetic acid can form ion pairs with peptide ions and improve their retention to the reverse phase column during desalting. It can, however, also cause ion suppression. For instance, if this is carried throughout the protocol and used in the elution of the peptides during desalting. We have previously associated such usage with the occurrence of ion suppression in the LC-MS/MS analyses.

4. *Alternative search and alignment strategies.* The previous examples have been described mostly in the context of using Progenesis. As an alternative, MaxQuant is an open-source quantitative proteomics software package that is built around the Andromeda search engine [7, 8]. The platform facilitates alignment and AUC quantification and has recently been developed to include visualization capabilities [9]. The output can be analyzed with the Perseus module that was developed for bioinformatics analysis of the MaxQuant and Andromeda proteomics data.

References

1. Ikonomou MG, Blades AT, Kebarle P (1991) Electrospray-ion spray: a comparison of mechanisms and performance. Anal Chem 63(18): 1989–1998

2. Abian J, Oosterkamp AJ, Gelpi E (1999) Comparison of conventional, narrow-bore and capillary liquid chromatography/mass spectrometry for electrospray ionization mass spectrometry: practical considerations. J Mass Spectrometry 34(4):244–254

3. Foss EJ, Radulovic D, Shaffer SA et al (2007) Genetic basis of proteome variation in yeast. Nat Genet 39(11):1369–1375

4. Nahnsen S, Bielow C, Reinert K et al (2013) Tools for label-free peptide quantification. Mol Cell Proteomics 12(3):549–556

5. Kalli A, Smith GT, Sweredoski MJ et al (2013) Evaluation and optimization of mass spectrometric settings during data-dependent acquisition mode: focus on LTQ-Orbitrap mass analyzers. J Proteome Res 12(7):3071–3086

6. Moulder R, Bhosale SD, Erkkila T et al (2015) Serum proteomes distinguish children developing type 1 diabetes in a cohort with HLA-conferred susceptibility. Diabetes 64(6): 2265–2278

7. Cox J, Mann M (2008) MaxQuant enables high peptide identification rates, individualized p.p.b.-range mass accuracies and proteome-wide protein quantification. Nat Biotechnol 26(12): 1367–1372

8. Cox J, Neuhauser N, Michalski A et al (2011) Andromeda: a peptide search engine integrated into the MaxQuant environment. J Proteome Res 10(4):1794–1805

9. Tyanova S, Temu T, Carlson A et al (2015) Visualization of LC-MS/MS proteomics data in MaxQuant. Proteomics 15(8):1453–1456

Chapter 5

Proteogenomic Methods to Improve Genome Annotation

Keshava K. Datta, Anil K. Madugundu, and Harsha Gowda

Abstract

Annotation of protein coding genes in sequenced genomes has been routinely carried out using gene prediction programs guided by available transcript data. The advent of mass spectrometry has enabled the identification of proteins in a high-throughput manner. In addition to searching proteins annotated in public databases, mass spectrometry data can also be searched against conceptually translated genome as well as transcriptome to identify novel protein coding regions. This proteogenomics approach has resulted in the identification of novel protein coding regions in both prokaryotic and eukaryotic genomes. These studies have also revealed that some of the annotated noncoding RNAs and pseudogenes code for proteins. This approach is likely to become a part of most genome annotation workflows in the future. Here we describe a general methodology and approach that can be used for proteogenomics.

Key words Mass spectrometry, Proteogenomics, Novel proteins, Pseudogenes, Noncoding RNAs

1 Introduction

The utility of whole genome sequence of an organism is determined based on the availability of accurate assembly and accompanying annotation. Human genome sequence and genome sequences of model organisms are among some of the well-annotated genomes in the public domain. These have played an important role in accelerating novel discoveries and scientific investigations. Traditionally, annotation of genes in newly sequenced genomes has been largely carried out using gene prediction programs supported by conservation and transcript evidence in these regions. Many genomes including human were annotated using this approach [1, 2]. This was the mainstay a decade ago as high-throughput methods to sequence or identify proteins were not as matured. The advent of mass spectrometry has transformed our ability to identify proteins in a high-throughput manner. Unlike antibodies, mass spectrometry has the potential to identify proteins in an unbiased manner without prior need to know all the proteins in a sample. However, the identification of proteins is limited by protein sequences that are in the protein database that is used to search mass spectrometry

Salvatore Sechi (ed.), *Quantitative Proteomics by Mass Spectrometry*, Methods in Molecular Biology, vol. 1410,
DOI 10.1007/978-1-4939-3524-6_5, © Springer Science+Business Media New York 2016

data. Traditionally, MS/MS searches have been carried out using publicly available databases including RefSeq [3], IPI [4], Uniprot [5], and NextProt [6].

Although mass spectrometers might routinely sample unknown proteins, they are not identified because MS/MS searches are only carried out against known protein databases. Proteogenomics is an emerging field where MS/MS searches are not limited to the available protein databases in public domain. In addition to known protein databases, the data is searched against putative protein databases developed by translating genomic regions that are currently not annotated as protein coding. These searches allow the identification of novel protein coding regions that have not been identified till now. This approach has now been used by several research groups resulting in the identification of novel protein coding regions in well-annotated genomes [7–10]. We have carried out proteogenomics in various organisms including microbes [11–13], protozoans [14, 15], insects [16], zebrafish [17], and humans [18]. We have been able to identify novel protein coding regions in all the organisms where we have carried out proteogenomics. These studies have also revealed that some of the annotated noncoding RNAs (ncRNAs) and pseudogenes are indeed coding for proteins. In addition, we have identified several examples where a single transcript is capable of coding for multiple proteins using alternative reading frames. Below, we provide a detailed workflow for carrying out proteogenomics using mass spectrometry.

2 Materials

2.1 Protein Extraction and Estimation

1. Lysis buffer: 4 % Sodium dodecyl sulfate (SDS), 100 mM Dithiothreitol (DTT), and 100 mM Tris pH 7.5.
2. BCA protein assay kit (Thermo Scientific Pierce).

2.2 SDS-PAGE

1. Resolving gel buffer: 1.5 M Tris–HCl, pH 8.8.
2. Stacking gel buffer: 1.0 M Tris–HCl, pH 6.8.
3. 30 % acrylamide/bis solution: Dissolve 29 g of acrylamide and 1 g of N,N'-Methylenebisacrylamide in 100 ml of water.
4. 10 % ammonium persulfate (APS).
5. N,N,N,N'-tetramethyl-ethylenediamine (TEMED).
6. Running buffer: 25 mM Tris–HCl, pH 8.3, 0.192 M glycine, 0.1 % SDS.
7. Loading buffer (6×): 375 mM Tris–HCl (pH 6.8), 6 % SDS, 25 % β-mercaptoethanol, 0.1 % bromophenol blue (BPB), 45 % glycerol.
8. Fixative: Methanol:Water:Glacial acetic acid (5:4:1).

9. Coomassie Brilliant Blue stain: Dissolve 3 g of CBB R-250 per liter of solution containing methanol, water, and glacial acetic acid in the ratio of 5:4:1.

2.3 Digestion

2.3.1 In-Gel Digestion

1. Destaining solution: 40 mM Ammonium Bicarbonate (ABC), 40 % Acetonitrile (ACN).

2. Reduction solution: 5 mM DTT in 40 mM ABC (prepare shortly before use).

3. Alkylation solution: 20 mM Iodoacetamide (IAA) in 40 mM ABC (prepare in the dark, shortly before use).

4. Trypsin (modified sequencing grade; Promega, Madison, WI, USA).

5. 5 % formic acid (aqueous).

6. Extraction buffer: 5 % formic acid, 40 % ACN.

2.3.2 In-Solution Digestion

1. Reduction solution: 100 mM DTT stock—prepare shortly before use.

2. Alkylation solution: 100 mM (IAA) stock—prepare in the dark, shortly before use.

3. Trypsin (modified sequencing grade; Promega, Madison, WI, USA).

4. 0.1 % formic acid.

2.3.3 Desalting

1. Sep-Pak C_{18} columns (Waters Corporation, Milford, MA, USA).

2. 100 % ACN.

3. Solvent A: 0.1 % formic acid.

4. Solvent B: 0.1 % formic acid, 30 % ACN.

2.4 Fractionation

2.4.1 Basic pH Reverse Phase Chromatography (bRPLC)

1. Solvent A: 10 mM Tetra Ethyl Ammonium Bicarbonate (TEABC) buffer, pH 9.5.

2. Solvent B: 10 mM TEABC buffer, 90 % ACN, pH 9.5.

3. Waters XBridge column (Waters Corporation, Milford, MA; 130 Å, 5 μm, 250×9.4 mm).

4. HPLC system.

2.4.2 Strong Cationic Exchange (SCX) Fractionation

1. Solvent A: 5 mM potassium phosphate (K_3PO_4) buffer containing 25 % ACN, pH 2.7.

2. Solvent B: 350 mM KCl in solvent A, pH 2.7.

3. PolySULFOETHYL A column (PolyLC, Columbia, MD; 200 Å, 5 μm, 200×4.6 mm).

4. HPLC system.

2.4.3 Desalting Using STAGE Tips	1. C_{18} STAGE tips.
	2. 100 % ACN.
	3. Solvent A: 0.1 % formic acid.
	4. Solvent B: 0.1 % formic acid, 30 % ACN.

2.5 LC-MS/MS Analysis

1. A high resolution mass spectrometer such as LTQ-Orbitrap series (Thermo Electron, Bremen, Germany).
2. A nanoflow HPLC such as Easy-nLCII (Thermo Scientific, Odense, Southern Denmark).
3. Solvent A: 0.1 % formic acid.
4. Solvent B: 0.1 % formic acid, 95 % ACN.

2.6 Protein Databases

2.6.1 Known Protein Database

Download the appropriate protein database for the organism of study. The sources of databases are listed in Table 3.

2.6.2 Custom Databases for Novel Peptide Identification

1. Sequences of mRNAs, noncoding RNAs (http://www.noncode.org/), and pseudogenes (http://pseudogene.org/).
2. Whole genome sequence of the organism under study.
3. Tools for translation of DNA or RNA sequence such as *transeq* from the EMBOSS package [19] or functional modules available in BioPerl, BioPython, and Biostrings (Bioconductor).

2.6.3 Contaminant Protein Sequences

Sequences of common contaminant proteins such as proteases and human keratins. They can be downloaded from www.thegpm.org/crap.

2.6.4 Decoy Database

Database of reversed sequences of all the proteins in the target database.

2.7 Search Engines

One or multiple search algorithms. Algorithms may be open source—X!Tandem, MS-GF+, OMSSA, MyriMatch, and MassWiz or commercially available search algorithms such as SEQUEST or Mascot.

2.8 Visualization Tools

They are required for proteogenomic analysis. Examples include Integrative Genome Viewer (IGV—https://www.broadinstitute.org/igv/) and genome browsers from Ensembl or UCSC.

3 Methods

3.1 Protein Extraction and Estimation

1. Homogenize cell/tissue samples using Dounce homogenizer/mortar and pestle/mechanical homogenizer in the presence of lysis buffer.

2. Sonicate the homogenized samples using a probe sonicator—12 pulses of 10 s each at an amplitude of 40 %.

3. Centrifuge the samples at high speed ($10,000 \times g$) for 10 min at 4 °C.

4. Collect the supernatant into a fresh tube and proceed to protein estimation.

5. Estimate protein concentration of cell/tissue lysate using BCA assay reagents.

3.2 Protein Digestion

3.2.1 SDS-PAGE and In-Gel Digestion

1. Resolve 50–200 μg of protein from each lysate on an SDS-PAGE.

2. Leave the gel in fixative solution for 10 min.

3. Stain the gel in Coomassie Brilliant Blue or Colloidal Coomassie for ~60 min.

4. Remove the stain and briefly destain for 15–20 min using fixative solution.

5. Remove the fixative and rinse the gel in water for 1 h.

6. Carry out all these steps in a clean environment wearing a lab coat to avoid keratin contamination.

7. Place the gel on a clean glass plate on a transilluminator.

8. Excise protein bands stained with Coomassie and cut the bands into 1×1 mm pieces.

9. Transfer the gel pieces to microfuge tubes.

10. Cover the gel pieces with sufficient destaining solution and place it on a rocker. Gently agitate to aid destaining.

11. Discard the destaining solution with a 200 μl pipette tip and repeat the procedure until the gel pieces are completely destained.

12. Once the gel pieces are completely destained, spin tubes and discard supernatant.

13. Add 0.5 ml 100 % ACN to each tube, incubate for 10–15 min until gel pieces dehydrate and become opaque. Spin tubes and discard the liquid.

14. Add sufficient reduction solution to completely cover the gel pieces. Incubate at 60 °C for 30 min.

15. Cool the tubes to room temperature; spin tubes and discard the supernatant. Add sufficient alkylation solution. Incubate the tubes for 10 min at room temperature in the dark.

16. Spin tubes and discard the supernatant and dehydrate gel pieces by adding 100 % ACN (sufficient volume to cover gel pieces). Remove ACN completely and place the tubes on ice.

17. Prepare sequencing grade trypsin at a concentration of 10 ng/μl in 40 mM ammonium bicarbonate (chilled). Prepare this solution on ice and shortly before use.

18. To the microfuge tubes incubated on ice, add enough trypsin solution to cover the dehydrated gel pieces and leave on ice for 45 min. Ensure the gel pieces are completely rehydrated (add more trypsin solution if necessary).

19. Once the gel pieces are completely rehydrated, remove excess trypsin and replace with sufficient 40 mM ABC to cover gel pieces. Incubate the tubes at 37 °C overnight.

20. Peptide extraction: Cool the tubes to room temperature, add 100 μl of 5 % formic acid (aqueous) to each tube and incubate for 10 min at 37 °C. Spin tubes and transfer supernatant into a fresh microfuge tube.

21. Add 100 μl extraction buffer to each tube and incubate for 10 min on a shaker. Spin tubes and pool the extract with the supernatant from earlier step. Repeat the procedure. For final extraction, add 100 % ACN sufficient to cover gel pieces, incubate for 10 min on a shaker. Spin tube and pool the extract with the supernatant from earlier steps.

22. Dry down the pooled supernatant by spinning tubes in a speedvac. Store the dried peptides at –80 °C until LC-MS/MS analysis.

3.2.2 In-Solution Digestion

1. Use about 2 mg of protein from each cell/tissue lysate.

2. Reduce proteins by adding 100 mM DTT to a final concentration of 5 mM and incubating at 60 °C for 60 min.

3. Add 200 mM IAA to a final concentration of 20 mM. Incubate the tubes for 10 min at room temperature in the dark. The concentration of SDS can be brought down to <0.05 % by carrying out buffer exchange using 3 kDa filters.

4. Remove 20 μg equivalent protein and store it at –20 °C (pre-digest).

5. To the remaining solution, add sequencing grade trypsin (Promega) at a ratio of 1:20 [enzyme:protein amount]. Incubate the tubes at 37 °C overnight.

6. Remove 20 μg equivalent protein (post-digest).

7. Acidify the digest with 0.1 % formic acid.

8. Check digestion efficiency by resolving pre-digest and post-digest on 10 % SDS-PAGE (optional).

9. Desalt the samples using SepPak columns.

3.2.3 Sep-Pak C_{18} Cleanup

1. Connect a 10-ml syringe (with plunger removed) to the shorter end of the column.

2. Pass 5 ml of 100 % ACN through the column.

3. Wash the column with 7 ml of solvent A (applied as 3.5 ml × 2).

4. Load the acidified protein digest.

5. Wash the column with 12 ml of solvent A (applied as 1 + 5 + 6 ml).

6. Elute peptides with 6 ml of solvent B (applied as 3 × 2 ml) into a 15 ml polypropylene tube.

7. Freeze the eluate and lyophilize it for 2 days.

8. Fractionate the lyophilized peptide mixture.

3.3 Fractionation

3.3.1 Basic pH Reverse Phase Liquid Chromatography (bRPLC)

1. Reconstitute the lyophilized samples in 1 ml of solvent A.

2. Load the reconstituted sample onto the XBridge C18 column.

3. Resolve the peptide mixture using a gradient of 0–100 % solvent B (Table 1) in 50 min at a flow rate of 1 ml/min.

4. Collect the fractions in a 96-well plate. Concatenate the fractions into a total of 12 fractions to reduce mass spec time.

5. Vacuum dry the pooled samples and store at −80 °C until LC-MS/MS analysis.

3.3.2 Strong Cationic Exchange (SCX) Chromatography

1. Reconstitute the lyophilized sample in 100 μl solvent A.

2. Inject 95 μl of the sample at a flow rate of 125 μl/min.

3. Resolve the peptide mixture using an increasing gradient of solvent B (Table 2) in 50 min.

4. Collect the fractions in a polypropylene 96-well plate (1 ml capacity, 3.1 mm height). Concatenate the fractions to reduce mass spec time.

5. Vacuum dry the fractions and desalt using STAGE tips to remove the excess salt.

3.3.3 Desalting Using STAGE Tips

1. Connect the STAGE tip to a 5-ml syringe. Pre-wet the tips with 30 μl of 100 % ACN.

2. Equilibrate the tip with 30 μl of 0.1 % formic acid twice.

Table 1
Liquid chromatography method for bRPLC fractionation

Time in minutes	0–5	5–10	10–40	40–45	45–46	46–50
Percentage of solvent B	1 %	1–10 %	10–35 %	35–100 %	100–1 %	1 %

Table 2
Liquid chromatography method for strong cationic exchange fractionation

Time in minutes	0–5	5–8	8–40	40–42	42–46	46–47	47–50
Percentage of solvent B	0 %	0–5 %	5–50 %	50–100 %	100 %	100–0 %	0 %

3. Reconstitute the sample in 30 μl of 0.1 % formic acid.

4. Load the sample onto the column. Collect the flow through and reload it.

5. Wash the column with 30 μl of 0.1 % formic acid twice.

6. Elute the peptides with 30 μl of 40 % ACN, 0.1 % formic acid twice.

7. Dry the eluate in a speed vac and store at –80 °C until LC-MS/MS analysis.

3.4 LC-MS/MS Analysis

1. The method of LC-MS/MS analysis for both in-gel and in-solution digested samples remains the same.

2. Reconstitute the dried samples in 20 μl of solvent A and transfer it into a 96-well plate

3. Load sample onto the enrichment column at a flow rate of 350 nl/min. Use a linear gradient of 7–35 % solvent B to separate the peptides on the analytical column.

4. Acquire MS and MS/MS data in high resolution mode (>30,000 for MS and >5000 for MS/MS).

5. Acquire MS/MS data for all the samples of all the fractionation methods and then proceed toward database searching.

3.5 Data Analysis

An overview of steps and outcomes in proteogenomic data analysis has been summarized in Fig. 1. In the following sections, we first describe the procedure for identifying known proteins and then the procedure to be followed for identifying novel protein coding regions

3.5.1 Database Searching for Identification of Known Proteins

1. Download the protein database for a given organism from resources such as RefSeq/UniProt (refer Table 3). Create a nonredundant database by merging the identical sequence records using sequence analysis tools or with custom scripts (*see* **Note 1**).

2. Include the contaminant protein sequences in the target database and prepare the decoy database by reversing the protein sequences (*see* **Note 2**).

3. Index the database by following instructions provided by respective search algorithms.

4. Carry out MS/MS search by defining appropriate parameters including choice of protease, number of missed cleavages, mass tolerance at precursor and fragment level, and modifications (*see* **Notes 3** and **4**).

5. Apply FDR threshold of ≤ 1 % at peptide and protein level [20, 21] to obtain the list of known proteins identified from these cell/tissue samples (*see* **Note 5**).

6. Filter out all the spectra that were not assigned to any peptides. These spectra will be used for proteogenomic analyses.

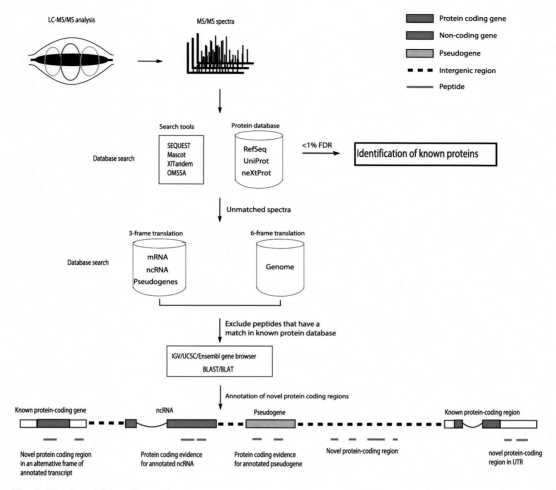

Fig. 1 Novel protein-coding region

3.5.2 Identification
of Novel Protein Coding
Regions

1. Custom databases have to be prepared for identification of novel protein coding regions in the genome. This includes three frame translated mRNAs, ncRNAs, and annotated pseudogene regions in addition to six frame translated genome.

2. These databases can be derived using *transeq* tool from the EMBOSS package [19] or sequence manipulation functions from BioPerl, BioPython, and Biostrings (Bioconductor).

3. After creating conceptually translated databases, filter and remove all the tryptic peptides that are less than seven amino acids. Further, remove all the stretches between two stop codons that do not contain methionine residue.

4. Carry out MS/MS searches of unassigned spectra as a two-step process.

Table 3
Sources of protein sequences

Database	Link
NCBI RefSeq	http://www.ncbi.nlm.nih.gov/refseq/
UniProtKB/Swiss-Prot	www.uniprot.org/uniprot/
neXtProt	www.nextprot.org/
IPI	ftp://ftp.ebi.ac.uk/pub/databases/IPI
Ensembl	http://www.ensembl.org
CCDS	http://www.ncbi.nlm.nih.gov/CCDS/
UCSC	https://genome.ucsc.edu
NCBI nr	ftp://ftp.ncbi.nlm.nih.gov/blast/db/
TAIR	https://www.arabidopsis.org/
FlyBase	http://flybase.org/
VectorBase	http://vectorbase.org/

5. Combine three frame translated mRNAs, ncRNAs, and pseudo-gene sequences as one database and carry out MS/MS search (*see* **Note 6**).

6. Unassigned spectra from the above step can then be searched against six frame translated genome database.

7. Apply FDR threshold of ≤1 % and filter the results.

8. From the list of peptides identified from the custom proteogenomic databases, filter the peptides that map to the known proteome by exact string match search.

3.6 Proteogenomic Analysis

1. The peptides that are uniquely identified from custom databases and could be mapped to single genome location via BLAST/BLAT are utilized to identify novel protein coding regions.

2. Map the novel peptides onto the genome of the organism under study using visualization tools such as IGV, UCSC, or Ensembl Genome Browser (*see* **Note 7**).

3. Based on the localization of these novel peptides on the genome, they can be classified as: (a) Intergenic and (b) Intragenic.

4. Peptides that map to intragenic regions serve as evidence of novel protein coding regions in UTRs, alternative reading frames of annotated mRNAs.

5. Peptides identified from annotated ncRNAs and pseudogenes serve as evidence for novel protein coding genes.

6. In addition, it is also possible to identify novel protein coding regions in intergenic regions with no annotation. This can be further corroborated with transcript evidence from ESTs and RNA-Seq datasets. UCSC browser also provides conservation plots that can serve as orthogonal evidence for protein coding potential of these regions.

4 Notes

1. Protein databases maintained by various repositories follow slightly different approaches to annotate protein sequences. There are differences in number of proteins annotated for a given organism because of this variability. To have a comprehensive list of annotated proteins for an organism, it is useful to combine data from these widely used resources. When a protein database for organism of study is not available or available partially, user may include protein sequences from taxonomically related organisms. The recommended method of searching the proteomic data in that case is through an iterative approach. The unmatched spectra from each step are further searched against the protein database of ancestral organisms. This step is repeated until an acceptable gain in protein number is observed.

2. The preferred method of creating decoy database is by reversing the protein sequences in the target database. It ensures that the composition and distribution of peptide length is similar to the target database and thus the statistical methods are valid. Commercial software are often equipped to automatically generate decoy databases.

3. It is best to acquire data on high resolution and high mass accuracy instruments. This allows defining narrow window for mass deviation during database search to identify accurate match. A wider window results in higher likelihood of false identifications due to increased search space. A recent article describes the effect of mass tolerance on FDR estimates and peptide identifications [22].

4. Defining appropriate modifications in a proteomics study is crucial for identification of peptides and proteins. For example, most proteomics workflows involve reduction and alkylation of cysteine residues during sample preparation. This modification has to be specified during database search in order to identify cysteine containing peptides. Similarly, oxidation of methionine and deamidation of asparagine and glutamine are commonly observed in vitro artifacts. These can be specified as variable modifications during database search. It is important to note that providing too many modifications increases the search space and adversely affects identification.

5. The standard FDR estimation methods work reasonably well for database search against known protein sequences. These are not optimal for proteogenomics searches as both forward and reverse databases are conceptual in nature. However, in the absence of optimal approaches, this approach is being employed in most studies. Decoy-free FDR estimation methods might prove valuable for proteogenomics in the future.

6. Six frame translation of genomes often result in large databases. Indexing such large databases is often difficult in many search algorithms. In such scenarios, one should choose algorithms that can handle such large databases. Further, it is recommended to carry out searches against six frame translated genome and three frame translated mRNA, ncRNA, and pseudogenes independently as a combined database would be too large and the FDR penalty would be high resulting in many false negatives.

7. Genome browsers are valuable in proteogenomics studies. They provide context as well as orthogonal lines of evidence including transcript and conservation in the same visual interface for annotating novel protein coding regions. UCSC and Ensembl browsers are some of the widely used genome browsers for these purposes. For organisms not available in above portals, users can configure the Integrative Genomics Viewer (IGV) available from Broad Institute with custom genome assembly and transcript annotations. IGV can be used offline and allows more customization.

Acknowledgements

We thank the Department of Biotechnology (DBT), Government of India, for research support to the Institute of Bioinformatics. Keshava K. Datta is a recipient of Research Fellowship from the University Grants Commission (UGC), Government of India. Anil K. Madugundu is a recipient of BINC-Research Fellowship from DBT.

References

1. Lander ES, Linton LM, Birren B et al (2001) Initial sequencing and analysis of the human genome. Nature 409(6822):860–921

2. Venter JC, Adams MD, Myers EW et al (2001) The sequence of the human genome. Science 291(5507):1304–1351

3. Pruitt KD, Tatusova T, Maglott DR (2007) NCBI reference sequences (RefSeq): a curated non-redundant sequence database of genomes, transcripts and proteins. Nucleic Acids Res 35(database issue):D61–65

4. Kersey PJ, Duarte J, Williams A et al (2004) The International Protein Index: an integrated database for proteomics experiments. Proteomics 4(7):1985–1988

5. UniProt: a hub for protein information (2015). Nucleic Acids Res 43(database issue): D204–D212

6. Gaudet P, Argoud-Puy G, Cusin I et al (2013) neXtProt: organizing protein knowledge in the context of human proteome projects. J Proteome Res 12(1):293–298

7. Brosch M, Saunders GI, Frankish A et al (2011) Shotgun proteomics aids discovery of novel protein-coding genes, alternative splicing, and "resurrected" pseudogenes in the mouse genome. Genome Res 21(5):756–767

8. Kumar D, Yadav AK, Kadimi PK et al (2013) Proteogenomic analysis of Bradyrhizobium japonicum USDA110 using GenoSuite, an automated multi-algorithmic pipeline. Mol Cell Proteomics 12(11):3388–3397

9. Gupta N, Benhamida J, Bhargava V et al (2008) Comparative proteogenomics: combining mass spectrometry and comparative genomics to analyze multiple genomes. Genome Res 18(7):1133–1142

10. Castellana NE, Payne SH, Shen Z et al (2008) Discovery and revision of Arabidopsis genes by proteogenomics. Proc Natl Acad Sci U S A 105(52):21034–21038

11. Kelkar DS, Kumar D, Kumar P et al (2011) Proteogenomic analysis of Mycobacterium tuberculosis by high resolution mass spectrometry. Mol Cell Proteomics 10(12):M111. 011627

12. Prasad TS, Harsha HC, Keerthikumar S et al (2012) Proteogenomic analysis of Candida glabrata using high resolution mass spectrometry. J Proteome Res 11(1):247–260

13. Nagarajha Selvan LD, Kaviyil JE, Nirujogi RS et al (2014) Proteogenomic analysis of pathogenic yeast Cryptococcus neoformans using high resolution mass spectrometry. Clin Proteomics 11(1):5

14. Pawar H, Sahasrabuddhe NA, Renuse S et al (2012) A proteogenomic approach to map the proteome of an unsequenced pathogen—Leishmania donovani. Proteomics 12(6): 832–844

15. Nirujogi RS, Pawar H, Renuse S et al (2014) Moving from unsequenced to sequenced genome: reanalysis of the proteome of Leishmania donovani. J Proteomics 97:48–61

16. Chaerkady R, Kelkar DS, Muthusamy B et al (2011) A proteogenomic analysis of Anopheles gambiae using high-resolution Fourier transform mass spectrometry. Genome Res 21(11): 1872–1881

17. Kelkar DS, Provost E, Chaerkady R et al (2014) Annotation of the zebrafish genome through an integrated transcriptomic and proteomic analysis. Mol Cell Proteomics 13(11):3184–3198

18. Kim MS, Pinto SM, Getnet D et al (2014) A draft map of the human proteome. Nature 509(7502):575–581

19. Rice P, Longden I, Bleasby A (2000) EMBOSS: the European Molecular Biology Open Software Suite. Trends Genet 16(6):276–277

20. Elias JE, Gygi SP (2007) Target-decoy search strategy for increased confidence in large-scale protein identifications by mass spectrometry. Nat Methods 4:207–214

21. Jeong K, Kim S, Bandeira N (2012) False discovery rates in spectral identification. BMC Bioinformatics 13:S2

22. Bonzon-Kulichenko E, Garcia-Marques F, Trevisan-Herraz M et al (2015) Revisiting peptide identification by high-accuracy mass spectrometry: problems associated with the use of narrow mass precursor windows. J Proteome Res 14(2):700–710

Chapter 6

Mass Spectrometry-Based Quantitative O-GlcNAcomic Analysis

Junfeng Ma and Gerald W. Hart

Abstract

The dynamic co- and post-translational modification (PTM) of proteins, O-linked β-D-N-acetylglucosamine modification (O-GlcNAcylation) of serine/threonine residues is critical in many cellular processes, contributing to multiple physiological and pathological events. The term "O-GlcNAcome" refers to not only the complete set of proteins that undergo O-GlcNAcylation but also the O-GlcNAc status at individual residues, as well as the dynamics of O-GlcNAcylation in response to various stimuli. O-GlcNAcomic analyses have been a challenge for many years. In this chapter, we describe a recently developed approach for the identification and quantification of O-GlcNAc proteins/peptides from complex samples.

Key words Chemoenzymatic labeling, Electron transfer dissociation (ETD), GalT1 labeling, O-GlcNAcylation, O-GlcNAcome, Photocleavage, Site mapping, SILAC, Quantitative mass spectrometry

1 Introduction

O-linked β-D-N-acetylglucosamine (O-GlcNAc) addition (O-GlcNAcylation) to serine/threonine residues is an important posttranslational modification on myriad proteins [1]. As a nutrient sensor, protein O-GlcNAcylation quickly responds to extracellular stimuli and nutrient status, regulating intracellular metabolic and signaling pathways [2, 3]. Aberrant O-GlcNAcylation contributes to the progression of multiple chronic diseases including diabetes [4–6], cancer [7–9], and neurodegenerative diseases [4, 10, 11].

Detection of protein O-GlcNAcylation has been a challenge since its discovery [12], largely due to the lack of sensitive tools [13–15]. Tritiated UDP-Galactose (i.e., UDP-[³H]-galactose) labeling has been used for the determination of O-GlcNAc status of proteins for over 30 years, and is still a commonly used approach. A number of pan-specific antibodies (e.g., CTD110.6 and RL2) have been exploited for probing O-GlcNAcylated proteins [14, 15].

Salvatore Sechi (ed.), *Quantitative Proteomics by Mass Spectrometry*, Methods in Molecular Biology, vol. 1410,
DOI 10.1007/978-1-4939-3524-6_6, © Springer Science+Business Media New York 2016

Although very useful for probing the O-GlcNAc status of proteins, these methods cannot provide accurate modification site information.

Newly emerging mass spectrometry techniques show tremendous advantages (e.g., high sensitivity, reliability, and throughput), substantially facilitating the discovery of O-GlcNAc sites on proteins and thus the elucidation of site-specific functions in diverse biological contexts. Due to the labile nature of the O-linked glycosidic bond between the GlcNAc moiety and its host peptides, traditional mass spectrometric approaches using collision induced dissociation (CID) and high-energy collision dissociation (HCD) often fail to directly assign the O-GlcNAc sites. However, those approaches are useful for indirect assignment of O-GlcNAc sites by combining with techniques that can convert the glycosidic bond to a CID/HCD-stable covalent bond (e.g., β-elimination followed by Michael addition with dithiothreitol (BEMAD)). As complementary fragmentation alternatives, electron-capture dissociation (ECD), and especially the recently introduced electron-transfer dissociation (ETD), show great promise for the detection of O-GlcNAc peptides by preserving the O-GlcNAc moiety on peptides, enabling facile and accurate assignment of modification sites. Another unique feature of these mass spectrometric approaches is their quantitative capacity by combining with isotopic labeling at the protein level (e.g., stable isotope labeling by amino acids in cell culture (SILAC)) or at the peptide level (e.g., isobaric tag for relative and absolute quantitation reagents (iTRAQ™; Applied Biosystems) and tandem mass tag reagents (TMT™; Thermo Fisher Scientific)) or label-free techniques.

As with other PTMs, enrichment is required for successful detection of O-GlcNAc peptides. To this end, a number of methods have been developed, including antibody-based immunocapture [16–18], lectin affinity [19–22], and chemical derivatization (e.g., metabolic labeling [23–27], chemoenzymatic labeling [28–34], and BEMAD [35–39]). Although relatively simple, the former two approaches often suffer from low affinity toward O-GlcNAc proteins/peptides. In contrast, chemical derivatization methods are generally unbiased for O-GlcNAc enrichment. One of the most recently developed chemical derivatization approaches is based upon chemical/enzymatic labeling (i.e., GalT1 labeling) followed by *Photo Cleavable*-PEG-*Biotin*-Alkyne (i.e., PC-Biotin)-based enrichment [31, 32, 34].

In this chapter, we focus on the analysis of O-GlcNAc proteins in complex samples. Specifically, we will mainly introduce a protocol by integrating SILAC-based protein labeling, GalT1 labeling, the subsequent PC-Biotin-based O-GlcNAc peptide enrichment, and the ETD-based mass spectrometric identification and quantification, as shown in Fig. 1a. Figure 1b illustrates a detailed scheme of GalT1 labeling and PC-Biotin-based enrichment, with the synthesis of PC-Biotin shown in Fig. 1c. It should be noted that

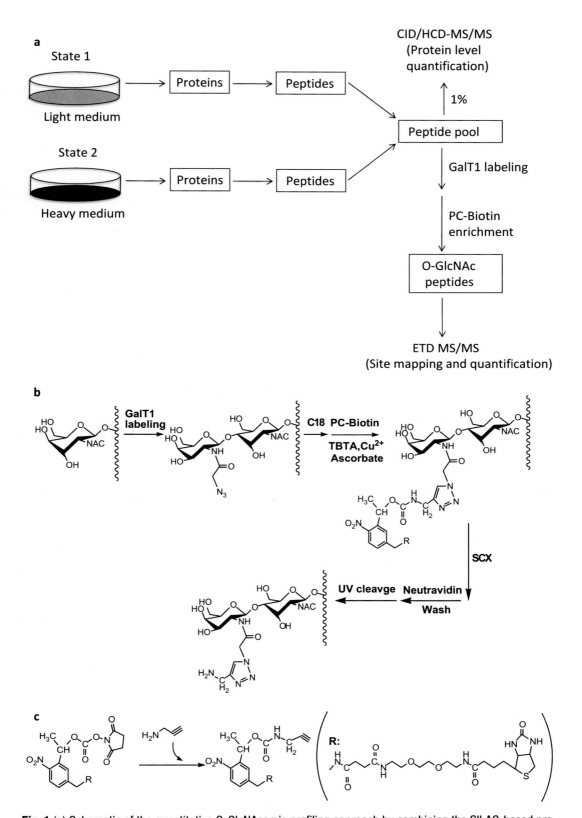

Fig. 1 (**a**) Schematic of the quantitative O-GlcNAcomic profiling approach by combining the SILAC-based protein labeling, the chemoenzymatic labeling and the subsequent PC-Biotin-based O-GlcNAc peptide enrichment, and the ETD-based mass spectrometric identification and quantification. (**b**) Schematic of the chemoenzymatic labeling and PC-Biotin-based O-GlcNAc peptide enrichment. (**c**) Synthesis of PC-Biotin

other isotopic labeling techniques (e.g., iTRAQ™, TMT™) and enrichment techniques could also be adapted and/or incorporated into this protocol for mass spectrometry-based O-GlcNAcomic profiling. These methodological developments are greatly facilitating the detection and quantification of protein O-GlcNAcylation, and will allow the dissection of the O-GlcNAc's biological roles at the individual site level on polypeptides.

---

2 Materials

2.1 SILAC Labeling

1. Heavy isotope-labeled amino acids (e.g., ^{13}C6-L-lysine and ^{13}C6-L-arginine for two-state SILAC experiment; Cambridge Isotope Laboratories).

2. Customized SILAC labeling medium: cell culture medium deficient in L-lysine and L-arginine, L-methionine, or other amino acids (AthenaES).

3. Supplements for a complete medium: e.g., dialyzed fetal bovine serum (Invitrogen or Quality Biological Inc.) and Penicillin-Streptomycin (Mediatech, Inc.).

4. 0.2-µm sterilization filter unit and sterile filter bottle (cell culture-grade).

5. Cell culture medium (DMEM or RPMI media containing the naturally abundant isotopic forms of amino acids).

6. Cells lines of interest.

7. Cell disassociation buffer (enzyme-free, PBS-based; Life Technologies).

8. Phosphate buffered saline (PBS).

9. PUGNAc (Sigma) and Thiamet G (Sigma or Cayman Chemical).

10. Protease inhibitor cocktail I (sigma).

11. Reagents and equipment for measuring total protein concentration.

2.2 GalT1 Labeling and PC-Biotin-Based O-GlcNAc Peptide Enrichment and Analysis

1. 100 % MeOH.

2. 100 % acetonitrile (ACN).

3. 10 mM Tris–HCl (pH 7.9).

4. 200 mM NaH_2PO_4/300 mM Sodium acetate (±400 mM KCl).

5. 5 mM KH_2PO_4, 25 % ACN (pH 3.0).

6. 80 % ACN/0.1 % Trifluoroacetic acid (TFA).

7. 0.1 % TFA.

8. 5 % Formic acid.

9. Chloroform ($CHCl_3$).

10. Urea.

11. Dithiothreitol (DTT).

12. Iodoacetamide.

13. Trypsin (Promega).

14. Calf intestinal phosphatase (CIP; New England Biolabs).

15. Peptide:N-glycosidase F (PNGase F; New England Biolabs).

16. 250 μm silica gel plate (Analtech).

17. TBTA (Tris-[(1-benzyl-1H-1,2,3-triazol-4-yl) methyl]amine, also known as tris-(benzyltriazolylmethyl)amine; Anaspec).

18. *tert*-Butanol (Sigma).

19. N-hydroxysuccinimidyl-PEG-biotin (Ambergen).

20. Propargylamine (Sigma).

21. 20 mM $CuSO_4$.

22. Sodium ascorbate (Sigma).

23. Click-iT *O*-GlcNAc Enzymatic Labeling System (Invitrogen).

24. Vydac C-18 Column (Nest Group #SUM SS18V).

25. SCX Column (Nest Group).

26. Empty Column (Nest Group).

27. High Capacity Neutravidin Agarose (Thermo Fisher Scientific).

28. UV lamp (Blak-Ray Lamp, Model XX-15; UVP, Upland, CA).

29. Speed-Vac concentrator.

30. LTQ Orbitrap XL™ ETD mass spectrometer, LTQ Orbitrap Velos™ ETD mass spectrometer, or other ETD-enabled mass spectrometers (Thermo Fisher Scientific).

3 Methods

3.1 SILAC Labeling

1. Weight out the appropriate amount of stable isotopes to customized SILAC media (as presented in Table 1 as a brief reference, *see* **Note 1**; SILAC Protein Quantitation Kits are also commercially available from several companies including Thermo Fisher Scientific and Life Technologies, which may not be that cost-effective).

2. Filter the reconstituted media with a 0.2-μm pore sterilization filter unit (*see* **Note 1**).

3. Add the required supplements to make the complete medium (for labeled and unlabeled medium). Generally, 10 % (v/v) dialyzed FBS and 100 U/mL Penicillin-Streptomycin are included, others are optional (*see* **Note 1**).

Table 1
Commonly used isotopic amino acids and their concentrations for the reconstitution of culture media for SILAC labeling

Amino acids (MW)	Amount in RPMI-1640 (Concentration)	Amount in DMEM (Concentration)
L-Arginine HCl (210.66)	242.3 mg/L (1.15 mM)	83.8 mg/L (0.398 mM)
L-Lysine HCl (182.65)	50.0 mg/L (0.274 mM)	145.8 mg/L (0.798 mM)
L-Tyrosine disodium salt (225.15)	25.0 mg/L (0.111 mM)	89.6 mg/L (0.398 mM)
$^{13}C_6$ L-Arginine HCl (216.62)	249.1 mg/L (1.15 mM)	86.2 mg/L (0.398 mM)
$^{13}C_6,^{15}N_4$ L-Arginine HCl (220.59)	253.7 mg/L (1.15 mM)	87.8 mg/L (0.398 mM)
$^{13}C_6$ L-Lysine HCl (188.60)	51.7 mg/L (0.274 mM)	150.5 mg/L (0.798 mM)
$^{13}C_6,^{15}N_2$ L-Lysine HCl (190.59)	52.2 mg/L (0.274 mM)	152.1 mg/L (0.798 mM)
$^{13}C_9$ L-Tyrosine (190.12)	21.1 mg/L (0.111 mM)	75.7 mg/L (0.398 mM)

4. Culture cells in SILAC heavy medium (i.e., isotopic amino acid-containing) and in normal medium for at least five passages (*see* **Notes 2** and **3**).

5. Harvest cells. Adherent cells from the dish can be released into ice-cold PBS using a cell scraper; suspension cells can be harvested by direct centrifugation.

6. Centrifuge cells at $500 \times g$ for 5 min at 4 °C. Wash once with ice-cold PBS.

3.2 Protein Extraction and Digestion

3.2.1 Extraction of Proteins from Cells

1. Resuspend the pellets in a cell lysis buffer (e.g., 50 mM Tris–HCl/pH 7.5, 150 mM NaCl, 1 mM EDTA, and 1 % NP40) supplemented with 2 μM PUGNAc and/or Thiamet G and 1× protease inhibitor cocktail I, incubated on ice for 30 min (brief sonication is recommended to achieve better yields of proteins) (*see* **Note 4**).

2. Centrifuge at $13,000 \times g$ for 10 min at 4 °C.

3. The concentration of the supernatant is determined by the Bradford assay or the bicinchoninic acid (BCA) assay according to manufacturer's instructions.

4. Mix equal amount of proteins from cell lysates from each state.

3.2.2 Digestion of Extracted Proteins

1. Add four volumes of cold acetone to the pooled proteins, keep at −80 °C for at least an hour.

2. Spin down the precipitates at $500 \times g$ for 5 min at 4 °C. Wash with cold acetone two more times. (Other protein precipitation approaches, e.g., the chloroform/methanol method, is optional.)

3. Carefully remove the liquid above the protein precipitates, leaving the cap open to allow air-dry for several minutes (*see* **Note 5**).

4. Resuspend the protein pellet in 8 M urea and 50 mM NH_4HCO_3 (pH 8.0).

5. Add freshly prepared DTT in 50 mM NH_4HCO_3 (final concn.: 10 mM), 37 °C for 30 min.

6. Add freshly prepared iodoacetamide (final concn.: 30 mM), RT for 30 min in dark.

7. Dilute with 100 mM NH_4HCO_3 (pH 8.0) to a final urea concentration <1 M, and then add trypsin (trypsin/protein = 1/50 (w/w)), with gentle shake at 37 °C overnight.

8. Acidify the peptide solution with 10 % TFA (final pH: ~3) and desalt with a Microspin c18 column, according to the manufacturer instructions.

9. Dry down with a SpeedVac.

3.3 GalT1 Labeling and PC-Biotin-Based O-GlcNAc Enrichment

3.3.1 Synthesis of PC-Biotin

1. Incubate 6 μmol of N-hydroxysuccinimidyl-PEG-biotin with 60 μmol of propargylamine in dry methanol at room temperature for 4 h in the dark (the reaction scheme is shown in Fig. 1c).

2. Run thin layer chromatography (250 μm silica gel plate) with methanol/chloroform (1:9, v/v) as the mobile phase to separate the reactants and the product (PC-Biotin).

3. Expose the thin layer plate briefly (<1 s) to 254 nm U.V. light, extract the product by scraping the thin-layer zone into dry methanol.

4. Remove silica gel by centrifugation.

5. Store the purified product in methanol at –20 °C until use (*see* **Note 6**).

3.3.2 O-GlcNAc Enrichment (Fig. 1b)

GalT1 Labeling

1. Resuspend the peptides (e.g., 20 μg of protein digest; from Subheading 3.2.2) in 100 μl of 10 mM HEPES with a pH of 7.9 (*see* **Note 7**).

2. Perform GalT1 labeling by sequentially adding 22 μl $MnCl_2$, 25 μl UDP-GalNAz (from the O-GlcNAc labeling kit from Life Technologies), mix well (*see* **Note 8**).

3. Add 15 μl GalT1 (from the O-GlcNAc labeling kit from Life Technologies; *see* **Note 9**) and PNGase F (500 U) into the reaction mixture, pipet up and down for ten times (*see* **Note 10**); leave at 4 °C overnight (*see* **Note 11**).

4. Add 20 U CIP (2 μl) and incubate at room temperature for 3 h on wheel.

5. Clean-up with a c18 Microspin column.

**Click Chemistry
and Neutravidin Capture**

1. Freshly prepare 50 mM sodium ascorbate in H_2O (*see* **Note 12**).

2. Add 8 µl of PC-PEG-Biotin-Alkyne, 8 µl 50 mM sodium ascorbate, 22 µl 1.7 mM TBTA (in 4:1 of *tert*-butanol:DMSO) sequentially into the peptides, vortex briefly, and spin down.

3. Add 4 µl freshly prepared 20 mM $CuSO_4$, mix (*see* **Note 13**).

4. Cover with aluminum foil and incubate overnight at room temperature (*see* **Note 14**).

5. Pre-wet the SCX column with 100 % MeOH and immerse in 200 mM NaH_2PO_4/300 mM sodium acetate; cover with Parafilm and let sit overnight at room temperature (*see* **Note 15**).

6. Spin the rest of the 0.2 M NaH_2PO_4/300 mM sodium acetate through.

7. Wash SCX column 3× with 200 µl 5 mM KH_2PO_4/25 % ACN (pH 3.0).

8. Add 80 µl 5 mM KH_2PO_4/25 % ACN (pH 3.0) to the peptide sample.

9. Load diluted sample (~100 µl) onto SCX column and spin through ($1000 \times g$)—repeat once.

10. Elute with 130 µl of 5 mM KH_2PO_4, 25 % ACN (pH 3.0)+400 mM KCl.

11. Neutralize sample with 0.2 % NH_4OH (final pH 7).

12. Put 500 µl high capacity neutralization agarose resin into a 15 mL conical tube; wash with 10 mL cold PBS for five times.

13. Load beads to the neutralized sample, cover with aluminum foil, and put on wheel for 2 h at room temperature.

14. Wash the beads 10× with cold PBS, 2× with H_2O, 1× with 20 % MeOH, 1× with 70 % MeOH (*see* **Note 16**).

15. Split the beads into ~4 PCR thin-wall tubes (with a final volume of ~100–150 µl in 70 % MeOH).

**UV-Cleavage of O-GlcNAc-
Tagged Peptides**

1. Put PCR tubes on a wheel ~2 in. from UV source (365 nm), rotate for 25 min (*see* **Note 17**).

2. Vortex briefly and pulse spin down beads, collect and pool supernatant from PCR tubes into one 1.5 mL tube.

3. Pulse spin down and collect supernatant a second time (no beads can be present; *see* **Note 18**).

4. Dry peptides to completion with a SpeedVac.

3.4 Mass Spectrometry and Data Analysis

1. Resuspend the peptides in 0.1 % formic acid, and load onto a 75 µm i.d. trap column packed with C18 particle (5 µm diameter, 120 Å) and an Eksigent nano-LC system (Dublin, CA) equipped with an integrated electrospray emitter tip.

2. Peptides can be gradient-eluted into the LTQ Orbitrap ETD mass spectrometer at a flow rate of 60–200 nL/min

(*see* **Note 19**). A quadrupole linear ion trap analyzer can be operated in a data-dependent mode to obtain ETD MS/MS spectra.

3. Peak lists can be generated from raw data files using Bioworks software (version 3.3.1 sp1). Open Mass Spectrometry Search Algorithm (OMSSA) (version 2.1.1) or others can be utilized to search c- and z-type fragment ions present in ETD MS/MS spectra against specific species (e.g., rat or mouse) in the NCBI nonredundant NR database. The general database searching parameters include the following: product ion mass tolerance, ±0.35 Da, up to three missed cleavages, in addition to variable modifications [e.g., oxidized methionine (+15.99), alkylated cysteine (+57.02 Da), tag on serine and threonine (+502.2024), light/heavy-labeled arginine/lysine] (*see* **Note 20**). A representative ETD spectrum of a synthetic O-GlcNAc peptide (YSPTgSPSK) is shown in Fig. 2 (as included is also a corresponding HCD spectrum showing the tags falling off from its peptide bone upon high-energy collision fragmentation conditions). The quantification of proteins/peptides can be obtained by using MaxQuant (http://141.61.102.17/maxquant_doku/doku.php?id=start) or other software available (*see* **Note 21**).

4   Notes

1. This step should be performed in a laminar flow hood.

2. Instead of the commonly used trypsin-EDTA solution, cell disassociation buffer (enzyme-free, PBS-based) is recommended to disassociate cells at all stages of passaging to avoid the introduction of naturally abundant amino acids into cells cultured in heavy-media.

3. Since dialyzed serum not only lacks free amino acids but also certain growth factors, some cell lines (e.g., nondividing cell lines and primary cells) might not grow well. Thus, the growth status of specific cell lines and the incorporation of heavy amino acids should be monitored prior to scaling up a SILAC experiment.

4. The inclusion of PUGNAc (Thiamet G is too specific for this purpose) into the lysis buffer is recommended to eliminate potential removal of GlcNAc from proteins mainly by lysosomal hexosaminidases during protein extraction.

5. Do not air-dry the protein precipitates too long, otherwise it would be difficult to re-dissolve them.

6. Mass spectrometry is recommended to confirm the identity of the product.

7. Peptides might be fractionated according to the complexity; standards (e.g., a synthetic O-GlcNAc peptide) should be included as a positive control.

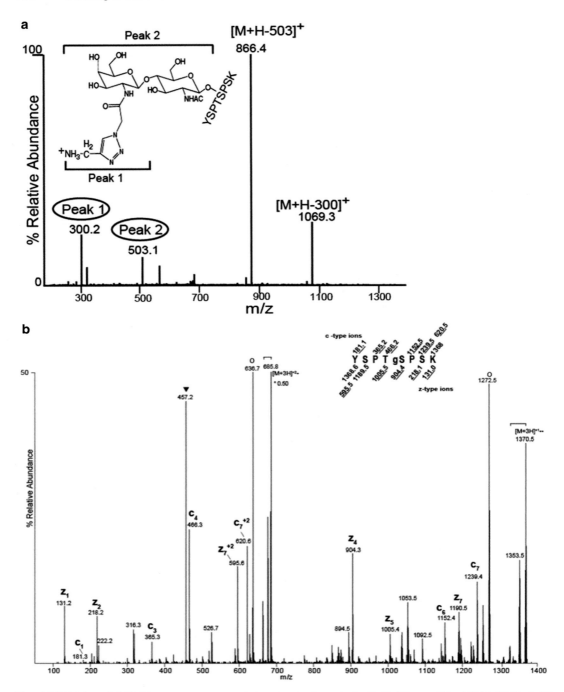

Fig. 2 HCD (**a**) and ETD (**b**) mass spectrum of the standard synthetic O-GlcNAc peptide (YSPTgSPSK; "gS" denotes an O-GlcNAc on Ser) after enrichment. (Reprinted from [31]; this figure was originally published in *Molecular and Cellular Proteomics* by Wang Z, Udeshi ND, O'Malley M, et al. Enrichment and site mapping of O-linked N-acetylglucosamine by a combination of chemical/enzymatic tagging, photochemical cleavage, and electron transfer dissociation mass spectrometry. 2010, 9, 153–160. © the American Society for Biochemistry and Molecular Biology.)

8. A 0.5 mM UDP-GalNAz solution is prepared by adding 144 mL of HEPES buffer (10 mM, pH 7.9) to component A. UDP-GalNAz solution should be aliquoted and stored in –80 °C freezer.

9. GalT1 should be kept at 4 °C. Do not freeze!

10. Do not vortex the reaction solution.

11. Make sure the final pH value of the reaction solution is ~7.9.

12. Sodium ascorbate should be freshly prepared to keep its reducing capacity.

13. $CuSO_4$ solution should be freshly prepared right before use.

14. The reaction tube should be covered with aluminum foil or kept in dark.

15. The SCX column should be well equilibrated before use.

16. Wash with 70 % MeOH is helpful to remove residual TBTA from the final product, otherwise it might interfere with the detection of some O-GlcNAc peptides by mass spectrometry.

17. It is recommended to run a positive control (e.g., cleavage of PC-Biotin) to check that photocleavage is working properly. If cleavage is not complete, run reaction longer or replace UV lamp bulbs.

18. Sample can be further filtered to remove the remaining beads that will clog the separation column.

19. LC conditions should be optimized according to the sample complexity.

20. Parameters, such as dynamic exclusion, MS/MS fragmentation, and collision energy, etc., should be optimized according to the specific instrument being used.

21. Manual inspection of the mass spectra is recommended for further confirmation of unambiguous site assignment and quantification.

Acknowledgments

Original research in this work was supported by NIH P01HL107153, R01DK61671, and NIH N01-HV-00240 (to G.W.H.). We appreciate Dr. Zihao Wang for his significant contribution for the initial development of this protocol and for his critical reading of the manuscript. Helpful discussion from the Hart laboratory is acknowledged. We also thank Drs. Feng Yang and Richard Smith and their coworkers at the Pacific Northwest National Laboratory (Richland, WA) for valuable comments of the enrichment protocol.

References

1. Hart GW, Housley MP, Slawson C (2007) Cycling of O-linked β-N-acetylglucosamine on nucleocytoplasmic proteins. Nature 446: 1017–1022

2. Hardivillé S, Hart GW (2014) Nutrient regulation of signaling, transcription, and cell physiology by O-GlcNAcylation. Cell Metab 20: 208–213

3. Harwood KR, Hanover JA (2014) Nutrient-driven O-GlcNAc cycling—think globally but act locally. J Cell Sci 127:1857–1867

4. Dias WB, Hart GW (2007) O-GlcNAc modification in diabetes and Alzheimer's disease. Mol Biosyst 3:766–772

5. Ma J, Hart GW (2013) Protein O-GlcNAcylation in diabetes and diabetic complications. Expert Rev Proteomics 10:365–380

6. Vaidyanathan K, Wells L (2014) Multiple tissue specific roles for the O-GlcNAc post-translational modification in the induction of and complications arising from type II diabetes. J Biol Chem 289:34466–34471

7. Slawson C, Hart GW (2011) O-GlcNAc signalling: implications for cancer cell biology. Nat Rev Cancer 11:678–684

8. Ma Z, Vosseller K (2014) Cancer metabolism and elevated O-GlcNAc in oncogenic signaling. J Biol Chem 289:34457–34465

9. Singh JP, Zhang K, Wu J, Yang X (2015) O-GlcNAc signaling in cancer metabolism and epigenetics. Cancer Lett 356:246–250

10. Lazarus BD, Love DC, Hanover JA (2009) O-GlcNAc cycling: implications for neurodegenerative disorders. Int J Biochem Cell Biol 41:2134–2146

11. Yuzwa SA, Vocadlo DJ (2014) O-GlcNAc and neurodegeneration: biochemical mechanisms and potential roles in Alzheimer's disease and beyond. Chem Soc Rev 43:6839–6858

12. Torres CR, Hart GW (1984) Topography and polypeptide distribution of terminal N-acetylglucosamine residues on the surfaces of intact lymphocytes: evidence for O-linked GlcNAc. J Biol Chem 259:3308–3317

13. Wang Z, Hart GW (2008) Glycomic approaches to study GlcNAcylation: protein identification, site-mapping, and site-specific O-GlcNAc quantitation. Clin Proteomics 4:5–13

14. Zachara NE (2009) Detecting the "O-GlcNAc-ome": detection, purification, and analysis of O-GlcNAc modified proteins. Methods Mol Biol 534:251–279

15. Ma J, Hart GW (2014) O-GlcNAc profiling: from proteins to proteomes. Clin Proteomics 11:8

16. Wang Z, Pandey A, Hart GW (2007) Dynamic interplay between O-linked N-acetylglucosaminylation and glycogen synthase kinase-3-dependent phosphorylation. Mol Cell Proteomics 6:1365–1379

17. Zachara NE, Molina H, Wong KY, Pandey A, Hart GW (2011) The dynamic stress-induced "O-GlcNAc-ome" highlights functions for O-GlcNAc in regulating DNA damage/repair and other cellular pathways. Amino Acids 40:793–808

18. Zhao P, Viner R, Teo CF et al (2011) Combining high-energy C-trap dissociation and electron transfer dissociation for protein O-GlcNAc modification site assignment. J Proteome Res 10:4088–4104

19. Vosseller K, Trinidad JC, Chalkley RJ et al (2006) O-Linked N-acetylglucosamine proteomics of postsynaptic density preparations using lectin weak affinity chromatography and mass spectrometry. Mol Cell Proteomics 5:923–934

20. Chalkley RJ, Thalhammer A, Schoepfer R et al (2009) Identification of protein O-GlcNAcylation sites using electron transfer dissociation mass spectrometry on native peptides. Proc Natl Acad Sci U S A 106:8894–8899

21. Trinidad JC, Barkan DT, Gulledge BF et al (2012) Global identification and characterization of both O-GlcNAcylation and phosphorylation at the murine synapse. Mol Cell Proteomics 11:215–229

22. Nagel AK, Schilling M, Comte-Walters S et al (2013) Identification of O-linked N-acetylglucosamine (O-GlcNAc)-modified osteoblast proteins by electron transfer dissociation tandem mass spectrometry reveals proteins critical for bone formation. Mol Cell Proteomics 12:945–955

23. Vocadlo DJ, Hang HC, Kim EJ et al (2003) A chemical approach for identifying O-GlcNAc-modified proteins in cells. Proc Natl Acad Sci U S A 100:9116–9121

24. Sprung R, Nandi A, Chen Y et al (2005) Tagging-via-substrate strategy for probing O-GlcNAc modified proteins. J Proteome Res 4:950–957

25. Hahne H, Sobotzki N, Tamara N et al (2013) Proteome wide purification and identification of O-GlcNAc-modified proteins using click chemistry and mass spectrometry. J Proteome Res 12:927–936

26. Zaro BW, Yang YY, Hang HC et al (2011) Chemical reporters for fluorescent detection and identification of O-GlcNAc-modified proteins reveal glycosylation of the ubiquitin ligase NEDD4-1. Proc Natl Acad Sci U S A 108:8146–8151

27. Boyce M, Carrico IS, Ganguli AS et al (2011) Metabolic cross-talk allows labeling of O-linked β-N-acetylglucosamine-modified proteins via the N-acetylgalactosamine salvage pathway. Proc Natl Acad Sci U S A 108:3141–3146

28. Khidekel N, Arndt S, Lamarre-Vincent N et al (2003) A chemoenzymatic approach toward the rapid and sensitive detection of O-GlcNAc posttranslational modifications. J Am Chem Soc 125:16162–16163

29. Khidekel N, Ficarro SB, Peters EC et al (2004) Exploring the O-GlcNAc proteome: direct identification of O-GlcNAc-modified proteins from the brain. Proc Natl Acad Sci U S A 101:13132–13137

30. Khidekel N, Ficarro SB, Clark MC et al (2007) Probing the dynamics of O-GlcNAc glycosylation in the brain using quantitative proteomics. Nat Chem Biol 3:339–348

31. Wang Z, Udeshi ND, O'Malley M et al (2010) Enrichment and site mapping of O-linked N-acetylglucosamine by a combination of chemical/enzymatic tagging, photochemical cleavage, and electron transfer dissociation mass spectrometry. Mol Cell Proteomics 9:153–160

32. Wang Z, Udeshi ND, Slawson C et al (2010) Extensive crosstalk between O-GlcNAcylation and phosphorylation regulates cytokinesis. Sci Signal 3:ra2

33. Parker BL, Gupta P, Cordwell SJ et al (2011) Purification and identification of O-GlcNAc-modified peptides using phosphate-based alkyne CLICK chemistry in combination with titanium dioxide chromatography and mass spectrometry. J Proteome Res 10:1449–1458

34. Alfaro JF, Gong CX, Monroe ME et al (2012) Tandem mass spectrometry identifies many mouse brain O-GlcNAcylated proteins including EGF domain-specific O-GlcNAc transferase targets. Proc Natl Acad Sci USA 109:7280–7285

35. Wells L, Vosseller K, Cole RN et al (2002) Mapping sites of O-GlcNAc modification using affinity tags for serine and threonine post-translational modifications. Mol Cell Proteomics 1:791–804

36. Vosseller K, Hansen KC, Chalkley RJ et al (2005) Quantitative analysis of both protein expression and serine/threonine post-translational modifications through stable isotope labeling with dithiothreitol. Proteomics 5:388–398

37. Ramirez-Correa GA, Jin W, Wang Z et al (2008) O-linked GlcNAc modification of cardiac myofilament proteins: a novel regulator of myocardial contractile function. Circ Res 103:1354–1358

38. Wang Z, Park K, Comer F et al (2009) Site-specific GlcNAcylation of human erythrocyte proteins: potential biomarker(s) for diabetes. Diabetes 58:309–317

39. Overath T, Kuckelkorn U, Henklein P et al (2012) Mapping of O-GlcNAc sites of 20S proteasome subunits and Hsp90 by a novel biotin-cystamine tag. Mol Cell Proteomics 11:467–477

Chapter 7

Isolating and Quantifying Plasma HDL Proteins by Sequential Density Gradient Ultracentrifugation and Targeted Proteomics

Clark M. Henderson, Tomas Vaisar, and Andrew N. Hoofnagle

Abstract

The sensitivity and specificity of tandem mass spectrometers have made targeted proteomics the method of choice for the precise simultaneous measurement of many proteins in complex mixtures. Its application to the relative quantification of proteins in high-density lipoproteins (HDL) that have been purified from human plasma has revealed potential mechanisms to explain the atheroprotective effects of HDL. We describe a moderate throughput method for isolating HDL from human plasma that uses sequential density gradient ultracentrifugation, the traditional method of HDL purification, and subsequent trypsin digestion and nanoflow liquid chromatography-tandem mass spectrometry to quantify 38 proteins in the HDL fraction of human plasma. To control for the variability associated with digestion, matrix effects, and instrument performance, we normalize the signal from endogenous HDL protein-associated peptides liberated during trypsin digestion to the signal from peptides liberated from stable isotope-labeled apolipoprotein A-I spiked in as an internal standard prior to digestion. The method has good reproducibility and other desirable characteristics for preclinical research.

Key words Multiple reaction monitoring, Tandem mass spectrometry, Nanoflow liquid chromatography, Stable isotope-labeled internal standard protein, Skyline, High-density lipoprotein, Protein quantification, Quantitative proteomics, Density gradient ultracentrifugation

1 Introduction

Elevated plasma concentrations of cholesterol bound to high-density lipoproteins (HDL-C) are associated with a reduced risk for developing cardiovascular disease, which has been well established in large epidemiological studies [1–3]. It has been proposed that HDL particles serve several anti-atherosclerotic functions, namely: reverse cholesterol transport from peripheral tissues, including from lipid plaques in the vascular intima of arteries [4, 5]; prevention or reduction of inflammation that mitigates endothelial cell activation [6, 7]; and prevention or reduction of the oxidation of low-density lipoproteins (LDL) [8, 9]. Analysis of the HDL proteome using

Salvatore Sechi (ed.), *Quantitative Proteomics by Mass Spectrometry*, Methods in Molecular Biology, vol. 1410,
DOI 10.1007/978-1-4939-3524-6_7, © Springer Science+Business Media New York 2016

shotgun mass spectrometry has demonstrated a complex heterogeneous composition of proteins of humoral and cellular origin with a vast array of functions [10–15]. The interplay of these proteins in the development of atherosclerosis and which proteins would serve as effective therapeutic targets is currently an intense area of research [14, 16]. Targeted mass spectrometry (MRM/SRM/PRM) has been demonstrated to be an effective method of quantifying apolipoproteins associated with HDL using stable isotope-labeled peptides and protein internal standards [17–19].

The protocol outlined here employs sequential potassium bromide density gradient ultracentrifugation to isolate the high-density lipoprotein ($\rho = 1.063$-1.210 g/mL) fraction of human plasma. Proteins are dialyzed to remove the potassium bromide and the protein concentration is determined using the Bradford method. Isolated HDL proteins are reduced, alkylated, and digested with trypsin. The resulting peptides are dried under vacuum and then reconstituted in an acidic slightly organic solvent for analysis using nanoflow liquid chromatography-tandem mass spectrometry. Peptides derived from the proteins of interest are identified by their retention time on the chromatographic column, their precursor mass (detected as the mass-to-charge ratio by the mass spectrometer), and the mass-to-charge ratio of specific fragments generated by collision-induced dissociation of the precursor peptide. Stable isotope-labeled internal standard protein (apolipoprotein A-I, apoA-I) spiked prior to digestion is used to control for digestion variability, matrix effects, and fluctuations in instrument performance. The chromatographic peak areas corresponding to the unlabeled endogenous peptides and the ^{15}N-labeled apoA-I peptides are determined in the Skyline software package [20]. The relative abundance of each protein in HDL is then calculated as sum of the peak area of two to six of the most intense precursor-fragment pairs (i.e., transitions) of the endogenous peptide normalized by the peak area of one of the stable isotope-labeled internal standard peptides.

2   Materials

All buffers used for HDL isolation should be prepared with ultrapure deionized water and analytical grade reagents. All solvents used in MS analysis should be LC-MS grade purity. All reagents and solvents should be stored at room temperature unless otherwise noted. Waste reagents and solvents should be appropriately disposed of according to local regulations.

2.1 Plasma HDL Isolation

1. Potassium bromide (KBr)-containing microcentrifuge tubes: Add 400 µL of 1.1863 g/mL (at 25 °C) KBr stock solution to 0.6 mL microcentrifuge tubes and dry down in vacuum

centrifuge. Each microcentrifuge tube should contain 108.7 mg of dry KBr (*see* **Note 1**).

2. Saline solution: 0.9 % (w/v) NaCl in H_2O with 500 mM EDTA.

3. Potassium phosphate/DTPA (K_2HPO_4/DTPA) buffer: 20 mM potassium phosphate, pH 7.4, 100 μM diethylene triamine pentaacetic acid (DTPA) in H_2O.

4. Potassium bromide (KBr) solution 1: density = 1.21 g/mL or 10.17 M in H_2O (at 25 °C).

5. Potassium bromide (KBr) solution 2: density = 1.063 g/mL or 8.93 M in H_2O (at 25 °C).

6. Dialysis cups: 3.5 kDa molecular weight cut-off (e.g., Thermo P/N 69552).

7. Microcentrifuge tube float, 500 mL beaker, and stir plate at 5 °C.

8. Ultracentrifuge with appropriate rotors and tubes (tube dimensions: 8 mm × 34 mm).

9. Metal (aluminum or stainless steel) ultracentrifuge tube rack.

10. Hamilton syringe with Chaney adapter to set volume at 125 μL.

11. Microcentrifuge capable of 15k × g.

12. Rocker table or similar agitation device.

2.2 Bradford Assay

1. Coomassie Plus Protein Assay Reagent (Thermo Pierce). Store at 4 °C. Warm to RT before use. The solution is light sensitive and exposure must be kept to a minimum.

2. Potassium phosphate/DTPA (K_2HPO_4/DTPA) buffer: 20 mM potassium phosphate, pH 7.4, 100 μM diethylene triamine pentaacetic acid (DTPA) in H_2O. Can use solution prepared in Subheading 2.1, **item 3**.

3. Bovine serum albumin (BSA) standard at 2 mg/mL in K_2HPO_4/DTPA buffer. Store at 4 °C. Warm to RT before use.

4. Clear flat-bottom 96-well plate.

5. Ultraviolet (UV) plate reader.

2.3 Stable Isotope-Labeled Proteins

1. ^{15}N-labeled apoA-I internal standard protein (e.g., Cambridge Isotopes Laboratories, Tewksbury, MA).

2.4 Trypsin Digestion

1. 100 mM Ammonium bicarbonate buffer (AmBic): Dissolve 395.3 mg AmBic in 50 mL H_2O. AmBic buffer should be prepared and used within 12 h.

2. 250 mM Dithiothreitol (DTT): Dissolve 50 mg of DTT in 648 mL of 100 mM AmBic. Prepare fresh solution every 12 h.

3. 0.2 % *Rapi*Gest SF Surfactant (Waters): Dissolve contents of 1 mg vial in 500 μL of 100 mM AmBic. Solution is stable for 1 week at 4 °C.

4. *Rapi*Gest/DTT solution: Combine 675 µL of 0.2 % *Rapi*Gest and 27 µL of 250 mM DTT. Sufficient for 24 digestions.

5. 500 mM Iodoacetamide (IAA): Dissolve 50 mg of IAA in 541 µL of 100 mM AmBic. Note: IAA is extremely light sensitive, prepare just prior to use ($\leq$2 min). Discard after use.

6. 1 mM Hydrochloric acid (HCl): Add 30 µL of 500 mM HCl stock to 15 mL of H_2O to obtain a 1 mM HCl solution to dissolve lyophilized trypsin.

7. 0.05 µg/µL Trypsin: Just prior to digestion, add 400 µL of 1 mM HCl to 20 µg of lyophilized trypsin (e.g., Promega Gold sequencing grade trypsin). Trypsin should be dissolved by gently re-pipetting solution until dissolved. Store on ice until use. Do not vortex.

8. All digestions should be performed in low protein binding microcentrifuge tubes, e.g., Eppendorf SafeLock LoBind 1.5 mL, in a Thermomixer that has been calibrated at 37 °C and speed of 1400 rpm.

9. 0.25 µg/µL ^{15}N-labeled apoaA-I IS$_{prot}$: Dissolve 250 µg of ^{15}N-labeled apoaA-I in 1000 µL of 100 mM AmBic. Keep on wet ice. Aliquot as necessary and store at –20 °C.

10. 10 % FA in H_2O (10 % FA): Add 455 µL of 88 % FA to a total volume of 4 mL in 15 mL conical tube using H_2O (*see* **Note 2**).

11. 1 % FA in Acetonitrile (1 % FA/ACN): Add 250 µL of 88 % FA to a total volume of 22 mL in a 50 mL conical tube using acetonitrile (ACN) (*see* **Note 2**).

12. Other equipment necessary for trypsin protein digestion includes: 1.5 mL low protein retention microcentrifuge tubes (e.g., Fisher Scientific P/N02-681-320), microcentrifuge capable of achieving 15k×g, speedvac, vortexer, phospholipid removal plate (e.g., Phenomenex Phree), and a positive pressure manifold, e.g., Biotage Pressure +96. The methods outlined here have been optimized for Phenomenex Phree 96-well plates.

2.5 Nano-LC-MRM-MS Solvents, Mobile Phases, and Columns

1. Sample suspension solvent: 95 % H_2O, 5 % acetonitrile, 0.1 % formic acid (FA).

2. Sample dilution solvent: 0.1 % FA in H_2O.

3. Mobile phase A: 98:1 H_2O to acetonitrile with 0.1 % FA. Mobile phase B: 98:1 acetonitrile to H_2O with 0.1 % FA.

4. Weak needle wash: H_2O with 0.1 % FA. Strong needle wash: 2-propanol with 0.1 % FA.

5. Trap column: XBridge BEH C18, 5 µm, 100 Å silica beads (Waters, MA) (or equivalent C18 packing material) are packed into a 0.1 mm i.d. fritted fused silica capillary (New Objective,

MA) (or equivalent) to 30×0.1 mm column and fused silica capillary is cut to length of ~15 cm.

6. Analytical column: XBridge BEH C18, 3.5 μm, 100 Å silica beads (Waters, MA) (or equivalent C18 packing material) are packed into a pulled tip fused silica capillary (75 μm i.d.) to 120 length.

7. Capillary column oven capable of maintaining temperature of 50 °C for analytical column.

3   Methods

All procedures should be performed at room temperature unless otherwise noted. When handling blood products, always take universal precautions such as appropriate PPE and proper use of microcentrifuges to minimize aerosols.

3.1 HDL Isolation from Human Plasma

1. Pre-chill ultracentrifuge and rotors to 5 °C.

2. If necessary, rapidly thaw plasma at 37 °C, then briefly centrifuge ($\leq$30 s at ~2000 $\times g$). Place plasma on wet ice.

3. Add 335 μL of plasma to microcentrifuge tubes containing 108.9 mg KBr, then place on rocker table to dissolve KBr. Briefly centrifuge and place tubes on wet ice.

4. Add 350 μL of plasma/KBr solution from **step 3** to ultracentrifuge tube using forward pipetting. Place tubes in metal rack (on wet ice) to keep the tubes chilled when not in the centrifuge.

5. Add 150 μL of KBr solution 1 (ρ = 1.21 g/mL at 25 °C) to ultracentrifuge using reverse pipetting.

6. Place ultracentrifuge tubes into chilled rotor. Spin samples at 120k rpm (625,698 $\times g$) for 4.5 h at 5 °C.

7. Remove 120 μL from meniscus. Dispense into the bottom of a new ultracentrifuge tube and rinse 1 time (*see* **Note 3**).

8. Add 239 μL saline solution to ultracentrifuge tube (*see* **Note 4**).

9. Add 140.8 μL KBr solution 2 (ρ = 1.063 g/mL at 25 °C) to the ultracentrifuge tube.

10. Place ultracentrifuge tube into chilled rotor. Spin samples at 120k rpm (625,698 $\times g$) for 2.5 h at 5 °C.

11. Using the Hamilton syringe with Chaney adapter set to 125 μL, carefully remove 125 μL from the bottom of the ultracentrifuge tube, wipe the tip to remove the lipids that became adhered from passing through the meniscus, and dispense into 0.6 mL microcentrifuge tube. Place samples on wet ice (~5 °C).

3.2 Dialysis of Isolated HDL	1. Test dialysis cups for leaks by placing a tube float with empty dialysis cups in beaker containing K_2HPO_4/DTPA buffer for 10 min at 5 °C with gentle stirring. Check for signs of leakage. No liquid should be in the dialysis cup when the membrane is intact, although the membrane will be wet.

2. Remove float with dialysis cups from beaker. Dispense 125 µL of HDL from **step 11** above onto the dialysis membrane. Carefully place the float back in the beaker and incubate for 3 h with stirring at 5 °C. Buffer should be changed three times during dialysis for a total of three, 3-h incubations.

3. Blot outer portion of dialysis membrane on a clean paper towel to remove excess dialysis buffer. Place the dialysis cup in a labeled 1.5 mL microcentrifuge tube, briefly spin to maximum rpm ($\sim$15k$\times g$) to rupture membranes and transfer dialyzed HDL to microcentrifuge tube.

4. Place tubes on wet ice ($\sim$5 °C). Once the protein concentration has been determined, the samples can be aliquoted as necessary.

3.3   Bradford Assay to Determine Protein Concentration of Isolated HDL

1. Dilute standard to make a 7-point calibration curve from 1000 to 15 µg/mL by making a 1:1 mixture of 2 mg/mL BSA standard with K_2HPO_4/DTPA buffer (1000 µg/mL) and serial dilute (twofold) to 15.6 µg/mL.

2. In a 0.6 mL tube containing 27 µL of the K_2HPO_4/DTPA buffer (1:10 dilution), add 3 µL of the dialyzed HDL from Subheading 3.2, **step 4**.

3. In duplicate, pipette 10 µL of standard or sample into the appropriate wells of the clear, flat-bottom 96-well plate. Do not allow bubbles to form in well when pipetting.

4. In each well, add 150 µL of the Coomassie Plus Protein assay reagent to each well. Again, do not introduce bubbles into the well.

5. Gently agitate the plate for 1 min and incubate for a total of 10 min (*see* **Note 5**).

6. Read plate at $\lambda = 595$ nm.

7. Once the protein concentrations of the HDL samples have been determined (be sure to account for 1:10 dilution), prepare aliquots containing 5 µg HDL protein in Eppendorf LoBind SafeLock Tubes.

3.4   Digestion of Isolated HDL Using Trypsin

1. To each 5 µg HDL protein aliquot, add 4 µL of the 0.25 µg/µL ^{15}N-labeled apoA-I IS$_{prot}$ working stock solution.

2. Denaturation and reduction of proteins is performed by adding 26 µL of *Rapi*Gest/DTT solution, vortex briefly and incubate for 1 h in Thermomixer at 37 °C and 1400 rpm. Then briefly centrifuge to collect condensation.

3. Prepare 500 mM IAA stock solution (*see* Subheading 2.4, **item 5**). Add 1.5 μL of the 500 mM IAA stock solution ($C_{final} = 15$ mM IAA) to each sample. Vortex briefly, then incubate in the dark for 15 min. During the last 5 min of the incubation, prepare the trypsin working stock as described above (*see* Subheading 2.4, **item 7**).

4. Add 5 μL of 0.05 μg/μL trypsin solution to each sample and incubate on Thermomixer at 37 °C and 1400 rpm for 3 h. Be sure to place trypsin back on ice.

5. Briefly vortex HDL digests to collect condensate. Then add another 5 μL of 0.05 μg/μL trypsin to each digestion and incubate for 17 h on Thermomixer at 37 °C and 1400 rpm.

6. After incubation, briefly vortex digests to collect condensate. Then add 62.4 μL of 10 % FA solution to cleave *Rapi*Gest and stop digestion. Incubate for 45 min in Thermomixer at 37 °C and 1400 rpm.

7. Centrifuge samples at ≥15k×*g* for 15 min. Then verify order of samples in regard to orientation on the SPE plate.

8. Add 390 μL of 1 % FA/ACN to each well. Then add 110 μL of HDL digest to each well. Seal plate with clear sealing tape and place on plate shaker for 5 min.

9. Orient SPE plate over collection plate, remove plate seal tape, and mount plates on positive pressure displacement manifold.

10. To collect eluent, set initial pressure at 4–5 psi for 5 min. Then increase pressure to 7–8 psi and hold for 5 min. Finally, increase pressure to 10–11 psi for 5 min.

11. Transfer eluent to 1.5 mL low protein binding microcentrifuge tubes and speed vac without heat (~RT) to dryness; approximately 12 h.

12. Store dried samples at –80 °C.

3.5 Nano-LC-MRM-MS Analysis of Tryptic Peptides

An example configuration of the nano-LC-MRM-MS system is a nanoACQUITY UPLC (Waters) coupled with TSQ Vantage (Thermo Scientific). However, the method can be adapted to other triple-quadrupole mass spectrometers. Alternatively, the method could also be used in the form of parallel reaction monitoring (PRM) on hybrid instruments (e.g., quadrupole time-of-flight or quadrupole-Orbitrap). Skyline could facilitate the transfer of methods described here to another instrument in a vendor-blind manner (*see* below). Capillary column and trap column are maintained at the temperature of 50 °C by means of a column heater.

3.5.1 Sample Suspension for Nano-LC-MRM Analysis

1. Remove samples from –80 °C and allow them to warm to room temperature.

2. Centrifuge samples at ≥15k×*g* for 5 min to ensure peptide pellet is in the bottom of the tube.

Table 1
Chromatography schedule and flow rate for nano-LC-MRM analysis of HDL peptides

Time (min)	Mobile phase A (%)	Mobile phase B (%)	Flow rate (μL/min)
0	99	1	0.6
2	93	7	0.6
17	75	25	0.6
20	65	35	0.6
22	20	80	0.6
25	20	80	0.6
26	99	1	0.8
37 (end)	99	1	0.0

3. Add 15 μL of 95:5 H_2O/ACN with 0.1 % FA sample suspension solvent to each sample and mix at 1400 rpm for 2 h at RT in a Thermomixer.

4. Then add 15 μL of 0.1 % FA dilution solvent to each sample and briefly vortex.

5. Centrifuge samples at $\geq 15 k \times g$ for 15 min, then transfer to compatible vials for the auto-sampler coupled to the nano-LC unit. Ensure that no bubbles are present in the solution.

3.5.2 Chromatographic Conditions

1. Injected sample is trapped and washed on the trapping column for 5 min with Mobile phase A at the flow rate 4 μL/min.

2. Peptides are eluted from the trap column onto the analytical column and separated by the gradient detailed in Table 1.

3.5.3 Mass Spectrometer Acquisition Conditions

Capillary temp = 325 °C.

Spray voltage = 2200 V.

Q1 peak width = 0.7.

Q3 peak width = 0.7.

Collision gas: 1.5 mTorr.

Collision energy: Calculated in Skyline.

Dwell time: 10 ms (except selected low abundance peptides where dwell set to 30 ms).

3.6 Develop MRM Transition Lists and Perform Data Analysis Using Skyline

Download and install Skyline from MacCoss laboratory Skyline webpage at: https://skyline.gs.washington.edu/labkey/project/home/software/Skyline/begin.view (*see* **Note 6**).

3.6.1 Skyline Settings for MRM Experiment

1. Under the Settings option, open the Peptide Settings window. Choose the Digestion tab and in the Enzyme dropdown menu, select Trypsin [KR|P]. Select 0 in the Max missed cleavages dropdown menu. The Background proteome should be set to none.

2. In the Filter tab, set the minimum length to 6 and maximum length to 25. Auto-select all matching peptides should be selected.

3. Under the Modifications tab, in the Structural modifications box, select Carbamidomethyl Cysteine and Oxidation (M). Then select OK.

4. Open Transition Settings and click on the Full-Scan tab. In the MS1 Filtering box, choose none from the Isotope peaks included dropdown menu. Below, in the MS/MS filtering box, choose none from the Acquisition method dropdown menu.

3.6.2 Preparing the Transition List

1. Copy and paste each peptide from Table 2 Skyline file Targets list on the left side of the window. Alternatively, import FASTA files for proteins being analyzed and cull the list of peptides to those listed in Table 2 by right-clicking on the protein name and selecting Pick Children (*see* **Note 7**).

2. For each apoA-I peptide that has a stable isotope-labeled internal standard counterpart, include the modified residues in the peptide by right-clicking on the precursor *m/z* and select Modify. In the Edit Modifications window, in the Isotope heavy column, select the dropdown menu that corresponds to each peptide and choose Label: 15N. Then select OK.

3. Select the optimized transitions for each peptide from Table 2. When selecting transitions for peptides that include an internal standard, be sure to check the box that synchronizes selection for the stable isotope-labeled peptide (*see* **Note 8**).

3.6.3 Export Transition List

1. From the menu bar, navigate to Export then choose Transition List. In the Export Transition List window, select the instrument from the Instrument type dropdown menu. Select Single Method with no optimization and a standard method type.

2. Import transition list directly into MS method software. Verify that all precursor ion, transition ion, declustering potentials, collision energy, and any other necessary method parameters required by the instrument were correctly imported into MS method (*see* **Note 9**).

3.6.4 Analyzing MRM Data in Skyline

1. Open the Skyline file that was used to prepare the transition list and save it as an analysis file specifically for these MS data and this experiment (*see* **Note 10**).

Table 2
List of HDL proteins, peptides, and corresponding transitions

Protein	Peptide	Precursor ion (*m/z*)	Fragment ion (*m/z*)					
			1	2	3	4	5	6
ALB	FQNALLVR	480.8	685.4	500.4				
ALB	LVNEVTEFAK	575.3	937.5	694.4				
APOA1	DLATVYVDVLK	618.3	936.5	736.4				
APOA1#	DLATVYVDVLK	624.3	945.5	743.4				
APOA1	DYVSQFEGSALGK	700.8	1023.5	808.4				
APOA1#	DYVSQFEGSALGK	708.3	1035.5	817.4				
APOA1	VQPYLDDFQK	626.8	1025.5	513.3				
APOA1#	VQPYLDDFQK	633.3	1035.5	518.2				
APOA2	EPC[+57.0] VESLVSQYFQTVTDYGK	1175.5	1436.7	583.3				
APOA2	EQLTPLIK	471.3	571.4	470.3				
APOA2	SPELQAEAK	486.8	788.4	443.2				
APOA4	LGEVNTYAGDLQK	704.4	794.4	631.3				
APOA4	LTPYADEFK	542.3	772.4	435.2				
APOA4	SELTQQLNALFQDK	817.9	835.4	537.3				
APOA5	LRPLSGSEAPR	591.8	913.5	816.4	703.3	616.3	559.3	
APOA5	SVAPHAPASPAR	580.8	903.5	669.4	598.3	487.8	452.2	403.7
APOB	IEIPLPFGGK	535.8	828.5	715.4				
APOB	SVSLPSLDPASAK	636.3	1085.6	885.5				
APOC1	EFGNTLEDK	526.7	605.3	504.3				
APOC1	EWFSETFQK	601.3	739.4	523.3				
APOC1	TPDVSSALDK	516.8	834.4	466.2				
APOC2	TAAQNLYEK	519.3	865.4	666.3				
APOC2	TYLPAVDEK	518.3	771.4	658.3				
APOC3	DALSSVQESQVAQQAR	858.9	1144.6	573.3				
APOC3	DYWSTVK	449.7	620.3	434.3				
APOC3	GWVTDGFSSLK	598.8	953.5	854.4				
APOC4	AWFLESK	440.7	623.3	476.3				
APOC4	ELLETVVNR	536.8	588.3	388.2				

(continued)

Table 2
(continued)

Protein	Peptide	Precursor ion (*m/z*)	Fragment ion (*m/z*)					
			1	2	3	4	5	6
APOD	IPTTFENGR	517.8	824.4	461.2				
APOD	NPNLPPETVDSLK	712.4	1098.6	985.5				
APOD	VLNQELR	436.3	659.3	545.3				
APOE	AATVGSLAGQPLQER	749.4	827.4	642.4				
APOE	SELEEQLTPVAEETR	865.9	902.5	801.4				
APOF	SGVQQLIQYYQDQK	849.4	1085.5	972.4				
APOF	SLPTEDC[+57.0]ENEK	661.3	923.3	561.2				
APOF	SYDLDPGAGSLEI	668.8	743.4	261.1				
APOH	ATVVYQGER	511.8	751.4	652.3				
APOH	TC[+57.0]PKPDDLPFSTVVPLK	638.7	743.5	357.2				
APOL1	LNILNNNYK	553.3	878.5	765.4				
APOL1	VTEPISAESGEQVER	815.9	1301.6	1091.5	804.4			
APOM	AFLLTPR	409.3	599.4	486.3				
APOM	DGLC[+57.0]VPR	408.7	531.3	371.2				
APOM	SLTSC[+57.0]LDSK	505.7	810.4	709.3				
C3	TGLQEVEVK	501.8	731.4	422.7				
C3	TIYTPGSTVLYR	685.9	1156.6	892.5				
C4A	GSFEFPVGDAVSK	670.3	919.5	772.4				
C4A	VFALDQK	410.7	721.4	574.3	503.3			
CETP	ASYPDITGEK	540.8	759.4	662.3				
CETP	GTSHEAGIVC[+57.0]R	593.8	804.4	675.4				
CETP	VIQTAFQR	481.8	750.4	622.3				
CLU	ASSIIDELFQDR	697.4	1035.5	922.4				
CLU	ELDESLQVAER	644.8	802.4	474.3	375.2			
CLU	LFDSDPITVTVPVEVSR	937.5	886.5	686.4				
HBB	LLVVYPWTQR	637.9	850.4	687.4				
HBB	SAVTALWGK	466.8	774.5	675.4				
HP	VTSIQDWVQK	602.3	1003.5	916.5	803.4			

(continued)

Table 2
(continued)

Protein	Peptide	Precursor ion (*m/z*)	Fragment ion (*m/z*)					
			1	**2**	**3**	**4**	**5**	**6**
HPR	GSFPWQAK	460.7	629.3	532.3	388.7			
HPR	LPEC[+57.0]EAVC[+57.0]GKPK	463.2	589.3	244.2				
HPR	TEGDGVYTLNDK	656.3	1081.5	909.5	590.3			
HPR	VGYVSGWGQSDNFK	772.4	1125.5	795.4				
IHH	AFQVIETQDPPR	700.9	955.5	369.2				
LCAT	LEPGQQEEYYR	706.3	1169.5	585.3				
LCAT	SSGLVSNAPGVQIR	692.9	941.5	669.4				
LCAT	STELC[+57.0]GLWQGR	653.8	876.4	716.4				
LPA	GTLSTTITGR	503.8	735.4	446.3				
LPA	TPAYYPNAGLIK	654.4	875.5	712.4				
LpPLA2	ASLAFLQK	439.3	719.4	606.4	535.3			
LpPLA2	IAVIGHSFGGATVIQTLSEDQR	767.1	634.3	547.2				
PCYOX1	LFLSYDYAVK	609.8	958.5	845.4				
PCYOX1	LVC[+57.0]SGLLQASK	588.3	963.5	482.3				
PLTP	AVEPQLQEEER	664.3	1028.5	514.8				
PLTP	FLEQELETITIPDLR	909	714.4	500.3				
PLTP	GAFFPLTER	519.3	762.4	615.3				
PON1	IFFYDSENPPASEVLR	942.5	982.5	868.5				
PON1	IQNILTEEPK	592.8	943.5	716.4				
PON1	STVELFK	412.2	635.4	536.3	294.2			
PON3	AQALEISGGFDK	618.3	1036.5	723.4				
PON3	LLNYNPEDPPGSEVLR	907	1195.6	854.5				
PON3	SVNDIVVLGPEQFYATR	954.5	1068.5	1011.5				
RBP4	LLNLDGTC[+57.0]ADSYSFVFSR	1033	742.4	409.2				
RBP4	YWGVASFLQK	599.8	849.5	350.1				
SAA1/2	DPNHFRPAGLPEKY	820.9	763.4	536.3				
SAA1/2	SFFSFLGEAFDGAR	775.9	935.5	822.4				
SAA4	FRPDGLPK	465.3	686.4	244.2				

(continued)

Table 2
(continued)

Protein	Peptide	Precursor ion (*m/z*)	Fragment ion (*m/z*)					
			1	**2**	**3**	**4**	**5**	**6**
SAA4	GPGGVWAAK	421.7	688.4	475.3				
SERPINA1	LSITGTYDLK	555.8	910.5	797.4				
SERPINA1	SVLGQLGITK	508.3	829.5	716.4				
SERPINA4	IAPANADFAFR	596.8	1008.5	504.7				
SERPINA4	VGSALFLSHNLK	643.4	971.6	711.4				
VDBP	THLPEVFLSK	390.9	494.3	352.2	234.1			
VDBP	VLEPTLK	400.2	700.4	587.3				
VTN	DVWGIEGPIDAAFTR	823.9	1076.5	947.5				
VTN	FEDGVLDPDYPR	711.8	875.4	647.3				
VTN	GQYC[+57.0]YELDEK	652.8	1119.5	956.4	796.4			

Note: # [15]N labeled peptides originating from ^{15}N-APOA-I. Protein abbreviation, protein name: ALB, serum albumin; APOA1, apolipoprotein A-I; APOA2, apolipoprotein A-II; APOA4, apolipoprotein A-IV; APOA5, apolipoprotein A-V; APOB, apolipoprotein B; APOC1, apolipoprotein C-I; APOC2, apolipoprotein C-II; APOC3, apolipoprotein C-III; APOC4, apolipoprotein C-IV; APOD, apolipoprotein D; APOE, apolipoprotein E; APOF, apolipoprotein F; APOH, apolipoprotein H; APOL1, apolipoprotein L-1; APOM, apolipoprotein M; C3, complement C3; C4A, complement C4A; CETP, cholesterol ester transfer protein; CLU, clusterin (apolipoprotein J); HBB, hemoglobin beta; HP, haptoglobin; HPR, haptoglobin-related protein; IHH, Indian hedgehog protein; LCAT, Phosphatidylcholine-sterol acyltransferase; LPA, lipoprotein a; LpPLA2, Lipoprotein-Associated Phospholipase A2; PCYOX1, prenylcysteine oxidase 1; PLTP, phospholipid transfer protein; PON1, paraoxonase 1; PON3, paraoxonase 3; RBP4, retinol binding protein 4; SAA1/2, serum amyloid A1/2; SAA4, serum amyloid A4; SERPINA1, serpin A1 (alpha-1-antitrypsin); SERPINA4, serpin A4 (kallistatin); VDBP, vitamin D binding protein; VTN, vitronectin

2. To import MS data into Skyline, from the menu bar, navigate to Import then choose Results. In the Import Results window, select Add single-injection replicates in files and none from the Optimizing dropdown menu. Then select OK.

3. Navigate to the data file containing the MS data to be analyzed. Select one or more MS spectra files for import into Skyline and select Open. Save the Skyline document after the spectral data has been imported (*see* **Note 11**).

4. Verify that the transitions assigned to each peptide by Skyline are correct. Adjust integration limits for the transition peak areas as necessary. Export the results in .csv file format for analysis in Microsoft Excel (or similar software) under File in the menu bar, then navigate to Export and select Report. In the Export Report window, select Transition Results, then Export. Name the file and select Save.

4   Notes

1. KBr stock solution at a density of 1.1863 g/mL is prepared by completely dissolving 124 g KBr in 400 mL of H_2O. Intermittent heat may be required to fully dissolve KBr. Allow temperature of KBr solution equilibrate to 25 °C, and then adjust density down to 1.1863 g/mL using water at 25 °C. Aliquot 400 μL of KBr stock solution into 0.6 mL microcentrifuge tubes. Dry down using vacuum centrifugation with heat at ~80 °C for 40 min with a total spin time of 3 h. Cap tubes immediately upon drying. We generally prepare 100 tubes per batch. To determine the variability of the amount of KBr added to the tubes, number and tare 10 of the tubes prior to adding KBr stock solution and weigh each tube again after drying down. Each tube should contain 108.7 mg of KBr. The variability of the amount of KBr added is typically <0.5 % CV.

2. Solutions are prepared by adding the total volume of H_2O (4.0 mL) or acetonitrile (22.0 mL) to the tube, then removing the volume of FA, 455 and 250 μL respectively, from the tube and discarding. Then add back the required volume of 88 % FA to H_2O or acetonitrile and mix well. For example, to make the 10 % FA solution, add 4.00 mL of H_2O to the tube. Then, using a micropipettor, remove 455 μL of H_2O and discard. Add back 455 μL of 88 % FA to the tube and mix well.

3. Dispense the 120 μL of lipoproteins into a 0.6 mL tube. Re-aspirate and dispense a second time to rinse the walls of the tip. Discard tip after second dispense and use a fresh tip for the next sample.

4. Dispense 239 μL normal saline near the top inside of the ultra-centrifuge tube to rinse the lipoproteins down off the side.

5. The Bradford reaction develops at different rates for BSA and for HDL proteins. Maintaining constant incubation times is vital for reproducible results between batches.

6. Detailed Skyline tutorials that outline MRM method development and other features of the software not included in this section can be downloaded from the MacCoss laboratory website at: https://skyline.gs.washington.edu/labkey/wiki/home/software/Skyline/page.view?name=tutorials.

7. The peptides and transitions in Table 2 represent proteotypic peptides for these 38 proteins. Proteotypic in this context means the peptide sequence has been determined to be unique in the human proteome (i.e., no splice isoforms, polymorphisms, or isobaric peptides being reported) and that it is readily detected using tandem MS. Furthermore, these peptide sequences: (1) do not contain methionine and (2) the peak area ratios of peptides from the same protein are highly correlated across a population of human samples.

8. If spectral libraries obtained from DDA data or another source are being used to aid in MS peak identification, please visit the Skyline software tutorial page (*see* **Note 5**) to find detailed tutorials that outline how to implement spectral libraries in Skyline.

9. Retention time scheduling for precursor ions can be added to the transition list .csv file in Microsoft Excel (or similar software) after exporting. Scheduled methods can be exported from Skyline using iRT retention time prediction. *See* **Note 5**.

10. Using a detailed file name that includes the acquisition data and pertinent information regarding the samples analyzed will make it easier to locate the Skyline data analysis file in the future.

11. If the MS file names have a common prefix, an Import Results pop-up window will appear indicating the common prefix and the option to retain or remove the prefix. Selecting to remove the prefix will not affect the original file name.

Acknowledgements

This work was supported by NIH grants: HL111375, DK035816, HL089504 and NIH training grant T32HL007028 and AHA grants 0830231N and 14GRNT18410022. We thank Jennifer Wallace for her contributions to this chapter.

References

1. Castelli WP, Anderson K, Wilson PWF et al (1992) Lipids and risk of coronary heart disease. The Framingham Study. Ann Epidemiol 2:23–28

2. Mahmood SS, Levy D, Vasan RS et al (2014) The Framingham Heart Study and the epidemiology of cardiovascular disease: a historical perspective. Lancet 383:999–1008

3. Boden WE (2000) High-density lipoprotein cholesterol as an independent risk factor in cardiovascular disease: assessing the data from Framingham to the veterans affairs high-density lipoprotein intervention trial. Am J Cardiol 86:19–22

4. Khera AV, Cuchel M, de la Llera-Moya M et al (2011) Cholesterol efflux capacity, high-density lipoprotein function, and atherosclerosis. N Engl J Med 364:127–135

5. Kontush A (2014) HDL-mediated mechanisms of protection in cardiovascular disease. Cardiovasc Res 103:341–349

6. Cheng AM, Rizzo-DeLeon N, Wilson CL et al (2014) Vasodilator-stimulated phosphoprotein protects against vascular inflammation and insulin resistance. Am J Physiol Endocrinol Metab 307:E571–E579

7. Cheng AM, Handa P, Tateya S et al (2012) Apolipoprotein A-I attenuates palmitate-mediated NF-κB activation by reducing toll-like receptor-4 recruitment into lipid rafts. PLoS One 7, e33917

8. Parthasarathy S, Raghavamenon A, Garelnabi M et al (2010) Oxidized low-density lipoprotein. In: Uppu RM, Murthy SN, Pryor WA et al (eds) Free radicals and antioxidant protocols SE-24. Humana Press, New York, pp 403–417

9. Navab M, Berliner JA, Subbanagounder G et al (2001) HDL and the inflammatory response induced by LDL-derived oxidized phospholipids. Arterioscler Thromb Vasc Biol 21:481–488

10. Karlsson H, Leanderson P, Tagesson C et al (2005) Lipoproteomics II: Mapping of proteins in high-density lipoprotein using two-dimensional gel electrophoresis and mass spectrometry. Proteomics 5:1431–1445

11. Vaisar T, Pennathur S, Green PS et al (2007) Shotgun proteomics implicates protease inhibition and complement activation in the antiinflammatory properties of HDL. J Clin Invest 117:746–756

12. Davidson WS, Silva RAGD, Chantepie S et al (2009) Proteomic analysis of defined HDL subpopulations reveals particle-specific protein clusters: relevance to antioxidative function. Arterioscler Thromb Vasc Biol 29:870–876

13. Gordon SM, Deng J, Lu LJ et al (2010) Proteomic characterization of human plasma high density lipoprotein fractionated by gel filtration chromatography. J Proteome Res 9:5239–5249

14. Hoofnagle AN, Heinecke JW (2009) Lipoproteomics: using mass spectrometry-based proteomics to explore the assembly, structure, and function of lipoproteins. J Lipid Res 50:1967–1975

15. Shah AS, Tan L, Long JL et al (2013) Proteomic diversity of high density lipoproteins: our emerging understanding of its importance in lipid transport and beyond. J Lipid Res 54:2575–2585

16. Birner-Gruenberger R, Schittmayer M, Holzer M et al (2014) Understanding high-density lipoprotein function in disease: recent advances in proteomics unravel the complexity of its composition and biology. Prog Lipid Res 56C:36–46

17. Agger SA, Marney LC, Hoofnagle AN (2010) Simultaneous quantification of apolipoprotein A-I and apolipoprotein B by liquid-chromatography-multiple-reaction-monitoring mass spectrometry. Clin Chem 56:1804–1813

18. Hoofnagle AN, Becker JO, Oda MN et al (2012) Multiple-reaction monitoring-mass spectrometric assays can accurately measure the relative protein abundance in complex mixtures. Clin Chem 58:777–781

19. Ronsein GE, Pamir N, von Haller PD et al (2015) Parallel reaction monitoring (PRM) and selected reaction monitoring (SRM) exhibit comparable linearity, dynamic range and precision for targeted quantitative HDL proteomics. J Proteomics 113:388–399

20. MacLean B, Tomazela DM, Shulman N et al (2010) Skyline: an open source document editor for creating and analyzing targeted proteomics experiments. Bioinformatics 26:966–968

Chapter 8

A Method for Label-Free, Differential Top-Down Proteomics

Ioanna Ntai, Timothy K. Toby, Richard D. LeDuc, and Neil L. Kelleher

Abstract

Biomarker discovery in the translational research has heavily relied on labeled and label-free quantitative bottom-up proteomics. Here, we describe a new approach to biomarker studies that utilizes high-throughput top-down proteomics and is the first to offer whole protein characterization and relative quantitation within the same experiment. Using yeast as a model, we report procedures for a label-free approach to quantify the relative abundance of intact proteins ranging from 0 to 30 kDa in two different states. In this chapter, we describe the integrated methodology for the large-scale profiling and quantitation of the intact proteome by liquid chromatography-mass spectrometry (LC-MS) without the need for metabolic or chemical labeling. This recent advance for quantitative top-down proteomics is best implemented with a robust and highly controlled sample preparation workflow before data acquisition on a high-resolution mass spectrometer, and the application of a hierarchical linear statistical model to account for the multiple levels of variance contained in quantitative proteomic comparisons of samples for basic and clinical research.

Key words Top-down proteomics, Top-down quantitation, Label-free quantitation, Quantitative mass spectrometry, Proteoform, Differential expression

1 Introduction

Mass spectrometry has emerged over the past few decades as a powerful tool for untargeted protein analysis in the clinical lab. The vast majority of proteomics research relies on the bottom-up approach, where proteins are digested with a protease, such as trypsin, prior to peptide detection and sequencing using tandem mass spectrometry [1, 2]. Precursor mass measurements, along with MS/MS fragmentation information, allow inference of the protein composition of the sample via these peptides. Top-down proteomics describes the process for identification and characterization of intact proteoforms [3] without the use of a protease. In doing so, top-down proteomics can fully characterize the composition of individual proteoforms, including proteolysis products, signal peptide cleavage, sequence variants, and PTMs co-occurring on the same molecule. A typical top-down workflow consists of

Salvatore Sechi (ed.), *Quantitative Proteomics by Mass Spectrometry*, Methods in Molecular Biology, vol. 1410,
DOI 10.1007/978-1-4939-3524-6_8, © Springer Science+Business Media New York 2016

single- or multi-step protein separations, such as RPLC [4] and GELFrEE [5], and the resulting protein fractions are then further separated by liquid chromatography in line with a mass spectrometer. Advances in MS instruments and protein separations have allowed top-down proteomics to become a robust technique for the identification and characterization of ~2000–3000 proteoforms [4–6]. Due to technical and analytical challenges inherent in the acquisition and bioinformatic analysis of top-down MS data, however, *quantitative* top-down methods analogous to those used by bottom-up proteomicists have lagged in terms of development and application.

While proteome-wide quantitation has remained elusive for proteoform-resolved measurements, several laboratories have published targeted studies that quantify whole proteins within a mixture of limited complexity. For example, two recent studies [7, 8] accomplished relative quantitation of multiple, coeluting proteoforms simply by measuring their intensity ratios in order to address relevant biomedical research questions. However, this kind of intraspectrum quantitation becomes less applicable to large multisample quantitative experiments in the face of ever growing biological variation among samples and technical variation among discrete data files as studies grow in scope. Several researchers have successfully applied various labeling techniques, like those often seen in quantitative bottom-up workflows, to intact protein analysis, including a study by our own group that quantified over 200 intact protein pairs by $^{14}N/^{15}N$ labeling in an anoxic yeast model [9]. In addition, studies have applied in vitro differential cysteine labeling [10] and tandem mass tag (TMT) workflows to perform MS1 and MS2-based quantitation of intact proteins [11]. However, these in vitro labeling methods are limited in their application to large-scale top-down quantitation studies, as differential labeling was found to alter chromatographic retention time and the existence of multiple precursor charge states in top-down measurements greatly complicates isobaric labeling experiments.

This chapter introduces a completely label-free method for high-throughput, quantitative profiling of intact proteins recently developed by our group [12]. We deferred from employing metabolic labeling, even though such an experiment circumvents the aforementioned chromatography challenges and requirement for single precursor ion selection. The preclusion of in vivo labeling techniques is necessary, as our method was developed to meet the growing demand for proteoform-resolved biomarker discovery in clinical research, where cells and tissue samples usually cannot be metabolically labeled. Statistical validity is also of prime importance in clinical studies, as they require great confidence in conclusions that may inspire expensive validation experiments and affect patient health in the future. While the label-free quantitation of proteoforms has been presented in various recent studies, including the introduction of "differential mass spectrometry" (dMS) by Yates and Hendrickson [13, 14], these analyses have been limited to just a few proteoforms and supporting statistical assessments were performed by the traditional Student's t-test. These statistics are insufficient to address the myriad, complex sources of technical variation implicit in large-scale, comparative proteomic workflows employed by clinical, quantitative, top-down proteomics in discovery mode.

The label-free method for the comparative quantitation of proteoforms described here relies on a robustly controlled sample processing, multiple levels of replication at the biological and technical level, reproducible high-performance nano-flow liquid chromatography online with high-resolution FTMS mass analysis, and a statistical platform based on a hierarchical linear model that considers multiple levels of variation in the ANOVA analysis (Fig. 1). While the experiment outlined in the following methods section was performed and validated in the model organism *Saccharomyces cerevisiae*, this proof-of-principle experiment can and has been carried out in mammalian cell lines and samples of peripheral blood mononuclear cells (PBMCs) isolated from human patients.

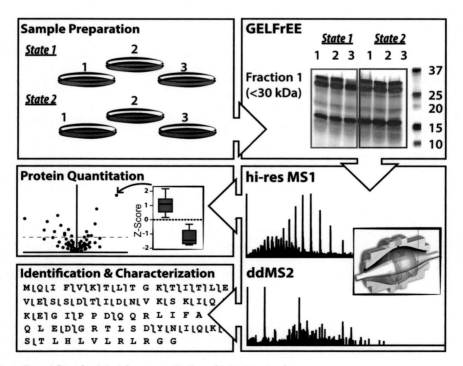

Fig. 1 Overall workflow for label-free quantitation of intact proteoforms

2   Materials

2.1   Cell Culture, Lysis, and Protein Fractionation

1. *Saccharomyces cerevisiae* S288c BY4742 and the *rpd3* (YNL330C) deletion mutant were obtained from ATCC (*see* Note 1).

2. Culture media: YPD broth (Sigma-Aldrich), with or without 0.2 g/L G-418 (Sigma-Aldrich).

3. Lysis Buffer: YPER (ThermoPierce, Rockford, IL) supplemented with 5 nM microcystin, 500 μM 4-(2-aminoethyl)-benzenesulfonyl fluoride (AEBSF), 100 mM sodium butyrate, and 100 mM dithiothreitol (DTT).

4. Bicinchoninic acid (BCA) assay kit (ThermoPierce, Rockford, IL) for protein concentration determination.

5. GELFREE 8100 device, 8 % GELFREE cartridge and GELFREE buffers (Expedeon, Cambridgeshire, UK), as supplied by the manufacturer.

2.2   LC-MS Data Acquisition

1. Dionex Ultimate 3000 RSLCnano system.

2. Mobile phase A: 0.2 % formic acid, 94.8 % water, 5 % acetonitrile.

3. Mobile phase B: 0.2 % formic acid, 4.8 % water, 95 % acetonitrile.

4. PLRP-S 1000 Å 5 μm bulk media (Agilent).

5. Orbitrap Elite mass spectrometer (Thermo Fisher, San Jose, CA) fitted with nanospray source.

6. PicoTip emitters (New Objective, Waltham, MA).

2.3   MS Data Analysis and Quantitation

1. Xtract: MS instrument vendor supplied software for deconvoluting spectra and obtaining intensities of masses.

2. ProSight PC: search engine for protein identification and characterization.

3. SAS: statistical analysis program.

3   Methods

The typical proteomic workflow for label-free top-down quantitation is depicted in Fig. 1. The experiment includes sample preparation, molecular-weight based protein fractionation, LC separation and MS data acquisition, differential quantitation, statistical analysis, and identification of quantified proteoforms.

The protocol described below is designed to compare the nuclear proteome of two *Saccharomyces cerevisiae* strains, the wild type versus the *rpd3* deletion mutant. Rpd3 is a histone deacetylase

"Control" (State A)			"Treatment" (State B)		
Bio Rep 1	Bio Rep 2	Bio Rep 3	Bio Rep 1	Bio Rep 2	Bio Rep 3
Tech Rep 1 / Tech Rep 2 / Tech Rep 3	Tech Rep 1 / Tech Rep 2 / Tech Rep 3	Tech Rep 1 / Tech Rep 2 / Tech Rep 3	Tech Rep 1 / Tech Rep 2 / Tech Rep 3	Tech Rep 1 / Tech Rep 2 / Tech Rep 3	Tech Rep 1 / Tech Rep 2 / Tech Rep 3

Fig. 2 Study design for comparing differential expression of proteins in two states with three biological replicates and four technical replicates ($2 \times 3 \times 3$ design)

and its deletion has been shown to increase acetylation levels of all core histones [15] and to have other global effects owing to a lack of epigenetic regulation [16].

3.1 Study Design

This is a protocol designed to compare the proteomes of two different biological samples (*see* **Note 2**). For each biological sample, it is recommended that three biological replicates and a minimum of three technical replicates are used to account for variability of biological and technical nature, respectively (Fig. 2).

3.2 Cell Culture, Lysis, and Protein Fractionation

3.2.1 Cell Culture

1. Single colonies of wild type *Saccharomyces cerevisiae* S288c BY4742 and the *rpd3* (YNL330C) deletion mutant (*rpd3Δ::KANMX*) were inoculated into 5 mL each of liquid YPD media without and with 0.2 g/L G-418, respectively. Three biological replicates of both WT and mutant strains were prepared.

2. Inoculated cultures were incubated overnight in a shaker at 250 rpm and 30 °C, and then centrifuged at $1,000 \times g$ for 10 min. Each cell pellet was gently resuspended in 1 mL YPD and then inoculated into 250 mL of YPD and YPD + G-418.

3. Cells were harvested at $OD_{600} = 0.7$ by centrifugation at $1,000 \times g$ for 20 min. The supernatant was discarded, pellets were washed once with distilled water, and pellet masses were measured before storage at −80 °C.

3.2.2 Cell Lysis

1. Thawed yeast cell pellets were lysed in 2.5 mL/g wet cell weight of Lysis Buffer. The mixtures were agitated at room temperature for 20 min, and then pelleted by centrifugation at $14,000 \times g$ for 10 min.

2. To isolate nuclear proteins, each cell pellet was first resuspended in 30 mL of water and centrifuged at $18{,}000 \times g$ for 10 min at 4 °C, to remove excess YPER.

3. Acid/urea extraction of the histone fraction was performed by adding 2.5 volumes of 8 M deionized urea (*see* **Note 3**) with 0.4 N sulfuric acid the pellets, vortexing for 5 min, and then incubating on ice for 30 min.

4. Sample fractions were loaded onto C4 solid phase extraction columns (Baker). 30 mL of 0.1 % trifluoroacetic acid (TFA) in water was used as the wash, and sample was eluted with 3 mL of 0.1 % TFA in 60 % acetonitrile. Each eluted fraction was dried and reconstituted with 1.0 % sodium dodecyl sulfate (SDS) solution.

5. Protein concentration was determined by BCA assay (Thermo Pierce, Rockford, IL).

3.2.3 Protein Fractionation

1. 400 µg of total protein as measured by BCA were prepared for GELFrEE according to the manufacturer's instructions. Up to 112 µL of each sample was combined with 30 µL sample buffer, 8 µL 1 M DTT, and distilled water to a final volume of 150 µL. The vial was heated for 5 min at 50 °C.

2. Each biological replicate (*see* **Note 4**) was separated on a single lane of an 8 % GELFREE cartridge (six lanes total) according to the manufacturer's instructions and the first fraction (0–30 kDa range of proteins) of each sample was collected.

3. SDS removal was done by methanol/chloroform/water precipitation [17].

   (a) Four volumes of methanol were added to each fraction, and vortexed vigorously for 30 s. One volume of chloroform was then added, and vortexed again. Three volumes of water were added to each fraction, and vortexed again. The fractions were centrifuged at 14,000 rpm for 10 min (*see* **Note 5**).

   (b) After centrifugation, the top aqueous/methanol layer was carefully pipetted off and discarded. Three volumes of methanol were slowly added, the vial was carefully inverted to mix, and then centrifuged again for 10 min at 14,000 rpm.

   (c) The supernatant was then completely removed, leaving the protein pellet in approximately 10 µL liquid (*see* **Note 6**). After the supernatant was removed, residual solvent was allowed to air dry.

4. Protein pellets were resuspended in 40 µL mobile phase A.

Table 1
LC gradient used for label-free quantitation of proteins

Time (min)	%B	Valve position
0	5	1_10
10	5	1_2
12	15	
62	55	
67	95	
72	95	
75	5	
90	5	

3.3 LC-MS Data Acquisition

It is important to randomize the order of data acquisition to reduce bias in results due to instrument performance decay during data collection (*see* **Note 4**).

1. Trap (150 μm × 2 cm) and analytical (75 μm × 15 cm) column were packed in-house with PLRP-S media into fused silica capillaries.

2. Proteins (5 μL) were injected onto a trap column using an autosampler (Thermo Dionex). The NC pump was operated at 300 nL/min and the loading pump at 3 μL/min. The trap and analytical were operated in a vented tee setup. Table 1 outlines the LC gradient used for protein separation and elution into the mass spectrometer.

3. The Orbitrap Elite was operated under standard operating conditions. More details can be found in Table 2.

3.4 MS Data Analysis and Quantitation

Our conventional top-down proteomics pipeline was used for proteoform identification and characterization, as previously described [5, 6, 12, 18].

3.4.1 Proteoform Identification and Characterization

1. ProSightHT within ProSightPC 3.0 was used to convert m/z data for each precursor/fragmentation scan pair to monoisotopic neutral mass values.

2. Neutral mass values were searched against a *S. cerevisiae* specific database built from UniProt release 2013_04.

3. Mass tolerance for precursors and fragments was set to 10 ppm.

4. Low confidence proteoform identifications were excluded by requiring those hits arising from an absolute mass search to have an *E*-value below 1×10^{-4} [5], and a more stringent

Table 2
MS method parameters for label-free quantitation of proteins

Parameter	Setting
Source	Nano-ESI
Capillary temperature (°C)	320
S-lens RF voltage	50 %
Source voltage (kV)	1.8
Full MS parameters	
Mass range (m/z)	500–2000
Resolution settings (FWHM at m/z *400*)	120,000
Target value	1×10^6
Max injection time (ms)	1000
Microscans	4
SID (V)	15
MS2 parameters	
Top-N MS^2	2
Resolution settings (FWHM at m/z 400)	60,000
Target value	1×10^6
Max injection time (ms)	2000
Isolation width (Da)	15
Minimum signal required	500
Normalized collision energy (HCD)	25
Default charge state	10
Activation time (ms)	0.1
Lowest m/z acquired	400
Charge state rejection on: 1+, 2+ and 3+	enabled
Dynamic exclusion	Repeat count 1 Repeat duration 240 s Exclusion list size 500 Exclusion duration 5000 Exclusion mass width by mass ±18 Da

E-value cutoff of 6×10^{-5} was applied to hits derived from a biomarker search as previously reported [19].

3.4.2 Proteoform Quantitation

All data files in the quantitation portion of the platform were processed using a collection of in-house tools to automate data analysis.

1. Files were analyzed for Quantitation Mass Targets (QMTs) using a moving spectral average and deisotoped with the Xtract algorithm (Thermo Fisher) at a signal to noise value of 6.

2. All QMTs were binned by mass (8 ppm) and retention time (8 min) to reduce data redundancy.

3. Intensities were normalized using the average total ion chromatogram intensity for each technical replicate.

4. The QMTs were grouped and artifactual ±1 Da deisotoping errors were removed. Final QMTs were stored within a SQLite reporting database.

5. Intensities were aggregated across all scans and charge states, to report one intensity value for each data file and QMT. These data were provided as a text file for further statistical processing (vide infra).

6. QMTs were matched to proteoforms identified (Subheading 3.4.1) using a 10 ppm tolerance.

3.4.3 Statistical Analysis

All statistical analyses were performed within SAS 9.4 (SAS Institute, Cary, NC).

1. Intensity data on the occurrence of putative proteoforms (QMTs) were tabulated and those not occurring in at least 50 % of all data files were excluded from further analysis (*see* **Note 7**).

2. Intensity values for the remaining QMTs were $\log_2$-transformed so that differences in estimated treatment-level intensities could be interpreted on a fold-change scale.

3. Two separate ANOVA analyses were performed; ANOVA-1 and -2. For the first analysis, ANOVA-1, intensity levels for each QMT were standardized to Z-scores across all samples. ANOVA-2 used unstandardized intensity values. ANOVA-1 was used to test the statistical significance of QMT intensity changes between the wild type and $\Delta rpd3$ mutant strains, while ANOVA-2 was used to estimate the size of the effect.

4. In both analyses, a hierarchical linear model was employed as the general statistical approach. The fixed effect hierarchical linear model can be expressed as

$$I_{ijk} = \mu + A_i + B_{j(ik)} + C_{k(ij)} + \varepsilon_{ijk}$$

5. In ANOVA-1, "I" represents the QMT intensity Z-score, while in ANOVA-2, this represents the $\log_2$-transformed intensity. In both models, μ is the true mean, A is the treatment factor levels (wild type and $\Delta rpd3$), B is the biological replicates' variance, C is the technical replicates' variance, and ε is the residual variance (*see* **Note 8**). In ANOVA-1, all *p-scores* were corrected for multiple testing at a false discovery rate of $\alpha = 0.05$ [20].

6. To visualize the results, we created a volcano plot which represented each proteoform (i.e., QMT) as a function of estimated effect size (in $\log_2$ fold-change) and the statistical confidence (the FDR) that there was a difference in normalized intensity between the two states "wild type" and "$\Delta rpd3$" (Fig. 3). Proteoforms identified to the right of the *y*-axis were upregulated in $\Delta rpd3$ as compared to WT.

7. Quantitation of individual proteoforms can be visualized with box and whisker plots, as seen in the case of Histone H4 in Fig. 4.

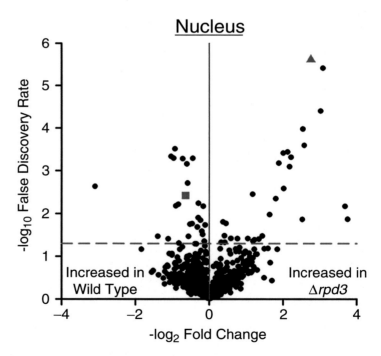

Fig. 3 Volcano plot generated to compare the WT vs. $\Delta rpd3$ strains of *S. cerevisiae* S288c for the nuclear proteins below 30 kDa having 54 differentially expressed proteoforms, above the 5 % FDR threshold (*dotted red line*). Datapoints highlighted above correspond to proteoforms of histone H4 (*red filled square* = diacetylated histone H4, *red filled triangle* = triacetylated histone H4) and their quantitation is explored further in Fig. 4. Reproduced from [12] with permission from the publisher

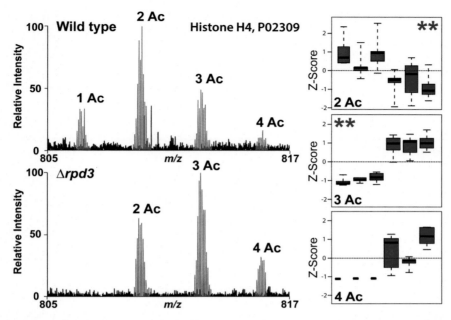

Fig. 4 Quantitation of histone H4 proteoforms. Box and whisker plots (panels at *right*) are presented for the histone H4 proteoforms with 2, 3, and 4 acetylations. The monoacetylated and tetraacetylated proteoforms were not considered within our analysis because they were observed in <50 % of the technical replicates. In all cases, mass spectra are the sums of individual scans across the full elution time of each proteoform within a single technical replicate. The symbol ** indicates significant proteoform abundance changes as reported by our platform. Reproduced from [12] with permission from the publisher

4   Notes

1. Yeast was used for the development of this protocol, but any protein source can be used instead. The lysis protocol would need to be modified but once the proteome is obtained the rest of the protocol can be followed.

2. It is possible to compare three or more states by adjusting the study design and statistical analysis.

3. Urea was deionized using AG 501-X8 resin (BioRad) per manufacturer's instructions.

4. Biological replicates were loaded on the cartridge in a randomized order. For a tutorial on how to randomize a list of items in an Excel spreadsheet go to http://www.excel-easy.com/examples/randomize-list.html.

5. The proteins will be at the interface between the two layers after this centrifugation step.

6. The methanol wash may be repeated once more. More methanol washes may lead to better removal of SDS, but also loss of protein.

7. This was done to address the problem of "missing values."

8. Notice that the technical replicates are nested within the biological replicates, and that each biological replicate is treated as a random effect, thus each biological preparation is allowed to have its own mean. Analysis was done using restricted maximum likelihood (REML) estimation.

Acknowledgement

The methods used in this article were developed with partial support from the National Institute of General Medical Sciences P41GM108569 for the National Resource for Translational and Developmental Proteomics (NRTDP) based at Northwestern University

References

1. Lu B, Motoyama A, Ruse C, Venable J, Yates JR 3rd (2008) Improving protein identification sensitivity by combining MS and MS/MS information for shotgun proteomics using LTQ-Orbitrap high mass accuracy data. Anal Chem 80(6):2018–2025. doi:10.1021/ac701697w

2. Wisniewski JR, Zougman A, Mann M (2009) Combination of FASP and StageTip-based fractionation allows in-depth analysis of the hippocampal membrane proteome. J Proteome Res 8(12):5674–5678. doi:10.1021/pr900748n

3. Smith LM, Kelleher NL (2013) Proteoform: a single term describing protein complexity. Nat Methods 10(3):186–187. doi:10.1038/nmeth.2369

4. Ansong C, Wu S, Meng D, Liu X, Brewer HM, Deatherage Kaiser BL, Nakayasu ES, Cort JR, Pevzner P, Smith RD, Heffron F, Adkins JN, Pasa-Tolic L (2013) Top-down proteomics reveals a unique protein S-thiolation switch in Salmonella Typhimurium in response to infection-like conditions. Proc Natl Acad Sci U S A 110(25):10153–10158. doi:10.1073/pnas.1221210110

5. Tran JC, Zamdborg L, Ahlf DR, Lee JE, Catherman AD, Durbin KR, Tipton JD, Vellaichamy A, Kellie JF, Li M, Wu C, Sweet SM, Early BP, Siuti N, LeDuc RD, Compton PD, Thomas PM, Kelleher NL (2011) Mapping intact protein isoforms in discovery mode using top-down proteomics. Nature 480(7376):254–258. doi:10.1038/nature10575

6. Catherman AD, Li M, Tran JC, Durbin KR, Compton PD, Early BP, Thomas PM, Kelleher NL (2013) Top down proteomics of human membrane proteins from enriched mitochondrial fractions. Anal Chem 85(3):1880–1888. doi:10.1021/ac3031527

7. Chamot-Rooke J, Mikaty G, Malosse C, Soyer M, Dumont A, Gault J, Imhaus AF, Martin P, Trellet M, Clary G, Chafey P, Camoin L, Nilges M, Nassif X, Dumenil G (2011) Posttranslational modification of pili upon cell contact triggers N. meningitidis dissemination. Science 331(6018):778–782. doi:10.1126/science.1200729

8. Dong X, Sumandea CA, Chen YC, Garcia-Cazarin ML, Zhang J, Balke CW, Sumandea MP, Ge Y (2012) Augmented phosphorylation of cardiac troponin I in hypertensive heart failure. J Biol Chem 287(2):848–857. doi:10.1074/jbc.M111.293258

9. Du Y, Parks BA, Sohn S, Kwast KE, Kelleher NL (2006) Top-down approaches for measuring expression ratios of intact yeast proteins using Fourier transform mass spectrometry. Anal Chem 78(3):686–694. doi:10.1021/ac050993p

10. Collier TS, Sarkar P, Franck WL, Rao BM, Dean RA, Muddiman DC (2010) Direct comparison of stable isotope labeling by amino acids in cell culture and spectral counting for quantitative proteomics. Anal Chem 82(20):8696–8702. doi:10.1021/ac101978b

11. Hung CW, Tholey A (2012) Tandem mass tag protein labeling for top-down identification and quantification. Anal Chem 84(1):161–170. doi:10.1021/ac202243r

12. Ntai I, Kim K, Fellers RT, Skinner OS, Smith AD, Early BP, Savaryn JP, LeDuc RD, Thomas PM,

Kelleher NL (2014) Applying label-free quantitation to top down proteomics. Anal Chem 86(10):4961–4968. doi:10.1021/ac500395k

13. Mazur MT, Cardasis HL, Spellman DS, Liaw A, Yates NA, Hendrickson RC (2010) Quantitative analysis of intact apolipoproteins in human HDL by top-down differential mass spectrometry. Proc Natl Acad Sci U S A 107(17):7728–7733. doi:10.1073/pnas.0910776107

14. Meng F, Wiener MC, Sachs JR, Burns C, Verma P, Paweletz CP, Mazur MT, Deyanova EG, Yates NA, Hendrickson RC (2007) Quantitative analysis of complex peptide mixtures using FTMS and differential mass spectrometry. J Am Soc Mass Spectrom 18(2):226–233. doi:10.1016/j.jasms.2006.09.014

15. Jiang L (2008) Analyzing post translational modifications on yeast core histones using Fourier transform mass spectrometry. University of Illinois at Urbana-Champaign, Ann Arbor, MI

16. Rundlett SE, Carmen AA, Kobayashi R, Bavykin S, Turner BM, Grunstein M (1996) HDA1 and RPD3 are members of distinct yeast histone deacetylase complexes that regulate silencing and transcription. Proc Natl Acad Sci U S A 93(25):14503–14508

17. Wessel D, Flugge UI (1984) A method for the quantitative recovery of protein in dilute solution in the presence of detergents and lipids. Anal Biochem 138(1):141–143

18. Ahlf DR, Compton PD, Tran JC, Early BP, Thomas PM, Kelleher NL (2012) Evaluation of the compact high-field orbitrap for top-down proteomics of human cells. J Proteome Res 11(8):4308–4314. doi:10.1021/pr3004216

19. Kellie JF, Catherman AD, Durbin KR, Tran JC, Tipton JD, Norris JL, Witkowski CE 2nd, Thomas PM, Kelleher NL (2012) Robust analysis of the yeast proteome under 50 kDa by molecular-mass-based fractionation and top-down mass spectrometry. Anal Chem 84(1):209–215. doi:10.1021/ac202384v

20. Benjamini Y, Hochberg Y (1995) Controlling the false discovery rate: a practical and powerful approach to multiple testing. J R Stat Soc Ser B 57(1):289–300. doi:10.2307/2346101

Chapter 9

Multiplexed Immunoaffinity Enrichment of Peptides with Anti-peptide Antibodies and Quantification by Stable Isotope Dilution Multiple Reaction Monitoring Mass Spectrometry

Eric Kuhn and Steven A. Carr

Abstract

Immunoaffinity enrichment of peptides using anti-peptide antibodies and their subsequent analysis by targeted mass spectrometry using stable isotope-labeled peptide standards is a sensitive and relatively high-throughput assay technology for unmodified and modified peptides in cells, tissues, and biofluids. Suppliers of antibodies and peptides are increasingly aware of this technique and have started incorporating customized quality measures and production protocols to increase the success rate, performance, and supply of the necessary reagents. Over the past decade, analytical biochemists, clinical diagnosticians, antibody experts, and mass spectrometry specialists have shared ideas, instrumentation, reagents, and protocols, to demonstrate that immuno-MRM-MS is reproducible across laboratories. Assay performance is now suitable for verification of candidate biomarkers from large scale discovery "omics" studies, measuring diagnostic proteins in plasma in the clinical laboratory, and for developing a companion assay for preclinical drug studies. Here we illustrate the process for developing these assays with a step-by-step guide for a 20-plex immuno-MRM-MS assay. We emphasize the need for analytical validation of the assay to insure that antibodies, peptides, and mass spectrometer are working as intended, in a multiplexed manner, with suitable assay performance (median values for 20 peptides: CV = 12.4 % at 740 amol/μL, LOD = 310 amol/μL) for applications in quantitative biology and candidate biomarker verification. The assays described conform to Tier 2 (of 3) level of analytical assay validation (1), meaning that the assays are capable of repeatedly measuring sets of analytes of interest within and across samples/experiments and employ internal standards for each analyte for confident detection and precise quantification.

Key words Anti-peptide antibody, Protein assay, Peptide assay, Multiplexed, Quantification, Mass spectrometry, Immunoaffinity enrichment, Reverse curves, Plasma, Biomarkers, Multiple reaction monitoring, Selected reaction monitoring, Parallel reaction monitoring

1 Introduction

Sensitive and selective detection and quantification of peptides using targeted mass spectrometry has become an essential component of verification studies of candidate disease biomarkers and is being

Salvatore Sechi (ed.), *Quantitative Proteomics by Mass Spectrometry*, Methods in Molecular Biology, vol. 1410, DOI 10.1007/978-1-4939-3524-6_9, © Springer Science+Business Media New York 2016

increasingly used in biology and clinical diagnostics (1–10 and elsewhere in this book). Historically these mass spectrometry-based peptide assays have been most widely developed and applied using triple quadrupole mass analyzers using a method known as multiple reaction monitoring (MRM) (also referred to as selected reaction monitoring, SRM) experiments [11–14], and assays multiplexed up to several hundred analytes are now achievable [15]. In these experiments a subset of sequence-defining fragment ions (usually 3–5) are selected from the precursor peptide and monitored to increase sensitivity and selectivity of analysis [16, 17]. With improvements in the sensitivity and data acquisition speed of mass spectrometers, these assays can now be robustly developed and applied using instruments that acquire full mass spectra at high resolution and mass accuracy, greatly increasing the selectivity and specificity of analysis, a method referred to as parallel reaction monitoring or PRM [18–20]. Adding stable isotope-labeled versions of the analyte peptides as internal standards [21–23], or labeled proteins when available [24, 25] is necessary to insure that the desired analyte is being measured and that the quantification is precise.

MRM assays can now be configured to quantitatively measure peptides and modified peptides from nearly any protein. However, sample complexity and the wide dynamic range of protein abundance in sample matrices like plasma and tissue require additional steps be taken besides assay development to insure detection of analytes that are present at low abundance in biological samples. Several approaches have become standard for plasma analysis, including the use of immunoaffinity depletion columns which remove the 6–60 most abundant proteins thereby facilitating detection of proteins present at lower abundances [26–30]. Fractionating peptide digests of depleted plasma by ion exchange [5, 8, 12] or high pH reversed-phase chromatography [7, 31–33] prior to targeted analysis by MS, a process referred to as fractionMRM (fMRM), reduces sample complexity and enhances sensitivity and specificity of analyte measurement. Combining immunoaffinity depletion with fMRM has resulted in robust, practical methods to quantitatively measure, in high multiplex, peptides from proteins that are present in the high picogram to low nanogram/mL concentration in plasma [5, 12, 15]. Even greater sensitivities can be achieved for small numbers of analytes by taking fMRM to extremes, isolating small numbers of peptides into very small volumes suitable for direct analysis using targeted MS [34].

Antibodies have been used by biologists and clinical laboratories for decades to enrich analytes of interest from biological samples by immunoprecipitation (IP) for detection and quantification [35, 36]. In 2004, Anderson et al. [37] described the use of antibodies raised against proteotypic tryptic peptides to immunoprecipitate analyte peptides from proteolytic digests of plasma. The enriched peptides were subsequently analyzed by LC-MRM-MS

iMRM-MS: immuno-multiple reaction monitoring mass spectrometry

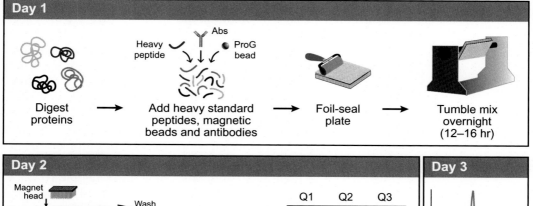

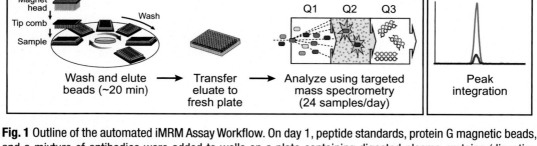

Fig. 1 Outline of the automated iMRM Assay Workflow. On day 1, peptide standards, protein G magnetic beads, and a mixture of antibodies were added to wells on a plate containing digested plasma proteins (digestion workflow not shown). After using a roller to firmly cover the plate with foil, samples were tumble-mixed 12–16 h overnight. On day 2, the Protein G beads, with antibodies and peptides bound, were washed and peptides eluted using a Kingfisher magnetic bead handler. Supernatants from the eluate plate were transferred to a fresh plate and analyzed by LC-MRM-MS. On day 3, data are analyzed. The heavy (*blue colored oval*) and light peptide (*red colored oval*) peak areas were integrated and the peak area ratios used to determine the molar concentration of the peptides in each sample

and quantified using stable isotope-labeled peptides added to the sample prior to IP and co-enriched with the analyte peptides (Fig. 1). This method was termed SISCAPA for Stable Isotope Standards and Capture by Anti-peptide Antibodies; more recently it has been referred to as immunoaffinity-MRM (iMRM). iMRM provides a one-step enrichment method capable of providing sufficient amounts of analyte peptides for MS analysis from even low abundance proteins. Prior removal of abundant proteins or fractionation at the protein or peptide level is not required [37–42]. Another advantage is that only a single capture Ab is required as the mass spectrometer substitutes for a secondary detection Ab, providing high sequence specificity and readily distinguishing the desired analyte from nonspecifically enriched peptides. The approach works equally well for modified peptides such as phosphopeptides [43], and it can adapted for and combined with capture at the protein level [44, 45]. Immunoaffinity enrichment of peptides requires generation of custom Abs for each peptide target. This can be a relatively lengthy and expensive process especially if the goal is to generate monoclonal Abs that can be distributed to labs throughout the world. However, the success rate for obtaining anti-peptide Abs useful in iMRM assays is substantially higher than for generating IP-competent anti-protein Abs [7, 37, 41, 46]. Highly purified peptide antigens are synthesized with ease and a single rabbit can be immunized in batches of up to five peptide antigens to yield mg quantities of IgG that IP sufficiently well and function in iMRM assays [7, 37, 41, 46]. Throughput can be significantly improved by using Protein G coated magnetic beads and bead-handling robotics to automate peptide capture, wash, and elution steps [38, 46]. iMRM assays can be multiplexed to as high as 50 antibodies in a single sample [6, 7, 41, 47–49]. Interlaboratory studies have shown that iMRM assays are robust and reproducible across laboratories, with detection limits approximating ca. 1 ng of protein per mL of plasma and assay CVs of 15 % or less [49]. The interested reader is directed to the growing body of literature describing configuration and use of iMRM assays for biology, preclinical and clinical measurements [6, 7, 40–50].

There are several distinct steps to the generation, analytical validation, and application of an iMRM assay. First, peptides for use as internal standards and for assay development are selected. This step is informed by what peptides have been previously observed for the proteins of interest. In the absence of experimental data, in silico methods have been developed and can be used. Peptides are examined for uniqueness to the candidate protein as well as to any other protein sequence in the sample to be studied, and nonspecific peptides discarded. When multiple peptides for a protein meet these criteria, those exhibiting the highest MS response, as well as those predicted to have good retention behavior on reversed-phase, are favored. Second, MRM transitions are selected and optimized for the heavy synthetic peptide standards to configure the LC-MRM-MS portion of the assay. Third, anti-peptide antibodies are made and the resulting Abs are evaluated for their ability to capture target peptides in a simplified iMRM assay in a matrix background that suitably mimics the matrix planned for final analysis (e.g. digested plasma for plasma assays, digested tissue from the same source of tissue, similar cell lysate backgrounds, etc). This step identifies which of the 2–5 immunogen peptides developed for each protein is efficiently captured and detected by iMRM, and is therefore suitable for full assay development. In addition, some evidence of how well the endogenous analyte is detected can also be derived at this step [46]. The performance (i.e., linearity, precision, LOD/LOQ) of the antibodies and selected peptides are then systematically evaluated. This is typically done using response curves generated by a method of standard addition in which increasing amounts of light peptide are added to the matrix while keeping the concentration of heavy peptide internal standard constant [51]. Alternatively, in cases where endogenous analyte was found or is expected to be present in the matrix, the heavy peptide may be added over a concentration range and a constant amount of light peptide (either added or endogenous) used as the internal standard. This approach is commonly referred to as "surrogate analyte" [52, 53]. Additional experiments may be used to further define the range and applicability of the iMRM assay, including repeatability, selectivity, stability, and reproducibility of endogenous detection [54]. In addition, experiments may be performed to optimize the amount of antibody per assay and determine the range of multiplexing (quantity of individual antibodies purified against separate peptide antigens used in a single capture) where performance is maintained [48].

Here we describe the generation of a 20-plex iMRM assay and the methods used to assess its performance in the context of a plasma matrix. The methods used are generalizable to smaller or larger multiplexes of Abs and are equally applicable to use in cell lines, tissues, or other biofluids like CSF.

2 Materials

1. Tryptic peptide standards (light versions): Amino acid sequences unique to a single protein (proteotypic) synthesized as free acids with unblocked termini (*see* **Note 1**), purified by RPLC, verified by MALDI and quantified by AAA. Light peptides are diluted, aliquoted, and formulated in 30 % acetonitrile/0.1 % formic acid at 100 pmol/μL. Refer to sequences and gene names in Table 1.

2. Tryptic peptide standards (heavy versions): Amino acid sequences that match the sequences of the light versions in 2.1 are synthesized with the C-terminal Arg or Lys residue labeled with heavy stable isotopes of carbon (^{13}C) or nitrogen (^{15}N) using $^{13}C_6$ l-Lysine, $^{13}C_6$ l-Arginine, or $^{13}C_6$, $^{15}N_4$ l-Arginine. The MRM-MS experiments rely on co-elution of the light and heavy versions of the peptides on RP-HPLC. Use of deuterium is not recommended for use in synthesis of heavy-labeled peptides as the isotope effect of deuterium (especially multiple deuterium atoms) can shift the retention time of the heavy versus the light version of the peptide on RP-HPLC. Heavy isotope-labeled peptides are diluted, aliquoted, and formulated identically to the light versions of the peptides as described, above. Refer to sequences and gene names in Table 1 (*see* **Note 2**).

3. Human plasma: plasma isolated from blood using potassium EDTA (purple tubes) from an individual or a pool of healthy individuals (Bioreclamation—K2EDTA), shipped in 1 mL aliquots and stored at −80 °C (*see* **Note 3**).

Table 1
Summary of peptides, proteins, and antibodies used in 20-plex iMRM assay evaluation

No.	mAb or pAb	Gene symbol	Uniprot protein name	Peptide sequence	Conc. (µg/µL)
1	pAb	TNNI	Troponin I	NITEIADLTQK	0.98
2	mAb	IL33	IL-33	TDPGVFIGVK	1
3	mAb	FTL	Ferritin light chain	LGGPEAGLGEYLFER	0.56
4	mAb	AFP	Alpha-fetoprotein	GYQELLEK	0.71
5	pAb	AFP	Alpha-fetoprotein	YIQESQALAK	0.12
6	mAb	ERBB2	Her-2	AVTSANIQEFAGCK	0.71
7	pAb	MUC1	Mucin-1	EGTINVHDVETQFNQYK	1.04
8	mAb	MUC16	Mucin-16	ELGPYTLDR	0.71
9	mAb	TG	Thyroglobulin	FSPDDSAGASALLR	0.8
10	pAb	TG	Thyroglobulin	VIFDANAPVAVR	1.39
11	pAb	ERBB2	Her-2	GLQSLPTHDPSPLQR	1.65
12	pAb	ERBB2	Her-2	VLGSGAFGTVYK	0.57
13	pAb	ANXA1	Annexin A1	GVDEATIIDILTK	0.68
14	pAb	CLIC1	Chloride intracellular channel 1	GFTIPEAFR	1.43
15	pAb	IL18	Interleukin 18	ISTLSCENK	2.22
16	pAb	NFKB2	Nuclear factor NF-kappa-B p100 subunit	IEVDLVTHSDPPR	0.82
17	pAb	FSCN1	Fascin	LSCFAQTVSPAEK	0.78
18	pAb	TAGLN	Transgelin	AAEDYGVIK	1.05
19	pAb	EZR	Ezrin	SQEQLAAELAEYTAK	0.94
20	pAb	PRDX4	Peroxiredoxin 4	QITLNDLPVGR	1.03

4. Polyclonal antibodies: polyclonal antibodies generated against target tryptic peptide sequences in rabbits (*see* **Note 4**), quantified by protein assay and formulated in 25 % glycerol/1× PBS/0.1 % sodium azide.

5. Monoclonal antibodies: monoclonal antibodies generated by clonal expansion of the rabbit immune cells isolated from the spleens harvested from the rabbits used in 4 (*see* **Note 5**), quantified by protein assay and formulated in 25 % glycerol/1× PBS pH 7.4/0.1 % sodium azide.

6. Peptide storage solvent: 30 % acetonitrile/0.1 % formic acid. 300 mL LC-MS grade acetonitrile, 700 mL HPLC grade water, 1 mL formic acid.

7. Antibody storage solution: 25 % glycerol/1× PBS pH 7.4/0.1 % sodium azide. 250 mL Glycerol, 1 packet PBS (Sigma), 1 g sodium azide dissolved into a final volume of 1 L HPLC grade water.

8. Sample diluent/Antibody Elution Solvent: 3 % Acetonitrile/5 % Acetic acid. 3 mL LC-MS grade acetonitrile, 5 mL acetic acid dissolved into a final volume of 100 mL HPLC grade water.

9. Trypsin, TPCK treated (Worthington).

10. TCEP solution: 0.5 M TCEP (BioRad).

11. Desalting Equilibration Solvent: 80 % Acetonitrile (ACN)/0.1 % trifluoroacetic acid (TFA). 800 mL of ACN, 1 mL TFA to a final volume of 1 L with HPLC grade water.

12. Desalting Load and Wash Solvent: 0.1 % TFA. Add 1 mL TFA to a final volume of 1 L with HPLC grade water.

13. Desalting Elution Solvent: 45 % ACN/0.1 % TFA. 450 mL of ACN, 1 mL TFA to a final volume of 1 L with HPLC grade water.

14. Antibody wash buffer 1: 1× PBS pH 7.4, 0.03 % CHAPS. 300 mg CHAPS, one packet of PBS (Sigma) dissolved in 1 L HPLC grade water.

15. Antibody wash buffer 2: 0.1× PBS pH 7.4, 0.03 % CHAPS. Add 100 mL of 1× PBS pH 7.4 and 300 mg CHAPS into 900 mL HPLC grade water.

16. Antibody storage buffer: 1× PBS/0.03 % CHAPS/0.1 % sodium azide. Dissolve 30 mg CHAPS and 1 g sodium azide in 1 L 1× PBS.

17. Antibody collection buffer: 1× PBS/100 mM Tris–HCl pH 8.1/0.03 % CHAPS/0.1 % sodium azide. Dissolve one packet of PBS (Sigma) and 28 g Tris–HCl pH 8.1 crystals (Sigma) 30 mg CHAPS and 1 g sodium azide in 1 L HPLC grade water.

18. Tris HCl solution: 200 mM Tris–HCl pH 8.1: Dissolve 14 g of Tris–HCl pH 8.1 crystals (Sigma) in 500 mL HPLC grade water.

19. 1 μm Protein G magnetic beads (Dynal) (NB: the 1 μm beads are no longer commercially available, but 2.8 μm beads are and can be used for the 20-plex level described here. An alternate source (Pierce/Life technologies/Thermo) of 1 μm Protein G magnetic beads may also be used).

20. KingFisher 96 magnetic particle processor (Thermo).

21. KingFisher 250 μL polypropylene 96-well plates (Thermo).

22. Barnstead Thermolyne Lab Quake Shaker (VWR).

23. Polypropylene 96-well hard-shell skirted PCR plates (BioRad).

24. Oasis HLB cartridges (Waters).

3   Methods

The key steps in developing and analytically validating iMRM assays are described below and illustrated in Fig. 1. Detailed descriptions of LC-MRM-MS data collection and analysis of MRM data can be found in references 3, 4, 12–17 and elsewhere in this volume.

3.1 Plasma Digestion and Desalt (Adapted from refs. 13, 49)

1. Remove 3×1 mL of plasma from the –80 °C freezer and thaw at ambient temperature (~30 min).

2. Turn on floor mixer incubator and set to 37 °C and rpm to 180. Configure with a 50 mL tube holder if necessary.

3. Add the following to one 50 mL Falcon tube, return any excess plasma to –80 °C freezer: 3 mL plasma, 2.73 g Urea, 1 mL 1 M Tris pH 8.0, 600 µL 0.5 M TCEP.

4. Mix briefly by gentle vortexing and place in incubator at 37 °C and 180 rpm. Once Urea has dissolved, incubate for additional 30 min.

5. Remove from incubator and cool to room temperature.

6. Weigh 462 mg of Iodoacetamide and dissolve in 5 mL 0.2 M Trizma pH 8.1 (500 mM IAA).

7. Add 2 mL of 500 mM IAA into 50 mL tube containing denatured plasma.

8. Mix briefly by gentle vortexing and let stand in the dark at room temperature for 30 min (*see* **Note 6**).

9. Add 40 mL 0.2 M Trizma pH 8.1. Total volume should be ~48 mL (*see* **Note 7**).

10. Verify pH ≥ 8.0 by pipetting 5 µL onto a 5–10 pH range pH test strip (EMD).

11. Carefully weigh 3 mg of TPCK-treated trypsin powder into a tared 15 mL Falcon tube (*see* **Note 8**).

12. Dissolve in 3 mL 0.2 M Trizma pH 8.1 and transfer to digestion mixture.

13. Incubate overnight (12–16 h) at 37 °C at 180 rpm.

14. Add 0.8 mL formic acid. Mix briefly by vortexing and verify pH < 3 by pipetting 5 µL of the digestion mixture onto a 2–5 pH range pH test strip.

15. Store at 4 °C until desalt step.

    If desalt is postponed until a later day, freeze digest mixture at –80 °C.

16. Prepare and label 3×1 g Oasis cartridges, one for each third of the total digest volume.

17. Install all three cartridges onto the vacuum manifold using pipet tip adaptors.

18. Condition each cartridge using 3×20 mL Desalting Equilibration Solvent (80 % ACN/0.1 % TFA).

19. Equilibrate each cartridge using 4×20 mL Desalting Wash and Loading Solvent (0.1 % TFA).

20. Add an additional 4 mL Desalting Wash and Loading Solvent (0.1 % TFA) to each cartridge but do not apply vacuum.

21. Divide the total digest volume into three equal volumes and add each third in 4 mL increments onto one of the three Oasis cartridges. Draw vacuum and load additional volume until the entire one-third of the total digest is loaded across the three cartridges.

22. Wash each cartridge using 3×20 mL Desalting Wash and Loading Solvent (0.1 % TFA).

23. Elute from each cartridge using into a fresh tube using 2×6 mL Desalting Elution Solvent (45 % ACN/0.1 % TFA).

24. Pool the eluates from all three Oasis cartridges into a single 50 mL Falcon tube.

25. Mix briefly by gentle vortexing and dispense an equivalent volume (e.g. 1.5 mL) into ten 2 mL polypropylene tubes (Sarstedt).

    Alternatively, the entire volume may be dried by lyophilization into a single tube.

26. Reduce volume in each tube to less than 0.5 mL/tube by rotary evaporation.

27. Add an additional equivalent volume (e.g. 1 mL) into each of the ten 2 mL tubes and dry to less than 0.5 mL by rotary evaporation. Continue until the remaining volume of the Oasis cartridge eluate is equally dispensed across all tubes.

28. Dry each tube completely by rotary evaporation.

29. Store at −80 °C until use.

3.2 Reverse Curve Preparation (Adapted from ref. 57)

1. Thaw peptide stock solutions on wet ice.

2. Combine 5 µL of each light peptide 100 pmol/µL solution into one tube (100 µL). Label as "Light Stock, 5 pmol/µL, 30 % ACN/0.1 % FA."

3. Combine 5 µL of each heavy peptide 100 pmol/µL solution into one tube (100 µL). Label as "Heavy Stock, 5 pmol/µL, 30 % ACN/0.1 % FA."

4. Resuspend 1×0.3 mL equivalent of digested lyophilized plasma into 270 µL 1× PBS, 0.03 % CHAPS and 30 µL 1 M Tris pH 8.0. Vortex and mix well for 30 min at RT.

5. Prepare 10 mL of peptide dilution buffer (1× PBS, 0.03 % CHAPS, 0.2 % digested plasma). Add 20 µL of resuspended digested plasma into 10 mL of 1× PBS, 0.03 % CHAPS.

6. Prepare 100 fmol/µL light peptide mix: add 10 µL of light peptide stock (5 pmol/µL) into 490 µL peptide dilution buffer (1× PBS, 0.03 % CHAPS, 0.2 % digested plasma).

7. Prepare Reverse Curve background plasma matrix: Pipet 5685 μL of 1× PBS, 0.03 % CHAPS pH 7.4 into a 15 mL Falcon tube. Add the resuspended 0.3 mL of digested plasma. Add 15 μL of 100 fmol/μL light peptide mix. Mix briefly by gentle vortexing.

8. Label 1.5 mL polypropylene centrifuge tubes No. 1–8.

9. Add 1045 μL of Reverse Curve Background Matrix to tube 8 and 700 μL into tubes 1–7. Keep tubes on wet ice.

10. Prepare 200 fmol/μL heavy peptide mix: add 20 μL of heavy peptide stock (5 pmol/μL) into 480 μL peptide dilution buffer (1× PBS, 0.03 % CHAPS, 0.2 % digested plasma).

11. Add 5 μL of 200 fmol/μL heavy peptide mix to tube 8. Mix briefly by gentle vortexing.

12. Transfer 350 μL of tube 8 into tube 7. Mix briefly by gentle vortexing.

13. Continue serial dilution repeating the process in similar manner transferring 350 μL from tube 7 to tube 6, tube 6 to tube 5 down to tube 2 remove 350 μL from tube 2 and discard (tube 1 is blank and contains light peptide only).

14. Freeze in –80 °C until next step (*see* **Note 9**).

3.3 Crosslinking Antibodies to Protein G Beads (Optional: See Note 10. Skip to Subheading 3.4 for Procedure Without Crosslinking Antibodies to Beads)

1. Prepare antibody crosslinking solutions:

   (a) Antibody equilibration solution: 200 mM triethanolamine (TEA) pH 8.5. Add 10 mL triethanolamine into 400 mL HPLC-grade water. Adjust pH to 8.5 using a target of 2 mL 5 N HCl. Add 1.8 mL of 5 N HCl, mix well and add the remaining 200 μL dropwise until pH is 8.5 (*see* **Note 11**).

   (b) Antibody crosslinking solution: 20 mM Dimethyl pimelimidate (DMP) in 200 mM TEA pH 8.5. Dissolve 1.03 g of DMP in 200 mL of Antibody Equilibration Solution.

   (c) Antibody quenching solution: 150 mM monoethanolamine (MEA) pH 9.0: Add 3.6 mL monoethanolamine in 400 mL HPLC-grade water. Adjust pH to 9.0 using a target of 7.5 mL 5 N HCl. Add 7.3 mL of 5 N HCl, mix well and add the remaining 200 μL dropwise until pH is 9.0.

   (d) Antibody wash solution: 5 % acetic acid/0.03 % CHAPS: Add 50 mL glacial acetic acid and 30 mg of CHAPS into a final volume of 1 L HPLC grade water.

   (e) Antibody Storage buffer: 1× PBS/0.03 % CHAPS/0.1 % sodium azide. Dissolve 30 mg CHAPS and 1 g sodium azide into a final volume of 1 L 1× PBS.

2. Add 1550 μL magnetic beads to volume containing the 775 μg required for this curve analysis in a 15 mL Falcon tube.

3. Tumble mix or rock mixture gently for 1–2 h at room temperature.

4. Place the magnet next to the tube and allow the beads to collect on the side of the tube, remove and discard supernatant.

5. Resuspend beads in 900 µL Antibody wash buffer 1 (1× PBS pH 7.4/0.03 % CHAPS). Mix briefly by gentle vortexing and store at 4 °C or on wet ice until use.

   The following crosslinking steps are performed at room temperature.

6. Place the magnet next to the tube and allow the beads to collect on the side of the tube.

7. Remove and discard supernatant. Add 1 mL Antibody equilibration solution and mix by gentle vortexing for 5 min.

8. Use magnet to collect beads on side of tube. Remove and discard supernatant and repeat equilibration with 1 mL Antibody equilibration solution.

9. Use magnet to collect beads on side of tube. Remove and discard supernatant. Add 1 mL Antibody crosslinking solution and mix by gentle vortexing. Continue tumble mixing for 30 min.

10. Use magnet to collect beads on side of tube. Remove and discard supernatant. Add 1 mL Antibody quenching solution and mix by gentle vortexing for 5 min. Continue tumble mixing for 60 min.

11. Use magnet to collect beads on side of tube. Remove and discard supernatant. Add 1 mL Antibody wash solution and mix by gentle vortexing for 5 min.

12. Use magnet to collect beads on side of tube. Remove and discard supernatant and repeat wash with 1 mL Antibody wash solution.

13. Use magnet to collect beads on side of tube. Remove and discard supernatant.

14. Add 1250 µL of Antibody storage buffer and mix by gentle vortexing. Store at 4 °C until use.

3.4 Antibody Affinity Enrichment (Adapted from ref. 49)

Day One

1. Thaw antibody stock solutions on wet ice.

2. Add 50 µg of each polyclonal antibody and 15 µg of each monoclonal antibody in a labeled 2 mL polypropylene centrifuge tube. Refer to Table 1 for antibody concentrations. Keep tubes on wet ice (*see* **Note 12**).

3. Bind antibodies to Protein G beads without crosslinking (optional—*see* **Note 10** and Subheading 3.3 for crosslinking antibodies to beads).

4. Add 1550 µL magnetic beads to volume containing the 775 µg required for this curve analysis in a 15 mL Falcon tube.

5. Tumble mix or rock mixture gently for 1–2 h at room temperature.

6. Place the magnet next to the tube and allow the beads to collect on the side of the tube.

7. Remove the supernatant and resuspend the beads in 1250 μL of 1× PBS, 0.03 % CHAPS pH 7.4. Mix briefly by gentle vortexing and store at 4 °C until use.

8. Thaw tubes containing the reverse curves prepared above.

9. Pipet 200 μL of tube 1 (blank sample, no heavy peptide added) into well A1. Pipet two more replicates of 200 μL of tube 1 into wells A2 and A3 of a Thermo 250 μL KF 96-well plate.

10. Repeat in series down the rows, pipetting three replicates of 200 μL of tube 2 into wells B1, B2, and B3 until three replicates of tube 8, which contains the highest concentration of heavy peptide (200 fmol total) are added into wells H1, H2, and H3. Refer to the plate maps in Table 2.

11. Add 50 μL of antibody bead mixture to each well, pipetting up and down 3–4 times to mix completely. Use a fresh pipet tip for each well.

12. Seal plate securely using a roller to press adhesive foil seal over all wells.

13. Place plate on Labquake mixer using rubber bands, Velcro strips, or ties and turn on to mix slowly inverting overnight (12–16 h) at 4 °C.

Day Two

14. Install the PCR magnet head on the Kingfisher bead handling platform.

15. Prepare and load the following plates on the Kingfisher:

    Plate 1: incubation plate (digested plasma, peptides, antibodies, and beads (~250 μL)).

    Plate 2: 250 μL Antibody wash buffer 1 (1× PBS/0.03 % CHAPS).

    Plate 3: 250 μL Antibody wash buffer 1 (1× PBS/0.03 % CHAPS).

    Plate 4: 250 μL Antibody wash buffer 2 (0.1× PBS/0.03 % CHAPS).

    Plate 5: 30 μL Antibody Elution Solvent: (3 % ACN/5 % acetic acid).

    Plate 6: 200 μL Antibody collection buffer: (1× PBS/100 mM Tris pH 8.0/0.03 % CHAPS/0.1 % sodium azide).

    Plate 7: tip comb.

    All solutions prepared for plates 1–4 and 6 are pipetted into KingFisher 250 μL wellplates. Solutions for elution (plate 5) are pipetted into a 96-well PCR plate. The tip comb is held in an empty 250 μL wellplate from plate 7. It is picked up at the beginning of the method to cover and protect the magnet and returned to plate 7 upon completion.

Table 2
Plate maps used for replicate samples of concentration points in the reverse curve by (A) concentration point and replicate and (B) by total heavy peptide amount

	1	2	3	4	5	6
(A) By concentration point and replicate						
A	pt1—1	pt1—2	pt1—3			
B	pt2—1	pt2—2	pt2—3			
C	pt3—1	pt3—2	pt3—3			
D	pt4—1	pt4—2	pt4—3			
E	pt5—1	pt5—2	pt5—3			
F	pt6—1	pt6—2	pt6—3			
G	pt7—1	pt7—2	pt7—3			
H	pt8—1	pt8—2	pt8—3			
(B) By total heavy peptide amount (fmol)						
A	0	0	0			
B	0.3	0.3	0.3			
C	0.8	0.8	0.8			
D	2.5	2.5	2.5			
E	7.4	7.4	7.4			
F	22.2	22.2	22.2			
G	66.7	66.7	66.7			
H	200	200	200			

16. Load the KingFisher program. Use the up ^ and down ˅ arrows to scroll through methods until the one described in Table 3 is displayed (*see* **Note 11**).

17. Remove plate from Labquake and centrifuge at 1400 RPM (130–400×*g* depending on the type of centrifuge and rotor) for 30–60 s to remove liquid that may be on the seal surface. Typically, a SpeedVac concentrator centrifuge equipped with a microplate rotor is used.

18. Remove the foil seal carefully (*see* **Note 13**).

19. Place the incubation plate in plate position 1 on the Kingfisher.

20. Press "start" to begin the method.

21. When the KingFisher method is finished (approximately 20 min (*see* **Note 14**))—seal the incubation plate (plate 1) and the collected antibody bead plate (plate 6) with adhesive foil and store at −80 °C and 4 °C, respectively.

Table 3
Description of the plate layouts and protocol steps in the KingFisher program

Instrument: KingFisher 96
Protocol template version: 2.6.0
Created: April 26, 2011

Plate layout:

No. and position	Type	Description	Contents	Volume (µL)
1	KingFisher 96–250 µL	Plate_1_beadAbPep	Plasma, ProG beads, Abs, peptides, 1× PBS pH 7.4, 0.03 % CHAPS	250
2	KingFisher 96–250 µL	Plate_2_wash_1	1× PBS pH 7.4, 0.03 % CHAPS	250
3	KingFisher 96–250 µL	Plate_3_wash_2	1× PBS pH 7.4, 0.03 % CHAPS	250
4	KingFisher 96–250 µL	Plate_4_wash_3	0.1× PBS pH 7.4, 0.03 % CHAPS	250
5	PCR—100 µL	Plate_5_elution	3 % Acetonitrile/5 % acetic acid	30
6	KingFisher 96–250 µL	Plate_6_collection	1× PBS pH 7.4, 0.03 % CHAPS, 0.1 % NaN$_3$	200
7	KingFisher 96–250 µL	Plate_7_tips	KF 96 tip comb	Empty

Protocol steps:

Step no.	Plate no.	Description	Beginning of step	Wash/elution parameters	End of step
1	1	Ab and Pep Capture	Release = yes Time = 10 s Speed = slow	Time = 5 m Speed = bottom slow	Collect = yes Count = 5
2	2	Wash 1	Release = yes Time = 10 s Speed = slow	Time = 1.5 m Speed = slow	Collect = yes Count = 5
3	3	Wash 2	Release = yes Time = 10 s Speed = slow	Time = 1.5 m Speed = slow	Collect = yes Count = 5
4	4	Wash 3	Release = yes Time = 10 s Speed = slow	Time = 1.5 m Speed = slow	Collect = yes Count = 5
5	5	Elution	Release = yes Time = 10 s Speed = bottom slow	Time = 5 m Speed = bottom slow Heating = no	Remove beads = Yes, collect count = 10, disposal plate = Plate 6

22. Remove elution plate (Plate 5) from the KingFisher and place onto the autosampler magnet plate on wet ice.

23. Label a fresh PCR plate "iMRM reverse curves" and add 5 µL of 3 % ACN/5 % HOAc to the wells designated in the plate map in Table 2.

24. Using a multichannel pipet set to 20 µL, draw the eluate supernatant without touching the bottom of the well and transfer into corresponding wells of a fresh PCR well plate (*see* **Note 15**).

25. Cover the PCR plates with silicon plate mats and transfer onto autosampler for LC-MRM-MS analysis (*see* **Note 16**).

3.5 LC-MRM-MS Analysis (Refer to refs. 3, 4, 12–17)

1. Inject one-third of the sample onto a triple quadrupole MS instrument configured with nanoflow liquid chromatograph and autosampler configured with a trap and analytical column and perform MRM-MS experiments, unscheduled or scheduled using the transition masses in Table 4 (*see* **Note 17**).

2. Verify the LC-MRM-MS system is ready for analysis by injecting and analyzing an appropriate system suitability standard [58], typically a mixture of peptide standards analyzed by MRM.

3. Inject samples in order from lowest concentration point to highest, replicates one, two and three for each point (e.g. pt1—0 fmol rep1, rep2, rep3) followed by an injection of blank (3 % ACN/5 % HOAc). Continue in sequence for the rest of the samples up to point 8 (200 fmol) (*see* **Note 18**).

4. Prepare a Skyline [59] document (version 3.1 https://brendanx-uw1.gs.washington.edu/labkey/project/home/software/Skyline/begin.view) that contains the peptide sequences with the light and heavy peptide masses of the peptides analyzed by MRM-MS (*see* **Note 19**).

5. Under Peptide Settings, confirm that the heavy label matches that heavy amino acid used for heavy peptide and select the light peptide as standard in the checkbox.

6. Import the reverse curve MS raw data in Skyline from the File, Import, Results drop-down window, selecting the appropriate data files.

7. Open and view the Result Grid, choose and add the columns "SampleGroup," "Concentration," and "IS Spike." Enter the curve designation (e.g. "pt1") and the concentration of heavy peptide (e.g. "0") in "Concentration" and concentration of light peptide (e.g. "20") in "IS Spike" (*see* **Note 20**).

8. Select "Integrate All" from the "Setting" drop-down window (a check mark will appear when selected). This makes sure that the integration for one version of the peptide (light or heavy) is applied to the other peptide (light or heavy) automatically.

Table 4
Unscheduled MRM method for 20 light and 20 heavy peptides, three transitions each (*n* = 120 transitions total)

Q1	Q3	Dwell	ID	DP	CE
623.3379	1018.542	10	TNNI3.NITEIADLTQK.+2y9.light	76.6	29.9
623.3379	788.4512	10	TNNI3.NITEIADLTQK.+2y7.light	76.6	29.9
623.3379	675.3672	10	TNNI3.NITEIADLTQK.+2y6.light	76.6	29.9
626.348	1024.562	10	TNNI3.NITEIADLTQK.+2y9.heavy	76.6	29.9
626.348	794.4714	10	TNNI3.NITEIADLTQK.+2y7.heavy	76.6	29.9
626.348	681.3873	10	TNNI3.NITEIADLTQK.+2y6.heavy	76.6	29.9
516.7898	816.4978	10	IL33.TDPGVFIGVK.+2y8.light	68.8	25.7
516.7898	719.445	10	IL33.TDPGVFIGVK.+2y7.light	68.8	25.7
516.7898	662.4236	10	IL33.TDPGVFIGVK.+2y6.light	68.8	25.7
519.7999	822.5179	10	IL33.TDPGVFIGVK.+2y8.heavy	68.8	25.7
519.7999	725.4652	10	IL33.TDPGVFIGVK.+2y7.heavy	68.8	25.7
519.7999	668.4437	10	IL33.TDPGVFIGVK.+2y6.heavy	68.8	25.7
804.4068	1154.584	10	FTL.LGGPEAGLGEYLFER.+2y10.light	89.8	37.2
804.4068	1083.547	10	FTL.LGGPEAGLGEYLFER.+2y9.light	89.8	37.2
804.4068	913.4414	10	FTL.LGGPEAGLGEYLFER.+2y7.light	89.8	37.2
807.4169	1160.604	10	FTL.LGGPEAGLGEYLFER.+2y10.heavy	89.8	37.2
807.4169	1089.567	10	FTL.LGGPEAGLGEYLFER.+2y9.heavy	89.8	37.2
807.4169	919.4615	10	FTL.LGGPEAGLGEYLFER.+2y7.heavy	89.8	37.2
490.2584	759.4247	10	AFP.GYQELLEK.+2y6.light	66.9	24.6
490.2584	631.3661	10	AFP.GYQELLEK.+2y5.light	66.9	24.6
490.2584	502.3235	10	AFP.GYQELLEK.+2y4.light	66.9	24.6
493.2684	765.4448	10	AFP.GYQELLEK.+2y6.heavy	66.9	24.6
493.2684	637.3862	10	AFP.GYQELLEK.+2y5.heavy	66.9	24.6
493.2684	508.3437	10	AFP.GYQELLEK.+2y4.heavy	66.9	24.6
575.8088	987.5469	10	AFP.YIQESQALAK.+2y9.light	73.1	28
575.8088	874.4629	10	AFP.YIQESQALAK.+2y8.light	73.1	28
575.8088	746.4043	10	AFP.YIQESQALAK.+2y7.light	73.1	28
578.8188	993.5671	10	AFP.YIQESQALAK.+2y9.heavy	73.1	28
578.8188	880.483	10	AFP.YIQESQALAK.+2y8.heavy	73.1	28

(continued)

Table 4
(continued)

Q1	Q3	Dwell	ID	DP	CE
578.8188	752.4244	10	AFP.YIQESQALAK.+2y7.heavy	73.1	28
549.2934	949.485	10	ERBB2.GLQSLPTHDPSPLQR.+3y8.light	71.2	31.5
549.2934	812.4261	10	ERBB2.GLQSLPTHDPSPLQR.+3y7.light	71.2	31.5
549.2934	697.3991	10	ERBB2.GLQSLPTHDPSPLQR.+3y6.light	71.2	31.5
552.6295	959.4933	10	ERBB2.GLQSLPTHDPSPLQR.+3y8.heavy	71.2	31.5
552.6295	822.4344	10	ERBB2.GLQSLPTHDPSPLQR.+3y7.heavy	71.2	31.5
552.6295	707.4074	10	ERBB2.GLQSLPTHDPSPLQR.+3y6.heavy	71.2	31.5
599.827	986.4942	10	ERBB2.VLGSGAFGTVYK.+2y10.light	74.8	29
599.827	842.4407	10	ERBB2.VLGSGAFGTVYK.+2y8.light	74.8	29
599.827	714.3821	10	ERBB2.VLGSGAFGTVYK.+2y6.light	74.8	29
603.8341	994.5084	10	ERBB2.VLGSGAFGTVYK.+2y10.heavy	74.8	29
603.8341	850.4549	10	ERBB2.VLGSGAFGTVYK.+2y8.heavy	74.8	29
603.8341	722.3963	10	ERBB2.VLGSGAFGTVYK.+2y6.heavy	74.8	29
748.3641	1325.615	10	ERBB2.AVTSANIQEFAGC[CAM]K.+2y12.light	85.7	34.9
748.3641	1224.568	10	ERBB2.AVTSANIQEFAGC[CAM]K.+2y11.light	85.7	34.9
748.3641	1066.499	10	ERBB2.AVTSANIQEFAGC[CAM]K.+2y9.light	85.7	34.9
748.3641	839.3716	10	ERBB2.AVTSANIQEFAGC[CAM]K.+2y7.light	85.7	34.9
751.3742	1331.636	10	ERBB2.AVTSANIQEFAGC[CAM]K.+2y12.heavy	85.7	34.9
751.3742	1230.588	10	ERBB2.AVTSANIQEFAGC[CAM]K.+2y11.heavy	85.7	34.9
751.3742	1072.519	10	ERBB2.AVTSANIQEFAGC[CAM]K.+2y9.heavy	85.7	34.9
751.3742	845.3917	10	ERBB2.AVTSANIQEFAGC[CAM]K.+2y7.heavy	85.7	34.9
674.657	928.4523	10	MUC1.EGTINVHDVETQFNQYK.+3y7.light	80.3	37.7
674.657	827.4046	10	MUC1.EGTINVHDVETQFNQYK.+3y6.light	80.3	37.7
674.657	699.3461	10	MUC1.EGTINVHDVETQFNQYK.+3y5.light	80.3	37.7
677.3284	936.4665	10	MUC1.EGTINVHDVETQFNQYK.+3y7.heavy	80.3	37.7
677.3284	835.4188	10	MUC1.EGTINVHDVETQFNQYK.+3y6.heavy	80.3	37.7

(continued)

Table 4
(continued)

Q1	Q3	Dwell	ID	DP	CE
677.3284	707.3603	10	MUC1.EGTINVHDVETQFNQYK.+3y5.heavy	80.3	37.7
532.2746	821.4152	10	MUC16.ELGPYTLDR.+2y7.light	69.9	26.3
532.2746	764.3937	10	MUC16.ELGPYTLDR.+2y6.light	69.9	26.3
532.2746	667.341	10	MUC16.ELGPYTLDR.+2y5.light	69.9	26.3
532.2746	382.7005	10	MUC16.ELGPYTLDR.+2y6+2.light	69.9	26.3
535.2846	827.4353	10	MUC16.ELGPYTLDR.+2y7.heavy	69.9	26.3
535.2846	770.4139	10	MUC16.ELGPYTLDR.+2y6.heavy	69.9	26.3
535.2846	673.3611	10	MUC16.ELGPYTLDR.+2y5.heavy	69.9	26.3
535.2846	385.7106	10	MUC16.ELGPYTLDR.+2y6+2.heavy	69.9	26.3
703.8492	960.5109	10	TG.FSPDDSAGASALLR.+2y10.light	82.4	33.2
703.8492	845.4839	10	TG.FSPDDSAGASALLR.+2y9.light	82.4	33.2
703.8492	687.4148	10	TG.FSPDDSAGASALLR.+2y7.light	82.4	33.2
706.8592	966.531	10	TG.FSPDDSAGASALLR.+2y10.heavy	82.4	33.2
706.8592	851.5041	10	TG.FSPDDSAGASALLR.+2y9.heavy	82.4	33.2
706.8592	693.4349	10	TG.FSPDDSAGASALLR.+2y7.heavy	82.4	33.2
636.359	1059.558	10	TG.VIFDANAPVAVR.+2y10.light	77.5	30.5
636.359	912.4898	10	TG.VIFDANAPVAVR.+2y9.light	77.5	30.5
636.359	726.4257	10	TG.VIFDANAPVAVR.+2y7.light	77.5	30.5
639.369	1065.578	10	TG.VIFDANAPVAVR.+2y10.heavy	77.5	30.5
639.369	918.5099	10	TG.VIFDANAPVAVR.+2y9.heavy	77.5	30.5
639.369	732.4458	10	TG.VIFDANAPVAVR.+2y7.heavy	77.5	30.5
694.3876	916.5714	10	ANXA1.GVDEATIIDILTK.+2y8.light	81.7	32.8
694.3876	815.5237	10	ANXA1.GVDEATIIDILTK.+2y7.light	81.7	32.8
694.3876	702.4396	10	ANXA1.GVDEATIIDILTK.+2y6.light	81.7	32.8
698.3947	924.5856	10	ANXA1.GVDEATIIDILTK.+2y8.heavy	81.7	32.8
698.3947	823.5379	10	ANXA1.GVDEATIIDILTK.+2y7.heavy	81.7	32.8
698.3947	710.4538	10	ANXA1.GVDEATIIDILTK.+2y6.heavy	81.7	32.8
519.2744	833.4516	10	CLIC1.GFTIPEAFR.+2y7.light	69	25.8
519.2744	732.4039	10	CLIC1.GFTIPEAFR.+2y6.light	69	25.8
519.2744	619.3198	10	CLIC1.GFTIPEAFR.+2y5.light	69	25.8

(continued)

Table 4
(continued)

Q1	Q3	Dwell	ID	DP	CE
524.2785	843.4598	10	CLIC1.GFTIPEAFR.+2y7.heavy	69	25.8
524.2785	742.4122	10	CLIC1.GFTIPEAFR.+2y6.heavy	69	25.8
524.2785	629.3281	10	CLIC1.GFTIPEAFR.+2y5.heavy	69	25.8
526.2581	938.4248	10	IL18.ISTLSC[CAM]ENK.+2y8.light	69.5	26.1
526.2581	851.3927	10	IL18.ISTLSC[CAM]ENK.+2y7.light	69.5	26.1
526.2581	637.261	10	IL18.ISTLSC[CAM]ENK.+2y5.light	69.5	26.1
530.2652	946.439	10	IL18.ISTLSC[CAM]ENK.+2y8.heavy	69.5	26.1
530.2652	859.4069	10	IL18.ISTLSC[CAM]ENK.+2y7.heavy	69.5	26.1
530.2652	645.2752	10	IL18.ISTLSC[CAM]ENK.+2y5.heavy	69.5	26.1
493.2597	809.39	10	NFKB2.IEVDLVTHSDPPR.+3y7.light	67.1	28.7
493.2597	708.3424	10	NFKB2.IEVDLVTHSDPPR.+3y6.light	67.1	28.7
493.2597	618.3226	10	NFKB2.IEVDLVTHSDPPR.+3y11+2.light	67.1	28.7
496.5958	819.3983	10	NFKB2.IEVDLVTHSDPPR.+3y7.heavy	67.1	28.7
496.5958	718.3506	10	NFKB2.IEVDLVTHSDPPR.+3y6.heavy	67.1	28.7
496.5958	623.3267	10	NFKB2.IEVDLVTHSDPPR.+3y11+2.heavy	67.1	28.7
719.3558	930.4891	10	FSCN1.LSC[CAM]FAQTVSPAEK.+2y9.light	83.6	33.8
719.3558	859.452	10	FSCN1.LSC[CAM]FAQTVSPAEK.+2y8.light	83.6	33.8
719.3558	731.3934	10	FSCN1.LSC[CAM]FAQTVSPAEK.+2y7.light	83.6	33.8
723.3629	938.5033	10	FSCN1.LSC[CAM]FAQTVSPAEK.+2y9.heavy	83.6	33.8
723.3629	867.4662	10	FSCN1.LSC[CAM]FAQTVSPAEK.+2y8.heavy	83.6	33.8
723.3629	739.4076	10	FSCN1.LSC[CAM]FAQTVSPAEK.+2y7.heavy	83.6	33.8
483.2506	823.4196	10	TAGLN.AAEDYGVIK.+2y7.light	66.3	24.3
483.2506	694.377	10	TAGLN.AAEDYGVIK.+2y6.light	66.3	24.3
483.2506	579.3501	10	TAGLN.AAEDYGVIK.+2y5.light	66.3	24.3
487.2577	831.4338	10	TAGLN.AAEDYGVIK.+2y7.heavy	66.3	24.3
487.2577	702.3912	10	TAGLN.AAEDYGVIK.+2y6.heavy	66.3	24.3
487.2577	587.3643	10	TAGLN.AAEDYGVIK.+2y5.heavy	66.3	24.3
826.4123	1066.542	10	EZR.SQEQLAAELAEYTAK.+2y10.light	91.4	38.1
826.4123	995.5044	10	EZR.SQEQLAAELAEYTAK.+2y9.light	91.4	38.1
826.4123	924.4673	10	EZR.SQEQLAAELAEYTAK.+2y8.light	91.4	38.1

(continued)

Table 4
(continued)

Q1	Q3	Dwell	ID	DP	CE
830.4194	1074.556	10	EZR.SQEQLAAELAEYTAK.+2y10.heavy	91.4	38.1
830.4194	1003.519	10	EZR.SQEQLAAELAEYTAK.+2y9.heavy	91.4	38.1
830.4194	932.4815	10	EZR.SQEQLAAELAEYTAK.+2y8.heavy	91.4	38.1
613.3486	984.5473	10	PRDX4.QITLNDLPVGR.+2y9.light	75.8	29.5
613.3486	770.4155	10	PRDX4.QITLNDLPVGR.+2y7.light	75.8	29.5
613.3486	656.3726	10	PRDX4.QITLNDLPVGR.+2y6.light	75.8	29.5
618.3527	994.5555	10	PRDX4.QITLNDLPVGR.+2y9.heavy	75.8	29.5
618.3527	780.4238	10	PRDX4.QITLNDLPVGR.+2y7.heavy	75.8	29.5
618.3527	666.3809	10	PRDX4.QITLNDLPVGR.+2y6.heavy	75.8	29.5

9. Confirm peak integration. Select "Retention Times, Replicate Comparison" under the "View" drop-down window and use the Retention Time plot to identify potential chromatograms requiring manual re-integration.

10. Activate QuaSAR from the "Tools" drop-down window [60].

11. Perform statistical analysis (LOD, LOQ, CV [61–64]) to assess assay performance. Check "plot each peptide," "CV table" and "LOD/LOQ table" under the "Generate" tab. Check "Standard present" and set "Analyte" and "Standard" fields as heavy area and light area, respectively. Accept default settings for AuDIT [63] and plot scales (*see* **Note 21**).

12. Evaluate analysis of the data and report assay performance using a combination of plots of the concentration curves, CV and LOD box and whisker plots for all peptides in the multiplex iMRM assay as shown in Fig. 2.

Fig. 2 (continued) the analysis. *QuaSAR implements a comprehensive and easy-to-use pipeline for the analysis of MRM-MS data and provides succinct visual summaries of various results including reproducibility, interferences, and detection limits. QuaSAR can be accessed at (http://genepattern.broadinstitute.org/gp/pages/index.jsf?lsid=QuaSAR). This link prompts the user to login at GenePattern, it also provides free registration at the GenePattern website upon choosing "click to register"; then under modules browse to "Proteomics" then to "Quasar" or search for the "Quasar" module directly. Transitions with interferences or high variability are detected using AuDIT [63], enabling focused reevaluation of the raw data and/or exclusion of erroneous transitions. Erroneous transitions are also visually marked in the data visualization plots

a Overall iMRM assay reproducibility

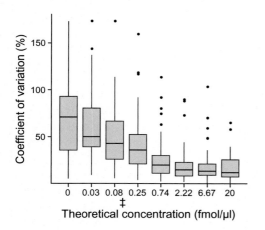

b Response curve analysis of ERRB2.AVTSANIQEFAGCK

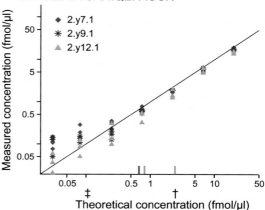

c AuDIT report

Peptide	Sample	Transition ID	Final p-value	Status	CV	CV status
AVTSANIQEFAGCK	iMRM-pt6	2.y11.1	0.6278	Good	0.086	Good
AVTSANIQEFAGCK	iMRM-pt6	2.y12.1	0.2670	Good	0.244 †	Bad
AVTSANIQEFAGCK	iMRM-pt6	2.y7.1	0.5893	Good	0.074	Good

d Assay performance summary

Peptide	Sample	Transition ID	LOD (fmol/µl)	LOQ (fmol/µl)	Slope	y-intercept
AVTSANIQEFAGCK	ERBB2	2.y9.1	0.258	0.773	0.894	−0.054
AVTSANIQEFAGCK	ERBB2	2.y12.1	0.111 ‡	0.332	0.712	0.011
AVTSANIQEFAGCK	ERBB2	2.y7.1	0.294	0.882	0.781	0.126

Fig. 2 Statistical analysis of assay performance. Key assay characteristics including LOD, LOQ, and CV [61, 62] as well as flagging of interferences observed in the specific transitions monitored were assessed using the tools QuaSAR* [64] and AuDIT [63]. (**a**) CVs of the heavy-to-light peptide peak area ratios at each theoretical concentration were calculated for every transition and plotted in the box and whisker format. Interquartile ranges are shaded in beige and outliers are displayed as *black dots*. The median CV for all measurements is represented by a *black line* within the box. (**b**) Example plot of observed vs. theoretical concentration (log scale) for each transition and each replicate of peptide AVTSANIQEFAGCK from the protein ERBB2. QuaSAR generates two separate plots (linear scale and log scale) for each peptide in the multiplex to evaluate individual peptide assay performance. *Color-keyed tick marks* on *x*-axis indicate specific transitions and the corresponding concentration points that are either inconsistent or more variable, and require manual inspection and interpretation of the integrated peak areas. Theoretical (*black solid*) line drawn with a slope = 1 for assessing the accuracy of the measurements. (**c**) *AuDIT Summary Report.* AuDIT determines whether the relative ratios for each transition for light and heavy peptide are consistent and flags those that are statistically inconsistent (*p*-value > 0.05). CV of the ratio of heavy to light peptide peak area (H/L) is used to filter transitions with unacceptably large variation (>20 %) between the replicates. CVs exceeding the threshold (transition ID 2.y12.1 (†)) are flagged in the table and designated by *color-keyed tick marks* along the *x*-axis on the plot. (**d**) *Assay Performance Summary Table.* The performance (LOD/LOQ, slope, *y*-intercept) is listed for each transition in QuaSAR summary tables to rank assay performance for each transition of each peptide. A second "CV final" table (not shown) reports CV for only the best transition, which is defined as the one with the lowest LOD. LOD is reported for each transition. The transition with the lowest LOD (‡) is consistent with the region of the curve in (**b**) where the curve begins to level off and a noticeable increase in replicate and transition variability is observed (**a**, **b**). A second "LOD final" table lists the transition that provides the lowest LOD for each peptide in

4 Notes

1. Prior experience using peptides synthesized with blocked termini (e.g. acetyl group on N-terminus, amide group on C-terminus) gave lower affinity anti-peptide Abs. Therefore we suggest not to use them. Trypsin treatment of proteins yields peptides with free termini. Peptide standards and the immunogen peptide sequence used for antibody generation, which may be synthesized with additional amino acids (e.g. Cys) for easier conjugation to KLH and purification media or spacer molecules (GSGS, or dPEG$_2$, dPEG$_4$) to increase immunogenicity, should be prepared as chemically similar to that expected from the native protein upon digestion with trypsin using amino acids with natural isotope (e.g. 12C > 13C > 14C) distribution and correspond to a tryptic sequence (R or K C-terminal amino acid).

   Peptides selected as target analytes should, in general, be unique to the protein of interest as well as being unique in the proteome (i.e. proteotypic). Evaluating if the sequence of a target peptide is unique can be determined by BLAST analysis http://www.uniprot.org/blast/ of each target peptide sequence or in batch mode for a group of peptides using the Peptide String Match utility in Spectrum Mill (http://proteomics.broadinstitute.org/). Species specificity depends on the source of samples intended for analysis. Peptide Selector, another utility in Spectrum Mill, provides an initial in silico ranking of peptides to monitor based on mass, sequence, and theoretical retention time.

   The best predictor of whether a given peptide will be useful for assay development is the observation of that peptide in your own data or public database containing high-quality MS and MS/MS data acquired on high-performance instrumentation at high mass accuracy and high resolution. Peptides can be ranked for assay development and synthesis based on frequency of detection, score (e.g., number and/or percentage of fragment ions in an observed spectrum that correspond to the target sequence) and retention time. Signal response by electrospray MS is another important parameter to consider, especially when deciding which of several peptides from a given protein to select for assay development. The response of tryptic peptides derived from the same protein can vary by up to 50-fold based on inherent physiochemical parameters [55]. In general, for peptides that are well recovered from sample processing, the peptides with the higher ESI-MS response will yield higher assay sensitivity. When this information is not readily obtained from available data, prediction tools can be used to algorithmically predict the highest responding peptides from a given protein using software tools such as ESP (http://www.broadinstitute.org/cancer/software/genepattern/esppredictor) or PeptideSieve [56] (http://www.systemsbiology.org/peptidesieve).

For iMRM assays designed to measure total protein, peptides containing amino acids susceptible to modification during biochemical processing (e.g. methionine oxidation, cyclization of N-terminal Glu or carboxamidomethyl Cys) should be avoided [12]. However, it is not always possible to do so, in which case multiple forms of the peptide may need to be synthesized and included in the final assay. Small biochemically introduced modifications such as phosphorylation or acetylation can be readily synthesized and assays constructed in the same manner as for unmodified peptides. Peptides predicted or known to contain large modifications such as N-linked carbohydrate should be avoided.

Synthesized peptides are purified by reversed-phase liquid chromatography (RPLC) and assessed for identity and purity by matrix-assisted laser desorption ionization (MALDI) and RPLC, respectively. Quantity and concentration of peptide are determined by Amino acid analysis (AAA).

2. Tryptic peptides will contain a C-terminal Lys or Arg unless derived from the C-terminus of the protein. These amino acids are preferred for stable isotopic incorporation for several reasons. The y-ion series of ions for tryptic peptides are usually among the most abundant ions in their MS/MS spectra. Each y-ion fragment will contain either the light version or the heavy version of Lys or Arg, depending on which version was selected for fragmentation in the MS. Several of these y-ions are monitored in the MRM-MS experiment and used for both identification and quantification. Use of common heavy amino acids at the C-termini of synthetic tryptic peptides also simplifies synthesis.

3. Plasma was used for this evaluation because it is the matrix in which the assays were designed to measure analytes. Plasma is readily available from healthy subjects collected under appropriate IRB collection protocols, and can be obtained commercially.

4. Immunogen peptides are conjugated to KLH, formulated with adjuvant and administered to New Zealand white rabbits following 77–120 day immunization schedules depending on vendor and protocol. Antibody titer is determined by peptide ELISA, coating free peptide on the plate. Rabbit sera with the highest titers are collected and purified by peptide affinity chromatography [46]. Glycerol and sodium azide are added to 25 % and 0.1 % respectively to aid antibody stability during freeze/thaw for antibodies stored at –20 °C and longer term storage at 4 °C [65, 66]

5. Lymphocytes isolated from the spleens harvested from the rabbits used to generate polyclonal antibodies were isolated and fused with partner cells to generate a mixture of hybridoma cell clones which are expanded. After expansion and growth, subclones are picked, expanded further and tested until a single cell population (monoclone) with a positive screen against the target peptide (by ELISA or by automated iMRM format [67, 68]) is isolated. IgG is purified from monoclonal cell cultures or is sequenced and expressed as recombinant protein that are purified by antibody affinity chromatography or protein G chromatography, respectively.

6. Dissolve IAA fresh for every digestion and do not reuse solutions. Continuous mixing is not necessary after the IAA is initially mixed. Since IAA breaks down when exposed to light, it is suggested to either use amber-tinted tubes that can be handled on the bench or to place the tubes in a light-tight enclosure during reaction.

7. Concentration of urea needs to be reduced from the initial concentration of greater than 6 M to less than 1 M for optimum trypsin efficiency.

8. This trypsin is purified from bovine pancreas, treated with l-(tosylamido 2-phenyl) ethyl chloromethyl ketone (TPCK) to inhibit chymotryptic activity [69] and lyophilized after dialysis with 1 mM HCl. Trypsin in powder form is very flaky and light and highly susceptible to static interactions. Use a static gun and weigh without wearing gloves to reduce static electricity. Be cautious when pipetting dissolution solvent (next step) to avoid aerating and forcing trypsin out of the tube. The optimum pH for trypsin is 8.3. Trypsin becomes active immediately upon dissolution of its powder form in 0.2 M Tris–HCl pH 8.1. To terminate digestion, trypsin enzymatic activity is reduced by the addition of acid to reduce the pH below 3.0.

9. Even if the samples are continuing directly to antibody affinity enrichment freeze these tubes at –80 °C. That way, freeze/thaw steps will be consistent between replicates, which is especially important when additional curve replicates are prepared for future studies.

10. Under acidic conditions, peptides as well as immunoglobulin protein will elute from the protein G beads. The large amount of desorbed protein will eventually overload the 75 μm ID × 10 cm analytical column (packed with 3 μm C18) that we estimate has a loading capacity of approximately 1 μg. Crosslinking antibodies to protein G beads via primary amines with dimethyl pimelimidate (DMP) allows the antibodies to be washed to reduce the amount of nonspecific background, the amount of bound passenger peptide (see Note 12) and retains the anti-

body on bead for the potential reuse. Antibodies cross-linked to beads should be tested by a capture efficiency test or mini-curve [46] prior to use.

11. Bead processing begins by selecting and starting the method using the dialog box of the KingFisher magnetic bead handler. The method outlined in Table 3 was programmed using KingFisher software version 2.6 and then sent to the instrument via a serial port connection. It is not necessary to have the computer on or running during the method once it is programmed and loaded onto the KingFisher. Method settings, such as duration and intensity of bead mixing or bead collection times, can be changed when the method is open on the computer, saved and then re-sent to the KingFisher. Outline of bead processing steps performed on the KingFisher:

   – Wash the beads twice with 250 µL PBS/0.03 % CHAPS (1.5 min per wash).

   – Wash the beads once with 250 µL 0.1× PBS/0.03 % CHAPS (1.5 min per wash).

   – Elute the peptides in 25 µL of 3 % acetonitrile/5 % acetic acid (5 min).

   – Collect used beads into a fresh collection plate (5 min).

   This method collects the magnetic beads after elution into a fresh collection plate (plate 6) containing Antibody collection buffer (1×PBS/100 mM Tris pH 8.0/0.03 % CHAPS/0.1 % sodium azide) and subsequently pooled, washed, and reused. When beads with crosslinked antibodies are used (*see* **Note 11**) these beads may be pooled, washed, and reused after re-equilibration in storage solution, 1×PBS pH 7.4/0.03 % CHAPS/0.1 % NaN3. They can be stored for longer terms (6 months or longer at 4 °C depending on the antibody), but should be retested prior to reuse.

12. Antibody mixtures may be made in advance of reverse curve preparations and stored at 4 °C for 3 months. Antibodies stored for longer than 3 month at 4 °C after thawing should be evaluated by a capture efficiency test or mini-curve [46] prior to use. Total mixture volume is dependent on the individual antibody concentrations. There are six monoclonal antibodies and 14 polyclonal antibodies in this 20-plex iMRM assay. Since 2 µg of each polyclonal antibody and 0.5 µg of each monoclonal antibody are used per enrichment, there will be 31 µg of total IgG added to each well (refer to Table 1). Protein G binding capacities depend on the type and size of bead and should be tested empirically for each batch and bead type. Protein G is a bacterial-derived protein that binds to the Fc portion of immunoglobulin heavy chain 66], and is the recommended

ligand for binding antibodies derived from rabbit. Protein G is commercially available conjugated on many bead types and sizes, from larger 20 μm porous beads to smaller 1 μm magnetic beads. Magnetic beads were chosen to make the process more amenable to automation. Here, we found 2 μL of 1 μm protein G magnetic beads sufficient to bind 1 μg of antibody. At this plex level, 62 μL of 1 μm beads are required per enrichment. If larger beads are used (e.g., 2.8 μm beads), more beads will be required leading to some increase in nonspecific binding. Binding capacities of alternate sources of protein G magnetic beads should be tested prior to use.

13. Removing the foil seal can be tricky, especially after being well applied and tumble mixed overnight. It may tear off in pieces, requiring multiple grip and tear motions to completely remove. The plate must be securely controlled in one hand to prevent it from tipping and mixing the contents of neighboring wells on the plate.

14. The duration and overall lapse time for the steps was optimized to keep the KingFisher method wash and elution time to less than 30 min. This upper time limit is based on the estimated Kd of these rabbit polyclonal antibodies calculated from the off-time measured under constant flow conditions in 1× PBS pH 7.4 [70].

15. Transferring 20 μL of eluate supernatant to a fresh PCR plate (or to fresh wells on the same plate) was found to increase the reproducibility of the subsequent LC-MRM-MS analysis by removing particulates and other precipitous solids may form during the wash and elution process. After placing the PCR plate on the autosampler magnet plate, wait 1–3 min for the magnet to draw and collect residual magnetic beads to the side before drawing up the eluate supernatant. However, it is not imperative to remove the entire volume from each well nor equivalent volumes from all wells since the heavy peptide standard have already been added to account for the variation that may occur during this process.

16. In cases where MS analysis may not occur right away, plates may be resealed with aluminum foil seals and stored at −80 °C until the instrument is ready. Prolonged freezer storage of digested plasma or enrichments from digested plasma may produce addition particulates and precipitate, which should be removed as described above using a magnet plate holder. Plates may be centrifuged briefly after thawing a frozen plate, for less than 2 min at 250 × g, to collect particulates in the bottom of the well. Although it is preferred to analyze a plate soon after processing, in some situations, laboratory workflows may be segregated between individuals or sites or labs, which may

make immediate analysis difficult. For these samples, processed iMRM samples may be sealed and frozen immediately after the automated wash and elution steps on the KingFisher and not transferred to a fresh plate until the day of MS analysis. Although a full stability study has not been conducted, to reduce nonspecific losses to plastic and evaporation, minimize the length of storage of antibody-enriched plates at −80 °C (preferably within a month of processing). This method was designed to prepare samples intended to be injected onto a LC configured for trap/elute [46, 49], i.e. samples are injected onto a trap column, washed to remove salts, then the valves are switched to put the trap in-line with an analytical column. Alternatively, samples may be desalted off-line using StageTips [71] or other SPE/C18 cartridges or columns and subsequently injected onto a MS equipped with a single analytical column format.

17. Instrument operating parameters should be optimized separately using mixtures of light and heavy peptides formulated in 3 % ACN/5 % HOAc containing 0.2 % digested plasma at a concentration suitable to inject 100 fmol in a background of ~100 ng plasma peptides to optimize instrument dependent conditions, such as source gas and collision energy. To prepare 0.2 % digested plasma, resuspend a 0.3 mL equivalent of digested plasma in 1× PBS/0.03 % CHAPS as described in Subheading 3.2, **step 4**. Pipet 10 μL into 10 mL of 3 % ACN/5 % HOAc and use this matrix to dilute the heavy and light peptide mixtures to 50 or 100 fmol/μL. Alternatively, 10 μL of the digested plasma resuspended as described in Subheading 3.2, **step 4** can be added to 10 mL of 3 % ACN/5 % HOAc prior to the preparation of the reverse curve background matrix described in Subheading 3.2, **step 7**.

Scheduled methods constrain the MRM scans to a defined a retention time window. Typically a retention time window of between 2 and 10 min is used when the number of peptides in an experiment exceeds 40 peptides. In this iMRM assay for example, MRM-MS analysis of 40 peptides, 20 light and 20 heavy, 3 transitions/peptide, requires a total of 120 scans for one measurement of each transition. Using a dwell time of 10 ms, it would take 1200 ms to complete one cycle. If the average peptide chromatographic peak width is 15 s (this may vary depending on the peptide retention time, LC, and column conditions) approximately 12.5 cycles or over 12 scans per transition could be acquired without scheduling. By only monitoring for peptides at their retention time (±2–5 min), scheduled MRM can be used to maintain the number of scans per transitions as the number of peptides (thus transitions) increases [15]. The number of peptides in this iMRM assay (20 peptide pairs, 40 total peptides, 3 transitions/peptide,

120 transitions total) can be acquired with over ten scans per transition, but the differences in precision offered by scheduled MRM can also be assessed. In our experience, ten or more scans over the elution profile of each peptide is needed for good inter- and intra-lab reproducibility [14, 49] although there are instances where this cannot be achieved and a lower number must be used to accommodate the plex size and chromatography conditions used. It is important for the publication of results and for method comparisons to state the number of scans used per peak area determination to compare results to across instrument platforms. These methods can be prepared, both for collision energy optimization and final data collection by exporting instrument-specific conditions from a Skyline document containing these peptides.

18. To minimize carryover, wash methods should be added in between each set of concentration points. Typically we insert two rapid reversed-phase gradients that cover the same or even a broader range of acetonitrile concentrations than used in the analytical gradient [58].

19. A Skyline document containing the peptide sequences and selected masses and transitions of the light and heavy peptides analyzed by MRM-MS may have been done earlier as part of iMRM methods generation. Spectral libraries for each peptide may have been generated and imported into the Skyline document for earlier data-dependent MS analysis. Spectral libraries displayed within Skyline are not required for integrating and processing these data, however, they provide helpful information troubleshooting data that are affected by interferences either from changes in matrix or chromatographic conditions.

20. Units of concentration are fmol/μL plasma (e.g. enter "20" for pt 8 which contains a total (*see* plate map in Table 2) of 200 fmol per well. Even though the total volume is 200 μL per well and the peptide concentration at the time of immunoaffinity capture is 1 fmol/μL, the concentration entered into the Results Grid is based on the starting amount of 10 μL of plasma per well. Concentrations are entered into the results grids as numbers without units.

21. When Quasar starts it will open an Immediate Window in Skyline where progress and any errors can be monitored. LOD, LOQ, and CVs for each transition will be calculated and summarized in two tables, one for all transitions and a second table for the performance of the best transition selected as the transition with the lowest CV. Box and whisker plots of these results as well as calibration or response curves will be generated as linear and log plots. Summary tables for the regression line, including R^2 and standard errors, slope and intercept will be generated and saved in the same Skyline directory.

22. Under acidic conditions, peptides as well as immunoglobulin protein will elute from the protein G beads. The large amount of desorbed protein will eventually overload the 75 μm ID × 10 cm analytical column (packed with 3 μm C18) that we estimate has a loading capacity of approximately 1 μg. Crosslinking antibodies to protein G beads via primary amines with dimethyl pimelimidate (DMP) allows the antibodies to be washed to reduce the amount of nonspecific background, the amount of bound passenger peptide (see Note 122) and retains the antibody on bead for the potential reuse. Antibodies cross-linked to beads should be tested by a capture efficiency test or mini-curve [46] prior to use.

Acknowledgments

This work was supported in part by grants from National Institutes of Health: HHSN268201000033C and R01HL096738 from NHLBI and Grants U24CA160034 from NCI Clinical Proteomics Tumor Analysis Consortium initiative and 5U01CA152990-05 from the NCI Early Detection Research Network program (to SAC).

References

1. Carr SA, Abbatiello SE, Ackermann BL, Borchers C, Domon B, Deutsch EW, Grant RP, Hoofnagle AN, Hüttenhain R, Koomen JM, Liebler DC, Liu T, Maclean B, Mani D, Mansfield E, Neubert H, Paulovich AG, Reiter L, Vitek O, Aebersold R, Anderson L, Bethem R, Blonder J, Boja E, Botelho J, Boyne M, Bradshaw RA, Burlingame AL, Chan D, Keshishian H, Kuhn E, Kinsinger C, Lee JS, Lee SW, Moritz R, Oses-Prieto J, Rifai N, Ritchie J, Rodriguez H, Srinivas PR, Townsend RR, Van Eyk J, Whiteley G, Wiita A, Weintraub S (2014) Targeted peptide measurements in biology and medicine: best practices for mass spectrometry-based assay development using a fit-for-purpose approach. Mol Cell Proteomics 13(3):907–917, PMCID: PMC3945918

2. Gillette MA, Carr SA (2013) Quantitative analysis of peptides and proteins in biomedicine by targeted mass spectrometry. Nat Methods 10:28–34

3. Picotti P, Aebersold R (2012) Selected reaction monitoring-based proteomics: workflows, potential, pitfalls and future directions. Nat Methods 9:555–566

4. Liebler DC, Zimmermann LJ (2013) Targeted quantitation of proteins by mass spectrometry. Biochemistry 52:3797–3806

5. Addona TA, Shi X, Keshishian H, Mani DR, Burgess M, Gillette MA, Clauser KR, Shen DX, Lewis GD, Farrell LA, Fifer MA, Sabatine MS, Gerszten RE, Carr SA (2011) A pipeline that integrates the discovery and verification of plasma protein biomarkers reveals candidate markers for cardiovascular disease. Nat Biotechnol 29:635–643

6. Whiteaker JR, Lin CW, Kennedy J, Hou LM, Trute M, Sokal I, Yan P, Schoenherr RM, Zhao L, Voytovich UJ, Kelly-Spratt KS, Krasnoselsky A, Gafken PR, Hogan JM, Jones LA, Wang P, Amon L, Chodosh LA, Nelson PS, McIntosh MW, Kemp CJ, Paulovich AG (2011) A targeted proteomics-based pipeline for verification of biomarkers in plasma. Nat Biotechnol 29:625–634

7. Keshishian H, Burgess MW, Gillette MA, Mertins P, Clauser KR, Mani DR, Kuhn EW, Farrell LA, Gerszten RE, Carr SA (2015) Multiplexed, quantitative workflow for sensitive biomarker discovery in plasma yields novel candidates for early myocardial injury. Mol Cell

Proteomics 14(9):2375–2393. doi:10.1074/mcp.M114.046813

8. Keshishian H, Addona T, Burgess M, Mani DR, Shi X, Kuhn E, Sabatine MS, Gerszten RE, Carr SA (2009) Quantification of cardiovascular biomarkers in patient plasma by targeted mass spectrometry and stable isotope dilution. Mol Cell Proteomics 8:2339–2349

9. Creech AL, Taylor JE, Maier VK, Wu X, Feeney CM, Udeshi ND, Peach SE, Boehm JS, Lee JT, Carr SA, Jaffe JD (2015) Building the Connectivity Map of epigenetics: chromatin profiling by quantitative targeted mass spectrometry. Methods 72:57–64. doi:10.1016/j.ymeth.2014.10.033

10. Yuan W, Sanda M, Wu J, Koomen J, Goldman R (2015) Quantitative analysis of immunoglobulin subclasses and subclass specific glycosylation by LC–MS–MRM in liver disease. Proteomics 116:24–33

11. Anderson L, Hunter CL (2006) Quantitative mass spectrometric multiple reaction monitoring assays for major plasma proteins. Mol Cell Proteomics 5:573–588

12. Keshishian H, Addona T, Burgess M, Kuhn E, Carr SA (2007) Quantitative, multiplexed assays for low abundance proteins in plasma by targeted mass spectrometry and stable isotope dilution. Mol Cell Proteomics 6:2212–2229

13. Addona TA, Abbatiello SE, Schilling B, Skates SJ, Mani DR, Bunk DM, Spiegelman CH, Zimmerman LJ, Ham AJL, Keshishian H, Hall SC, Allen S, Blackman RK, Borchers CH, Buck C, Cardasis HL, Cusack MP, Dodder NG, Gibson BW, Held JM, Hiltke T, Jackson A, Johansen EB, Kinsinger CR, Li J, Mesri M, Neubert TA, Niles RK, Pulsipher TC, Ransohoff D, Rodriguez H, Rudnick PA, Smith D, Tabb DL, Tegeler TJ, Variyath AM, Vega-Montoto LJ, Wahlander A, Waldemarson S, Wang M, Whiteaker JR, Zhao L, Anderson NL, Fisher SJ, Liebler DC, Paulovich AG, Regnier FE, Tempst P, Carr SA (2009) Multisite assessment of the precision and reproducibility of multiple reaction monitoring-based measurements of proteins in plasma. Nat Biotechnol 27:633–641

14. Abbatiello SE, Schilling B, Mani DR, Zimmermann LJ, Hall SC, MacLean B, Albertolle M, Allen S, Burgess M, Cusack MP, Ghosh M, Hedrick V, Held JM, Inerowicz HD, Jackson A, Keshishian H, Kinsinger CR, Lyssand J, Makowski L, Mesri M, Rodriguez H, Rudnick P, Sadowski P, Sedransk N, Shaddox K, Skates SJ, Kuhn E, Smith D, Whiteaker JR, Whitwell C, Zhang S, Borchers CH, Fisher SJ, Gibson BW, Liebler DC, MacCoss MJ, Neubert TA, Paulovich AG, Regnier FE, Tempst P, Carr SA (2015) Large-scale inter-laboratory study to develop, analytically validate and apply highly multiplexed, quantitative peptide assays to measure cancer-relevant proteins in plasma. Mol Cell Proteomics 14(9):2357–2374. doi:10.1074/mcp.M114.047050

15. Burgess MW, Keshishian H, Mani DR, Gillette MA, Carr SA (2014) Simplified and efficient quantification of low abundance proteins at very high multiplex by targeted mass spectrometry. Mol Cell Proteomics 13(4):1137–1149

16. Picotti P, Rinner O, Stallmach R, Dautel F, Farrah T, Domon B, Wenschuh H, Aebersold R (2010) High-throughput generation of selected reaction-monitoring assays for proteins and proteomes. Nat Methods 7:43–46. doi:10.1038/nmeth.1408

17. Ebhardt HA (2014) Selected reaction monitoring mass spectrometry: a methodology overview. Methods Mol Biol 1072:209–222

18. Peterson AC, Russell JD, Bailey DJ, Westphall MS, Coon JJ (2012) Parallel reaction monitoring for high resolution and high mass accuracy quantitative, targeted proteomics. Mol Cell Proteomics 11:1475–1488. doi:10.1074/mcp.O112.020131

19. Gallien S, Duriez E, Crone C, Kellmann M, Moehring T, Domon B (2012) Targeted proteomic quantification on quadrupole-orbitrap mass spectrometer. Mol Cell Proteomics 11:12. doi:10.1074/mcp.O112.019802

20. Gallien S, Kim SY, Domon B (2015) Large-scale targeted proteomics using internal standard triggered-parallel reaction monitoring. Mol Cell Proteomics 14(6):1630–1644. doi:10.1074/mcp.O114.043968

21. Barnidge DR, Dratz EA, Martin T, Bonilla LE, Moran LB, Lindall A (2003) Absolute quantification of the G protein-coupled receptor rhodopsin by LC/MS/MS using proteolysis product peptides and synthetic peptide standards. Anal Chem 75:445–451

22. Gerber SA, Rush J, Stemman O, Kirschner MW, Gygi SP (2003) Absolute quantification of proteins and phosphoproteins from cell lysates by tandem MS. Proc Natl Acad Sci U S A 100:6940–6945

23. Kuhn E, Wu J, Karl J, Liao H, Zolg W, Guild B (2004) Quantification of C-reactive protein in the serum of patients with rheumatoid arthritis using multiple reaction monitoring mass spectrometry and 13C-labeled peptide standards. Proteomics 4:1175–1186

24. Brun V, Dupuis A, Adrait A, Marcellin M, Thomas D, Court M, Vandenesch F, Garin J (2007) Isotope-labeled protein standards. Mol Cell Proteomics 6:2139–2149

25. Singh R, Crow FW, Babic N, Lutz WH, Lieske JC, Larson TS, Kumar R (2007) A liquid chromatography-mass spectrometry method for the quantification of urinary albumin using a novel N-15-isotopically labeled albumin internal standard. Clin Chem 53:540–542

26. Echan LA, Hsin-Yao Tang HY, Nadeem Ali-Khan N, KiBeom Lee K, Speicher DW (2005) Depletion of multiple high-abundance proteins improves protein profiling capacities of human serum and plasma. Proteomics 5(13):3292–3303. doi:10.1002/pmic.200401228

27. Hinerfeld D, Innamorati D, Pirro J, Tam SW (2004) Serum/plasma depletion with chicken immunoglobulin Y antibodies for proteomic analysis from multiple mammalian species. J Biomol Tech 15(3):184–190

28. Liu T, Qian WJ, Mottaz HM, Gritsenko MA, Norbeck AD, Moore RJ, Purvine SO, Camp DG 2nd, Smith RD (2006) Evaluation of multiprotein immunoaffinity subtraction for plasma proteomics and candidate biomarker discovery using mass spectrometry. Mol Cell Proteomics 5(11):2167–2174

29. Qian WJ, Kaleta DT, Petritis BO, Jiang H, Liu T, Zhang X, Mottaz HM, Varnum SM, Camp DG 2nd, Huang L, Fang X, Zhang WW, Smith RD (2008) Enhanced detection of low abundance human plasma proteins using a tandem IgY12-SuperMix immunoaffinity separation strategy. Mol Cell Proteomics 7(10):1963–1973

30. Tu C, Rudnick RA, Martinez MY, Cheek KL, Stein SE, Slebos RJC, Liebler DC (2010) Depletion of abundant plasma proteins and limitations of plasma proteomics. J Proteome Res 9(10):4982–4991

31. Yang F, Shen Y, Camp DG II, Smith RD (2012) High pH reversed-phase chromatography with fraction concatenation as an alternative to strong-cation exchange chromatography for two-dimensional proteomic analysis. Expert Rev Proteomics 9(2):129–134. doi:10.1586/epr.12.15

32. Batth TS, Francavilla C, Jesper V, Olsen JV (2014) Off-line high-pH reversed-phase fractionation for in-depth phosphoproteomics. J Proteome Res 13:6176–6186. doi:10.1021/pr500893m

33. Mertins P, Yang F, Liu T, Mani DR, Petyuk VA, Gillette MA, Clauser KR, Qiao JW, Gritsenko MA, Moore RJ, Levine DA, Townsend R, Erdmann-Gilmore P, Snider JE, Davies SR, Ruggles KV, Fenyo D, Kitchens RT, Li S, Olvera N, Dao F, Rodriguez H, Chan DW, Liebler D, White F, Rodland KD, Mills GB, Smith RD, Paulovich AG, Ellis M, Carr SA (2014) Ischemia in tumors induces early and sustained phosphorylation changes in stress

kinase pathways but does not affect global protein levels. Mol Cell Proteomics 13:1690–1704. doi:10.1074/mcp.M113.036392, First published on April 9, 2014

34. Shi T, Fillmore TL, Sun X, Zhao R, Schepmoes AA, Hossain M, Xie F, Wu S, Kim JS, Jones N, Moore RJ, Paša-Tolić L, Kagan J, Rodland KD, Liu T, Tang K, Camp DG II, Smith RD, Qian WJ (2012) Antibody-free, targeted mass-spectrometric approach for quantification of proteins at low picogram per milliliter levels in human plasma/serum. Proc Natl Acad Sci U S A 109(38):15395–15400

35. Dickson C (2008) Protein techniques: immunoprecipitation, in vitro kinase assays, and Western blotting. Methods Mol Biol 461:735–744. doi:10.1007/978-1-60327-483-8_53

36. Kaboord B, Perr M (2008) Isolation of proteins and protein complexes by immunoprecipitation. Methods Mol Biol 424:349–364. doi:10.1007/978-1-60327-064-9_27

37. Anderson NL, Anderson NG, Haines LR, Hardie DB, Olafson RW, Pearson TW (2004) Mass spectrometric quantitation of peptides and proteins using Stable Isotope Standards and Capture by Anti-Peptide Antibodies (SISCAPA). J Proteome Res 3:235–244

38. Whiteaker JR, Zhao L, Zhang HY, Feng LC, Piening BD, Anderson L, Paulovich AG (2007) Antibody-based enrichment of peptides on magnetic beads for mass-spectrometry-based quantification of serum biomarkers. Anal Biochem 362:44–54

39. Berna MJ, Zhen Y, Watson DE, Hale JE, Ackermann BL (2007) Strategic use of immunoprecipitation and LC/MS/MS for trace-level protein quantification: myosin light chain 1, a biomarker of cardiac necrosis. Anal Chem 79:4199–4205

40. Hoofnagle AN, Becker JO, Wener MH, Heinecke JW (2008) Quantification of thyroglobulin, a low-abundance serum protein, by immunoaffinity peptide enrichment and tandem mass spectrometry. Clin Chem 54:1796–1804

41. Kuhn E, Addona T, Keshishian H, Burgess M, Mani DR, Lee RT, Sabatine MS, Gerszten RE, Carr SA (2009) Developing multiplexed assays for troponin I and interleukin-33 in plasma by peptide immunoaffinity enrichment and targeted mass spectrometry. Clin Chem 55:1108–1117

42. Ocana MF, Neubert H (2010) An immunoaffinity liquid chromatography-tandem mass spectrometry assay for the quantitation of matrix metalloproteinase 9 in mouse serum. Anal Biochem 399:202–210

43. Whiteaker JR, Zhao L, Yan P, Ivey RG, Voytovich UJ, Moore HD, Lin C, Paulovich

AG (2015) Peptide immunoaffinity enrichment and targeted mass spectrometry enables multiplex, quantitative pharmacodynamic studies of phospho-signaling. Mol Cell Proteomics 14(8):2261–2273. doi:10.1074/mcp.O115.050351

44. Palandra J, Finelli A, Zhu M, Masferrer J, Neubert H (2013) Highly specific and sensitive measurements of human and monkey interleukin 21 using sequential protein and tryptic peptide immunoaffinity LC-MS/MS. Anal Chem 85(11):5522–5529. doi:10.1021/ac4006765

45. Neubert H, Muirhead D, Kabir M, Grace C, Cleton A, Arends R (2013) Sequential protein and peptide immunoaffinity capture for mass spectrometry-based quantification of total human β-nerve growth factor. Anal Chem 85(3):1719–1726. doi:10.1021/ac303031q

46. Whiteaker JR, Zhao L, Abbatiello SE, Burgess M, Kuhn E, Lin CW, Pope ME, Razavi M, Anderson NL, Pearson TW, Carr SA, Paulovich AG (2011) Evaluation of large scale quantitative proteomic assay development using peptide affinity-based mass spectrometry. Mol Cell Proteomics 10(4):M110.005645

47. Whiteaker JR, Zhao L, Anderson L, Paulovich AG (2010) An automated and multiplexed method for high throughput peptide immunoaffinity enrichment and multiple reaction monitoring mass spectrometry-based quantification of protein biomarkers. Mol Cell Proteomics 9:184–196

48. Whiteaker JR, Zhao L, Lin C, Yan P, Wang P, Paulovich AG (2012) Sequential multiplexed analyte quantification using peptide immunoaffinity enrichment coupled to mass spectrometry. Mol Cell Proteomics 11(6):M111.015347. doi:10.1074/mcp.M111.015347

49. Kuhn E, Whiteaker JR, Mani DR, Jackson AM, Zhao L, Pope ME, Smith D, Rivera KD, Anderson NL, Skates SJ, Pearson TW, Paulovich AG, Carr SA (2012) Inter-laboratory evaluation of automated, multiplexed peptide immunoaffinity enrichment coupled to multiple reaction monitoring mass spectrometry for quantifying proteins in plasma. Mol Cell Proteomics 11(6):M111.013854, PMCID: PMC3433918

50. Kushnir MM, Rockwood AL, Roberts WL, Abraham D, Hoofnagle AN, Meikle AW (2013) Measurement of thyroglobulin by liquid chromatography–tandem mass spectrometry in serum and plasma in the presence of antithyroglobulin autoantibodies. Clin Chem 59(6):982–990

51. Harris DC (2003) Quantitative chemical analysis, 6th edn. W.H. Freeman, New York

52. Li W, Cohen LH (2003) Quantitation of endogenous analytes in biofluid without a true blank matrix. Anal Chem 75(21):5854–5859

53. Jones BR, Schultz GA, Eckstein JA, Ackermann BL (2012) Surrogate matrix and surrogate analyte approaches for definitive quantitation of endogenous biomolecules. Bioanalysis 4(19):2343–2356. doi:10.4155/bio.12.200, PMID: 23088461

54. Whiteaker JR, Halusa GN, Hoofnagle AN, Sharma V, MacLean B, Yan P, Wrobel JA, Kennedy J, Mani DR, Zimmerman LJ, Meyer MR, Mesri M, Rodriguez H, Clinical Proteomic Tumor Analysis Consortium (2014) CPTAC Assay Portal: a repository of targeted proteomic assays. Nat Methods 11(7):703–704. doi:10.1038/nmeth.3002

55. Fusaro VA, Mani DR, Mesirov JP, Carr SA (2009) Computational prediction of high responding peptides for development of targeted protein assays by mass spectrometry. Nat Biotechnol 27(2):190–198

56. Mallick P, Schirle M, Chen SS, Flory MR, Hookeun Lee H, Martin D, Ranish J, Raught B, Schmitt R, Werner T, Kuster B, Aebersold R (2007) Computational prediction of proteotypic peptides for quantitative proteomics. Nat Biotechnol 25:125–1314. doi:10.1038/nbt1275

57. Kuhn E, Ross J, Abbatiello SE, Mani DR, Carr SA (2012) Reversing the curve: determining LOD in the presence of endogenous signal using SID-MRM-MS. Presented at the 60th annual conference on mass spectrometry, Poster MP01-004

58. Abbatiello SE, Mani DR, Schilling B, Maclean B, Zimmerman LJ, Feng X, Cusack MP, Sedransk N, Hall SC, Addona T, Allen S, Dodder NG, Ghosh M, Held JM, Hedrick V, Inerowicz HD, Jackson A, Keshishian H, Kim JW, Lyssand JS, Riley CP, Rudnick P, Sadowski P, Shaddox K, Smith D, Tomazela D, Wahlander A, Waldemarson S, Whitwell CA, You J, Zhang S, Kinsinger CR, Mesri M, Rodriguez H, Borchers CH, Buck C, Fisher SJ, Gibson BW, Liebler D, MacCoss M, Neubert TA, Paulovich AG, Regnier F, Skates SJ, Tempst P, Wang M, Carr SA (2013) Design, implementation, and multi-site evaluation of a system suitability protocol for the quantitative assessment of instrument performance in LC-MRM-MS. Mol Cell Proteomics 12:2623–2639. doi:10.1074/mcp.M112.027078

59. MacLean B, Tomazela DM, Shulman N, Chambers M, Finney GL, Frewen B, Kern R, Tabb DL, Liebler DC, MacCoss MJ (2010) Skyline: an open source document editor for

creating and analyzing targeted proteomics experiments. Bioinformatics 26(7):966–968. doi:10.1093/bioinformatics/btq054

60. Broudy D, Killeen T, Choi M, Shulman N, Mani DR, Abbatiello SE, Mani D, Ahmad R, Sahu AK, Schilling B, Tamura K, Boss Y, Sharma V, Gibson BW, Carr SA, Vitek O, MacCoss MJ, MacLean B (2014) A framework for installable external tools in Skyline. Bioinformatics 30(17):2521–2523. doi:10.1093/bioinformatics/btu148

61. Currie LA (1968) Limits for qualitative detection and quantitative determination. Anal Chem 40:586–593

62. Linnet K, Kondratovich M (2004) Partly nonparametric approach for determining the limit of detection. Clin Chem 50(4):732–740

63. Abbatiello SE, Mani DR, Keshishian H, Carr SA (2010) Automated detection of inaccurate and imprecise transitions in quantitative assays of peptides by multiple monitoring mass spectrometry. Clin Chem 56(2):291–305, PMCID: PMC2851178

64. Mani DR, Abbatiello SE, Carr SA (2012) Statistical characterization of multiple-reaction monitoring mass spectrometry (MRM-MS) assays for quantitative proteomics. BMC Bioinformatics 13(Suppl 16):S9. doi:10.1186/1471-2105-13-S16-S9

65. Daugherty AL, Mrsny RJ (2006) Formulation and delivery issues for monoclonal antibody therapeutics. Adv Drug Deliv Rev 58(5–6):686–706

66. Harlow E, Lane D (1999) Using antibodies: a laboratory manual. Cold Spring Harbor Laboratory Press, USA

67. Schoenherr RM, Zhao L, Whiteaker JR, Feng L, Li L, Lina L, Liu X, Paulovich AG (2010) Automated screening of monoclonal antibodies for SISCAPA assays using a magnetic bead processor and liquid chromatography-selected reaction monitoring-mass spectrometry. J Immunol Methods 353(1–2):49–61

68. Razavi M, Frick LE, LaMarr WA, Pope ME, Miller CA, Anderson NL, Pearson TW (2012) High-throughput SISCAPA quantitation of peptides from human plasma digests by ultrafast, liquid chromatography-free mass spectrometry. J Proteome Res 11(12):5642–5649. doi:10.1021/pr300652v

69. Kostka V, Carpenter FH (1964) Inhibition of chymotrypsin activity in crystalline trypsin preparations. J Biol Chem 239(6):1799–1803

70. Pope ME, Soste MV, Eyford BA, Anderson NL, Pearson TW (2009) Anti-peptide antibody screening: selection of high affinity monoclonal reagents by a refined surface plasmon resonance technique. J Immunol Methods 341(1–2):86–96

71. Rappsilber J, Ishihama Y, Mann M (2003) Stop and go extraction tips for matrix-assisted laser desorption/ionization, nanoelectrospray, and LC/MS sample pretreatment in proteomics. Anal Chem 75:663–670

Chapter 10

High-Throughput Quantitative Proteomics Enabled by Mass Defect-Based 12-Plex DiLeu Isobaric Tags

Dustin C. Frost and Lingjun Li

Abstract

Isobaric labeling has become a popular technique for high-throughput, mass spectrometry (MS)-based relative quantification of peptides and proteins. However, widespread use of the approach for large-scale proteomics applications has been limited by the high cost of commercial isobaric tags. To address this, we have developed our own N,N-dimethyl leucine (DiLeu) multiplex isobaric tags as a cost-effective alternative that can be synthesized with ease using readily available isotopic reagents. When paired with high-resolution tandem mass (MSn) acquisition, mass defect-based DiLeu isobaric tags allow relative quantification of up to twelve samples in a single liquid chromatography (LC)–MS^2 experiment. Herein, we present detailed methods for synthesis of 12-plex DiLeu isobaric tags, labeling of complex protein digest samples, analysis by high-resolution nanoLC–MSn, and processing of acquired data.

Key words Quantitative proteomics, Isobaric labeling, DiLeu, High-resolution mass spectrometry, Mass defect, Isobaric tag synthesis, Multiplexed quantitation

1 Introduction

The introduction of multiplexed isobaric tags has been a pivotal advancement for mass spectrometry-based quantification approaches by permitting analysis of many complex biological samples in a single experiment. In contrast to mass difference strategies such as stable isotope labeling by amino acids in cell culture (SILAC) [1, 2] and reductive dimethylation [3–6] that increase mass spectral complexity as a result of increasing the number of quantitative channels, isobaric tags allow greater levels of multiplexing without affecting mass spectral complexity. Every tag in the multiplexed set is identical in mass and contains the same number of stable isotopes, but each has a unique arrangement of isotopes between the reporter group and balance group of the chemical structure. When peptide samples labeled with each of the tags are pooled and analyzed by LC–MS, a single precursor is detected for each peptide in the MS^1 scan, but upon MS^2

Salvatore Sechi (ed.), *Quantitative Proteomics by Mass Spectrometry*, Methods in Molecular Biology, vol. 1410,
DOI 10.1007/978-1-4939-3524-6_10, © Springer Science+Business Media New York 2016

fragmentation, distinct reporter ions are generated in the low-mass region which can be compared to allow relative quantification between the pooled samples. Initial commercial isobaric tag offerings included isobaric tags for relative and absolute quantification (iTRAQ) [7, 8] in 4-plex and 8-plex configurations and tandem mass tags (TMT) [9, 10] in a 6-plex configuration. Recently, TMT reagents were expanded to 10-plex configuration, by the addition of mass defect-based variants [11–13], for use with high-resolution MS platforms. While these products have established themselves as powerful tools for high-throughput quantitative proteomics studies, the steep financial investment required to purchase the reagents has thus far been a significant barrier to making isobaric labeling a routine approach.

We have since developed our own custom N,N-dimethyl leucine (DiLeu) isobaric tags as a cost-effective alternative to expensive commercial isobaric tags [14]. DiLeu tags can be synthesized with ease in just two or three steps using established and relatively simple chemistry. The straightforward synthetic procedure means that even those with little to no synthetic chemistry experience can synthesize any particular tag at high yield (~80 %) with a high rate of success in a period of just 1 or 2 days. The required equipment, materials, and isobaric reagents are readily available to any lab. DiLeu tags can be synthesized in bulk amounts at scales appropriate for the needs of any research plan and stored safely for months or even years without degrading.

DiLeu isobaric tags resemble commercial isobaric tags in that they are composed of a reporter group, a balance group, and an amine-reactive group for targeting for selective modification of peptide N-termini and lysine side chains (Fig. 1a). Incorporation of the DiLeu tag onto peptides adds a modest mass of 145 Da per label and enhances their electrospray ionization efficiency by increasing hydrophobicity. Since the DiLeu tag is just a single dimethylated amino acid, no abnormal cleavages or artifacts are produced during fragmentation of labeled peptides that could complicate peptide sequence identification. On the contrary, the dimethyl leucine tag increases peptide fragmentation efficiency by increasing proton affinity at N-termini and lysine side chains, which can improve peptide sequence identification [15–17]. The first generation of DiLeu isobaric tags was developed as a 4-plex set with reporter ions at m/z 115, 116, 117, and 118. We recently increased the multiplexing capacity threefold by adding mass defect-based isotopologues that differ from the originals by ~6 mDa to create an isobaric 12-plex set suitable for use with high-resolution MS platforms [18]. The 12-plex set of DiLeu isobaric tags is composed of two 115 variants (a-b), three 116 variants (a-c), three 117 variants (a-c), and four 118 variants (a-d) (Fig. 1b). Since no additional synthetic steps or custom isotopic reagents are necessary to synthesize the mass defect-based variants,

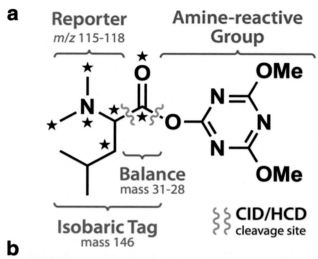

| DiLeu | Stable Isotopes | | Reporter |
Label	Reporter	Balance	Mass
115a	●	● ●	115.12476
115b	●	● ●	115.13108
116a	● ●	●	116.12812
116b	● ●	●	116.13444
116c	● ●	●	116.14028
117a	● ● ●	●	117.13147
117b	● ● ●	●	117.13731
117c	● ● ●	●	117.14363
118a	● ● ● ●		118.13483
118b	● ● ● ●		118.14067
118c	● ● ● ●		118.14699
118d	● ● ● ●		118.15283

● ^{13}C ● ^{2}H ● ^{15}N ● ^{18}O

Fig. 1 The 12-plex DiLeu general structure. *(a)* The DiLeu isobaric tagging reagent consists of a reporter group, balance group, and amine-reactive triazine ester group. Stars indicate positions of isotopic substitution. *(b)* Stable isotopes (^{13}C, ^{2}H, and ^{15}N) incorporated into the reporter group are mass-balanced by stable isotopes (^{13}C, ^{18}O) in the carbonyl balance group. Unique combinations of isotopes incorporated into the reporter group yield two 115 variants, three 116 variants, three 117 variants, and four 118 variants whose isotopologues differ in mass by approximately 6 mDa. Reprinted with permission from Frost et al. [18]. Copyright 2015 American Chemical Society

synthesis of 12-plex DiLeu is equally approachable as 4-plex DiLeu. Resolving each of the 12-plex DiLeu reporter ions requires an MSn resolving power of at least 30 k (@ m/z 400). Alternatively, 7-plex quantitation is possible at a resolving power of 15 k by using channels 115a, 116a, 116c, 117a, 117c, 118a, and 118c, for which the smallest spacing between two reporter ions is ~12 mDa. If high-resolution instrumentation is not available, 4-plex quantitation using channels 115a, 116b, 117a, and 118a is recommended.

Herein, we present detail protocols for synthesis of 12-plex DiLeu isobaric tags, labeling of complex protein digest samples, analysis by high-resolution nanoLC–MSn, and processing of acquired data. DiLeu tag synthesis begins with ^{18}O exchange of leucines for the 115 and 116 variants, followed by dimethylation of leucines for all tags and purification via flash column chromatography. Protein samples are enzymatically digested in solution overnight and desalted prior to labeling. DiLeu labels undergo a brief activation step and are used immediately for protein digest labeling. Labeled samples are pooled and fractionated using strong cation exchange (SCX) chromatography, and the fractions are desalted in preparation for nanoLC–MSn analysis. Online reversed-phase nanoLC separation is performed using a Waters nanoAcquity UPLC system, and eluted peptides are analyzed by high-resolution HCD MSn on an Orbitrap Elite mass spectrometer. Acquired spectra are processed using Proteome Discoverer for identification of proteins and extraction of reporter ion intensities.

2   Materials

2.1   12-Plex DiLeu Isobaric Tag Synthesis

1. Leucines: L-leucine (Sigma), L-leucine-1-^{13}C,^{15}N (Sigma), L-leucine-3-^{13}C,^{15}N (Sigma), L-leucine-1,2-^{13}C (Sigma).

2. ^{18}O Exchange: ^{18}O water (H$_2$^{18}O) (Sigma), HCl (g) (Sigma), small bore PTFE tubing, pH strips, StratoSpheres PL-HCO$_3$ MP resin (3.59 mmol/g, 100 Å, 500–600 μm) (Agilent Technologies, Santa Clara, CA), Pasteur pipettes, ACS grade ethanol (EtOH) (Fisher Scientific, Pittsburgh, PA), ACS grade methanol (MeOH) (Fisher Scientific).

3. Dimethylation: sodium cyanoborohydride (NaBH$_3$CN) (Sigma), sodium cyanoborodeuteride (NaBD$_3$CN) (Sigma), formaldehyde (CH$_2$O, 37 % w/w in H$_2$O) (Sigma), formaldehyde-^{13}C (^{13}CH$_2$O, 20 % w/w in H$_2$O) (Sigma), formaldehyde-d$_2$ (CD$_2$O, 20 % w/w in D$_2$O) (Sigma), distilled water (H$_2$O), deuterium oxide (D$_2$O) (Sigma).

4. Thin layer chromatography (TLC): glass TLC plates, capillary micropipets (Wiretrol II, Drummond Scientific Company, Broomall, PA), forceps, heat gun, ninhydrin stain (1.5 g ninhydrin, 100 mL of n-butanol, 3 mL glacial AcOH; light sensitive).

5. Glass dram vials with caps (10 dram, 1 dram, ½ dram).

6. Ice water bath.

7. Ring stand with 3-pronged clamp.

8. Hot plate stirrer and micro stir bars.

9. Industrial grade nitrogen (N_2) gas, pressure regulator, PVC or PTFE tubing.

2.2 Flash Column Purification

1. Glassware: round-bottom flasks (100 mL, 250 mL, 500 mL), beakers (250 mL, 500 mL), Erlenmeyer flask (1 L), stir rod, test tubes (16 × 150 mm), Pasteur pipettes.

2. Flash column materials: glass gravity column equipped with Teflon stopcock (26 × 305 mm, ST 24/40 joint) (Synthware C184323, Kemtech America Inc, Pleasant Prairie, WI), glass 90° inner joint inlet adapter (ST 24/40 joint) (Chemglass CG-1014-01, Chemglass, Vineland, NJ), 500 mL reservoir (ST 24/40 joint) (Synthware C182403, Kemtech America Inc), silica gel (SiliaFlash P60, 230–400 mesh), cotton wadding, sand, ring stand with Burette clamp, air line, PVC tubing.

3. Flash column solvents: ACS grade dichloromethane (DCM, CH_2Cl_2), ACS grade MeOH.

4. Thin layer chromatography (TLC): glass TLC plates, capillary micropipets, forceps, heat gun, $KMnO_4$ stain (1.5 g $KMnO_4$, 10 g K_2CO_3, 1.25 mL 10 % NaOH dissolved in 200 mL distilled water; light sensitive).

5. Rotary evaporator (Büchi Rotavapor, Büchi Labortechnik AG, Switzerland), faucet vacuum aspirator pump (Nalgene Polypropylene Vacuum Pump Aspirator 6140-0010, Thermo Scientific) with attached PVC tubing or mechanical vacuum pump with attached rubber tubing.

6. Centrifugal vacuum concentrator (SpeedVac, Thermo Scientific, or equivalent).

7. Desiccator cabinet (Nalgene 5317-0120, Thermo Scientific).

2.3 Protein Digestion and Desalting

1. Denaturation: 7 M urea in 50 mM Tris–HCl pH 8.

2. Reduction: 500 mM dithiothreitol (DTT).

3. Alkylation: 500 mM iodoacetamide (IAA).

4. Protein digestion: Trypsin/Lys-C mix, Mass Spec Grade (Promega, Madison, WI), 50 mM Tris–HCl (pH 8), 10 % trifluoroacetic acid (TFA).

5. Desalting: Sep-Pak C18 cartridge (Waters, Milford, MA) or Bond Elut OMIX C18 pipette tips (100 µL) (Agilent Technologies, Santa Clara, CA), HPLC grade acetonitrile (ACN), HPLC grade water, 1 % TFA or 1 % heptafluorobu-

tyric acid (HFBA), 0.1 % FA or 0.1 % HFBA, 60 % HPLC grade ACN in 0.1 % FA.

6. Centrifugal vacuum concentrator.

2.4   DiLeu Isobaric Labeling

1. DiLeu activation: 4-(4,6-Dimethoxy-1,3,5-triazin-2-yl)-4-methylmorpholinium tetrafluoroborate (DMTMM BF$_4$) (Sigma), *N*-Methylmorpholine (NMM) (Sigma), anhydrous dimethyl formamide (DMF) (Sigma), needle ($\leq$18 G $\times$ 1½ in.) (BD, Franklin Lakes, NJ), syringe (1 mL) (BD), glass dram vial, rubber septum, Bunsen burner, benchtop microcentrifuge (Microcentrifuge 5424, Eppendorf, Hauppauge, NY) (or equivalent); Optional: dry solvent system, round-bottom flask (25 mL), rubber septum.

2. DiLeu labeling: 0.5 M triethylammonium bicarbonate (TEAB) buffer pH 8.5, ACS grade ACN, 5 % hydroxylamine (NH$_2$OH) in water.

3. Centrifugal vacuum concentrator.

2.5   Strong Cation Exchange SPE and Desalting

1. SCX SPE: SCX SpinTips sample prep kit (SP-155-24kit, Protea Biosciences, Morgantown, WV), benchtop microcentrifuge.

2. Peptide desalting: Bond Elut OMIX C18 pipette tips (100 μL), HPLC grade ACN, 1 % FA, 0.1 % FA, 60 % HPLC grade ACN in 0.1 % FA.

3. Centrifugal vacuum concentrator.

2.6   Nano Liquid Chromatography–Tandem Mass Spectrometry

1. Orbitrap Elite mass spectrometer (Thermo Scientific, San Jose, CA) or Orbitrap Fusion mass spectrometer (Thermo Scientific) (recommended for most accurate and precise quantification).

2. nanoAcquity nanoflow UPLC (Waters, Milford, MA).

3. Analytical column: 15 cm $\times$ 75 μm microcapillary column fabricated in-house with integrated emitter tip and packed with Bridged Ethylene Hybrid C18 particles (1.7 μm, 130 Å, Waters).

4. Sample dissolution: 3 % Optima LC/MS grade ACN (Fisher Scientific) in Optima LC/MS grade 0.1 % FA in water (Fisher Scientific).

5. Mobile phase A: 5 % ACS grade dimethyl sulfoxide (DMSO) in Optima LC/MS grade 0.1 % FA in water.

6. Mobile phase B: 5 % ACS grade DMSO in Optima LC/MS grade 0.1 % FA in ACN.

2.7   Data Analysis

1. Software: Proteome Discoverer (version 1.4 or later, Thermo Scientific), Microsoft Excel (Microsoft, Redmond, CA).

3   Methods

3.1   12-Plex DiLeu Isobaric Tag Synthesis

The reaction schemes and specific isotopic reagents required for each of the 12-plex DiLeu isobaric tags are outlined in Fig. 2 (*see* **Note 1**).

3.1.1 ^{18}O Water Acidification

Prepare saturated HCl H$_2$^{18}O pH 1 solution in a fume hood and wear gloves and safety glasses. Hydrogen chloride gas is toxic, corrosive, and hazardous to eyes, skin, and respiratory tract.

1. Transfer 2.5 mL ^{18}O water to a small glass vessel and chill in an ice bath.

2. Connect the small bore PTFE tubing to the HCl gas cylinder and slowly bubble HCl gas into the ^{18}O water for several minutes until it turns bright green in color. Ensure that the pH is ≤1 using a pH strip.

3. Store HCl H$_2$^{18}O solution tightly sealed with a PTFE-lined cap at −20 °C.

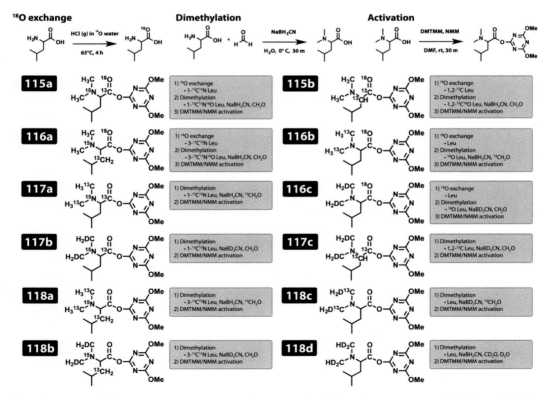

Fig. 2 Overview of DiLeu reaction schemes, synthetic steps, and isotopic reagents required for each of the 12-plex DiLeu isobaric tags. Adapted with permission from Frost et al. [18]. Copyright 2015 American Chemical Society

3.1.2 *^{18}O Exchange* ^{18}O exchange of L-leucine or heavy L-leucine is required for tags 115a, 115b, 116a, 116b, and 116c. Saturated HCl solution is corrosive and hazardous to eyes, skin, and respiratory tract. Handle with gloves and wear safety glasses.

1. To a small glass vial with flea micro stir bar and PTFE snap cap or PTFE-lined septum screw cap (*see* **Note 2**), add 100 mg L-leucine or heavy L-leucine.

2. Dissolve leucine in 500 μL HCl $H_2{}^{18}O$ solution (pH 1) and seal the reaction vessel securely (*see* **Note 3**).

3. Stir on a hotplate at 65 °C for 4 h or longer.

4. The reaction progress may be monitored by removing a 1 μL aliquot, diluting 1000-fold in ACN:H₂O, and analyzing by direct infusion MS (*see* **Notes 4** and **5**).

5. Transfer the reaction mixture to a 250 mL round-bottom flask, dilute to 100 mL with EtOH, and dry on the rotovap. Perform several EtOH washes (50–100 mL) until the presence of HCl is no longer detected by smell and a white solid, ^{18}O-exchanged leucine, is obtained.

6. Remove trace amounts of acid from the ^{18}O-exchanged leucine using StratoSpheres PL-HCO₃ MP acid-scavenging beads. Transfer 400 mg StratoSpheres PL-HCO₃ beads to a tall glass 10-dram vial with plastic snap cap. Wash beads three times with 5 mL portions of MeOH.

7. Dissolve ^{18}O-exchanged leucine in MeOH and transfer to beads for a total volume of 15 mL. Vortex or agitate gently for 30 min; white solid will precipitate out of solution as acid is removed.

8. Transfer the suspension to a 250 mL round-bottom flask and wash beads with several 5–10 mL MeOH washes until the washes run clear. Dry the combined washes on the rotovap to obtain ^{18}O-exchanged leucine, in free base form, as a white solid.

9. Transfer the ^{18}O-exchanged leucine to a weighed glass vial with MeOH and dry under an N_2 stream to determine yield (~95 %).

10. Store in desiccator at room temperature.

3.1.3 *N,N-dimethylation*   Perform this reaction in a fume hood and wear gloves when handling sodium cyanoborohydride and formaldehyde solution. Refer to Fig. 2 for specific isotopic reagents required for each of the 12-plex DiLeu isobaric tags.

1. To a small glass ½-dram vial with flea micro stir bar and snap cap, add 50 mg of L-leucine or heavy L-leucine.

2. Add 60 mg NaBH₃CN or NaBD₃CN (~2.5× molar excess).

3. Add 1 mL H₂O or D₂O and stir the suspension in an ice water bath.

4. Add dropwise 150 μL formaldehyde solution (37 % w/w) or 275 μL formaldehyde-d_2 or formaldehyde-^{13}C solution (20 % w/w) (~2.5× molar excess) while stirring, and cap vial. Evolution of gas will be evident.

5. Stir in the ice water bath for 30 min. Leucine will dissolve as the reaction progresses; a small volume of formaldehyde can be added to help drive the reaction to completion. Vortex briefly to wash unreacted material from the sides of the vial if necessary.

6. Monitor the reaction progress by spotting a few μL on a TLC plate, staining with ninhydrin solution, and heating gently with a heat gun to detect any unreacted amines. Reaction is complete when the spot shows no color.

7. Upon completion, the reaction mixture may be clear or cloudy depending on the purity of the isotopic reagents. Transfer the reaction mixture to a 100 mL round-bottom flask with several small MeOH washes and dry on the rotovap to yield a gel.

3.2   Flash Column Purification

Perform the flash column cleanup in a fume hood equipped with an air line. Silica gel is an irritant to the respiratory tract and may also irritate eyes and skin. Handle with gloves and wear safety glasses.

1. In a 250 mL beaker, suspend silica gel in DCM and mix with a glass stir rod to obtain a free-flowing slurry line.

2. Mount the glass chromatography column to a ring stand and pack a small amount of cotton wadding into the bottom of the column. Wet the cotton plug with DCM and drain the solvent into a beaker.

3. Pour silica gel suspension into the column and apply pressure from the air line via PVC tubing and glass vacuum adapter to compress the column and drain the solvent. Add silica gel and apply pressure until the column is 12–13 cm in height.

4. Equilibrate the column by adding one column volume of DCM, draining into a 250 mL beaker, and adding this volume back to the column. Repeat twice, then drain solvent to within a few mm of the silica surface.

5. Dissolve the reaction mixture in 5–10 mL 15:1 DCM:MeOH (*see* **Note 6**) and apply uniformly over the top of the silica surface using a long Pasteur pipette, without disturbing the surface. Drain solvent to within a few mm of the silica surface. Wash the round-bottom flask with several small washes and apply each wash to the top of the column to ensure a complete transfer. Drain, gently wash the sides of the column with DCM, and drain again.

6. Gently add fresh DCM to the column without disrupting the silica surface until the volume reaches the top. Pour sea sand into the column to form a layer atop the silica surface; this will help protect it from disruption when adding eluting solvent.

7. Attach the reservoir and elute with the following solvent gradient:

	DCM (mL)	MeOH (mL)
15:1	150	10
10:1	400	40
8:1	160	20
5:1	250	50
1:1	300	300

Fractions 15:1 through 5:1 contain contaminating reaction byproducts. Collect these volumes as waste in 1 L Erlenmeyer flasks. Drain solvent to within a few mm of the silica surface before adding the 1:1 DCM:MeOH solvent. Collect the 1:1 volume in fractions of 22 mL in test tubes (16 × 150 mm)—the dimethyl leucine product elutes during this part of the gradient.

8. Spot every other 1:1 fraction tube on a TLC plate, stain with $KMNO_4$, and heat gently with a heat gun. Spots that develop yellowish in color contain the dimethyl leucine product. If the last collected fractions show up as colored spots, continue eluting with 1:1 DCM:MeOH and spot fraction tubes until no color is shown by TLC to ensure complete recovery.

9. Separation of dimethyl leucine from contaminating compounds in the 1:1 fraction tubes can be more confidently determined by removing 5 μL aliquots, diluting to 1 mL with ACN:H$_2$O, and analyzing by direct infusion MS (*see* **Note 7**). Combine the fractions that contain the isolated dimethyl leucine compound in a 500 mL round-bottom flask. For the column set-up and fraction collection described here, this is often around fraction tube 10 and later. Avoid including the 1:1 fraction tubes 1–8 as these may contain contaminating compounds. Dry on the roto-vap to obtain dimethyl leucine as a white solid.

10. Transfer the dimethyl leucine to a weighed glass vial with 1:1 DCM:MeOH and dry under an N_2 stream to determine yield (~80 %).

11. Divide into single-use 1 mg aliquots by dissolving to 10 mg/mL with 1:1 DCM:MeOH and pipetting 100 μL into 0.6 mL microcentrifuge tubes. Dry aliquots in the SpeedVac.

12. Store aliquots in desiccator at –20 °C.

3.3 Protein Digestion and Desalting

3.3.1 Protein Reduction, Alkylation, and Digestion

1. Solubilize protein samples in 7 M urea in 50 mM Tris–HCl pH 8.
2. Add DTT solution to a final concentration of 5 mM and incubate at 37 °C for 30 min.
3. Add IAA solution to a final concentration of 15 mM and incubate in the dark at room temperature for 30 min.
4. Add a 6× volume of 50 mM Tris–HCl pH 8 buffer to reduce the concentration of urea to 1 M or less.
5. Dissolve Trypsin/Lys-C mix to 1 µg/µL with the included resuspension buffer or with 50 mM Tris–HCl pH 8 buffer.
6. Add Trypsin/Lys-C mix at a 25:1 protein:protease (w/w) ratio and incubate overnight at 37 °C.
7. Add 10 % TFA solution to a final concentration of 1 % to quench the digestion.
8. Centrifuge at 14,000×*g* for 10 min to pellet particulate material.
9. Transfer the supernatant to a separate microcentrifuge tube and concentrate to approximately 100 µL in the SpeedVac.
10. Store –20 °C or proceed to desalting step.

3.3.2 Desalting

C18 solid phase extraction is necessary to remove urea salts and the amine-containing Tris–HCl buffer from the sample prior to labeling. The following procedure describes desalting a sample with a 100 µL Bond Elut OMIX C18 SPE pipette tip (*see* **Note 8**).

1. Ensure that the protein digest sample is pH ≤3 with a pH strip, adjusting with TFA or HFBA if necessary, and is at a volume of approximately 100 µL.
2. To a separate, labeled 0.6 mL microcentrifuge tube, aliquot 100 µL 60 % ACN in 0.1 % FA. The bound, desalted protein digest sample will be eluted from the C18 SPE pipette tip into the tube.
3. Securely, attach the 100 µL Bond Elut OMIX C18 SPE pipette tip to the pipettor and set it to 100 µL. For the following steps, keep the plunger depressed between each iteration or step so as to not introduce air through the sorbent material.
4. Wet the tip by aspirating 100 µL 50:50 ACN:H_2O and discard solvent. Repeat.
5. Equilibrate the tip by aspirating 100 µL 1 % TFA or HFBA and discard solvent. Repeat.
6. Bind the sample by aspirating 100 µL of the protein digest sample into the tip. Dispense back into the tube and aspirate up to 10 cycles. Dispense the sample solution back into the original tube.
7. Wash the bound protein digest sample by aspirating 100 µL 0.1 % TFA or HFBA and discard the solvent. Repeat.

8. Elute the bound, desalted protein digest sample into the previously aliquoted elution tube by aspirating 100 μL 60 % ACN in 0.1 % FA and dispensing.

9. Dry in the SpeedVac.

10. Store at −20 °C.

3.4 DiLeu Isobaric Labeling

DiLeu isobaric tags require a brief activation step, in which the amine-reactive triazine ester group is incorporated, just prior to peptide labeling. It is critical to use anhydrous solvent during this step; any moisture present in the reaction mixture will reduce activation efficiency by hydrolyzing the amine-reactive group. The activation solution consists of DMTMM and NMM at equimolar amounts in dry DMF and is used immediately upon preparation (*see* **Note 9**). Active DiLeu should be used immediately for optimal peptide labeling efficiency (*see* **Note 10**). Samples must be free of any amine-containing buffers or non-peptide compounds, as these will interfere with peptide labeling. The peptide labeling mixture—containing TEAB buffer, ACN, and active DiLeu in DMF—should be at pH ~8 with an organic solvent (ACN and DMF) concentration of around 70 % v/v. The 1 mg aliquots of DiLeu prepared in Subheading 3.2 are sufficient for labeling approximately 50–75 μg of protein digest.

3.4.1 Protein Digest Sample Preparation

1. To a 1.5 mL microcentrifuge tube, combine 600 μL ACN and 400 μL 0.5 M TEAB pH 8.5 and vortex to make 1 mL 60:40 ACN:0.5 M TEAB pH 8.5 solution.

2. Reconstitute the protein digest samples in 70 μL 60:40 ACN:0.5 M TEAB pH 8.5 solution.

3. Chill on ice or store at 4 °C until the DiLeu labeling step.

3.4.2 DiLeu Activation

1. Ensure that each of the 12-plex DiLeu aliquots to be activated are free of moisture by drying them in the SpeedVac for 15 min.

2. Flame a 25 mL round-bottom flask over a Bunsen burner to evaporate any adsorbed moisture and let the flask cool. Attach the flask to the dry solvent system and apply vacuum. Evacuate air from the flask using several cycles and dispense at least 1 mL dry DMF. Remove the flask and seal with rubber septum. Use this solvent within 30 min to minimize risk of moisture intake. If a dry solvent system is not available, *see* **Note 11**.

3. To a 1.5 mL microcentrifuge tube, add 31 mg DMTMM BF$_4$.

4. Add 989.6 μL dry DMF and vortex briefly to dissolve.

5. Add 10.4 μL NMM and vortex thoroughly.

6. To each of the twelve 1 mg aliquots of DiLeu, immediately add 50 μL of the freshly prepared activation solution (equating to

a 0.7× molar ratio of DMTMM/NMM to DiLeu). Sonicate for 10 s to free solid material from the sides of the tube.

7. Vortex for 30 min at room temperature. DiLeu will dissolve as the reaction progresses, but the solution may remain slightly cloudy since DiLeu is in molar excess to the activation reagents.

8. Centrifuge at $14,000 \times g$ for 1 min to pellet any unreacted material. Use only the supernatant, which contains 0.7 mg active DiLeu, for labeling.

9. Proceed immediately to the labeling step.

3.4.3 DiLeu Labeling

1. To each protein digest sample in 70 µL 60:40 ACN:0.5 M TEAB pH 8.5 solution, add 45–50 µL of freshly activated DiLeu solution (*see* **Note 12**).

2. Vortex for 2 h at room temperature.

3. Add 6.5 µL 5 % hydroxylamine to a final concentration of ~0.25 % and incubate for 15 min to quench the labeling reaction.

4. Combine samples in equal amounts and dry in the SpeedVac.

3.5 Strong Cation Exchange SPE and Desalting

3.5.1 Strong Cation Exchange SPE

Strong cation exchange is necessary to remove unreacted DiLeu and activation byproducts from the labeled sample (*see* **Note 13**). The following procedure describes SCX SPE fractionation of a labeled sample with SCX SpinTips by eluting via cation displacement with a stepwise addition of ammonium format in 10 % ACN pH 3 (*see* **Note 14**).

1. Reconstitute the labeled protein digest sample in 200 µL reconstitution solution. Ensure that the pH is ≤3 with a pH strip, adjusting with formic acid if necessary.

2. Place a SpinTip microcentrifuge tube adapter into a 2 mL microcentrifuge tube and insert a SpinTip into the adapter.

3. Rinse the SpinTip by adding 50 µL of the reconstitution solution to the top and centrifuging at $4000 \times g$ for 2 min. Repeat, then discard the rinsing solution.

4. Bind the sample to the SpinTip by adding it to the top and centrifuging at $4000 \times g$ for 2 min. Cycle the flow-through back through the SpinTip with more centrifuge cycles twice. Retain the flow-through.

5. Wash the sample by adding 150 µL of the wash solution to the top and centrifuging at $4000 \times g$ for 2 min. Discard the wash flow-through and repeat twice for a total of three wash cycles.

6. Transfer the SpinTip to a new 2 mL microcentrifuge tube. Elute the sample by adding 200 µL of 20 mM ammonium formate in 10 % ACN pH 3 solution to the top and centrifuging at $4000 \times g$ for 2 min. Retain the elution fraction.

7. Transfer the SpinTip to a new 2 mL microcentrifuge tube. Elute the sample by adding 200 μL of 40 mM ammonium formate in 10 % ACN pH 3 solution to the top and centrifuging at $4000 \times g$ for 2 min. Retain the elution fraction.

8. Continue eluting the sample into new 2 mL microcentrifuge tubes with 200 μL each of 60, 80, 100, 150, 250, and 500 mM ammonium formate in 10 % ACN pH 3 solution and retain all elution fractions.

9. Dry the eight elution fractions in the SpeedVac.

10. Store at −20 °C.

3.5.2 Desalting

C18 solid phase extraction is necessary to remove salts and other ion-suppressing interferences prior to LC–MS analysis (*see* **Note 8**).

1. Reconstitute the SCX fractionated labeled samples in 100 μL 0.5 % TFA or HFBA. Ensure that the pH is ≤3 with a pH strip, adjusting with TFA or HFBA if necessary.

2. Follow the procedure described in Subheading 3.3.2, **steps 2–9**, to desalt the samples.

3. Store at −20 °C.

3.6 Nano Liquid Chromatography–Tandem Mass Spectrometry

The following procedure details high-resolution nano-MS/MS acquisition on a system consisting of a nanoflow UPLC and an Orbitrap Elite mass spectrometer (*see* **Note 15**).

1. Reconstitute each sample in 3 % Optima LC/MS grade ACN in Optima LC/MS grade 0.1 % FA in water (*see* **Note 16**).

2. Inject 4 μL onto the analytical column in 97 % mobile phase (MP) A, 3 % mobile phase B.

3. Elute peptides over a 120 min gradient of 5–35 % mobile phase B at a flow rate of 300 nL/min. Follow this with 50 % mobile phase B for 10 min to elute hydrophilic peptides, then with 95 % mobile phase B for 10 min to wash the column. Finish with 15 min of 3 % mobile phase B to re-equilibrate the column.

4. Acquire spectra in the Orbitrap mass analyzer in data-dependent mode in the m/z range of 380–1600 with an MS resolving power of 120 k (at m/z 400), an automatic gain control (AGC) target of 1×10^6, and a maximum injection time of 150 ms. Select the top 10 most abundant precursors with charge 2+ or greater for MS^2 acquisition in the Orbitrap with an MS/MS resolving power of 60 k using higher energy C-trap dissociation (HCD), an isolation width of 2.0 Da, a normalized collision energy (NCE) of 27 (*see* **Note 17**), an AGC target of 3×10^4, a maximum injection time of 250 ms, and a lower mass limit of 110 m/z. Enable dynamic exclusion of precursors for 30 s with a 10 ppm m/z tolerance. Optionally, CID MS^2

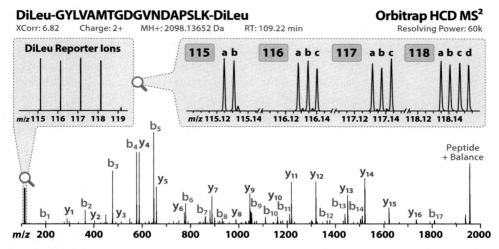

Fig. 3 A representative MS² spectrum of a 12-plex DiLeu-labeled yeast tryptic peptide acquired in the Orbitrap at 60 k resolving power following HCD fragmentation (NCE 29). DiLeu reporter ion signals (1:1 ratio) are fully resolved in the low-mass region for relative quantification, and high coverage of b- and y-ions provides confident peptide sequence identification. Reprinted with permission from Frost et al. [18]. Copyright 2015 American Chemical Society

followed by HCD MS³ acquisition of the top MS² fragment can be used to greatly improve reporter ion quantitative accuracy and precision, but this will also greatly reduce the number of quantified peptides (*see* **Note 18**).

3.7 Data Analysis

Raw mass spectral data is processed with Proteome Discoverer to identify peptides and proteins via Sequest HT database search and to extract reporter ion intensities for manual isotopic interference correction. Other software suites compatible with Thermo raw files and capable of processing reporter ion peaks may be used as alternatives. Figure 3 shows a representative MS² spectrum of a peptide identified following analysis of a 12-plex DiLeu-labeled yeast digest sample.

3.7.1 Proteome Discoverer Quantification Method

1. In the Administration menu, choose Maintain Chemical Modifications.

2. Add DiLeu 12-plex as a modification with Delta Mass: 145.12801, Substitution: H(13) C(6) ¹³C(2) N ¹⁸O, Position: Any_N_Terminus. Expand the entry with the [+] and select Amino Acid Name: N-Terminus, Classification: Isotopic Label.

3. Add a duplicate DiLeu 12-plex with Position: Any. Expand the entry with the [+] and select Amino Acid Name: Lysine, One Letter Code: K, Classification: Isotopic Label.

4. In the Administration menu, choose Maintain Quantification Methods.

5. Click Add, and name the quantification method 12-plex DiLeu as a new reporter ion quan method.

6. In the Quan Channels tab, choose 12-plex DiLeu as the residue modification on K and as the N-terminal modification. Add each of the 12-plex DiLeu reporter ion channels and their monoisotopic m/z values (*see* Fig. 1). Reporter Ion Isotopic Distribution may be entered for each channel but will not be used since interference corrections will be applied manually.

7. In the Ratio Reporting tab, add the desired ratios based on the quantitative comparisons to be made.

8. In the Ratio Calculation tab, check Show the Raw Quan Values and Reject All Quan Values if not All Quan Channels are Present. Leave Apply Quan Value Corrections unchecked.

9. In the Protein Quantification tab, check Show Peptide Ratio Counts, Show Protein Ratio Variabilities, Consider Proteins Groups for Peptide Uniqueness, and Use Only Unique Peptides.

3.7.2 Proteome Discoverer Workflow

1. The Spectrum Files node is used to specify raw files for processing and database search. If replicate LC–MS runs were acquired for a single sample or fraction, choose the set of replicates in the Parameters pane as the Input Data to process the combined data set. Process non-replicate samples or fractions separately.

2. The Reporter Ions Quantifier node (Proteome Discoverer version 1.4 and later) is used to extract reporter ion intensities from raw files specified in the connected Spectrum Files node. In the Parameters pane, choose the following:

Quantification Method	*Quantification Method:* 12-plex DiLeu
Peak Integration	*Integration Tolerance:* 10 ppm
	Integration Method: Most Confident Centroid
Scan Event Filters	*Mass Analyzer:* FTMS
	MS Order: MS^2 or MS^3 (*see* **Note 19**)
	Activation Type: HCD

3. The Spectrum Selector node is used to filter spectra from the raw files specified in the connected Spectrum Files node. In the Parameters pane, choose the following:

General Settings	*Precursor Selection:* Use MS1 Precursor
Scan Event Filters	*Mass Analyzer:* Is FTMS
	MS Order: Is MS²
	Activation Type: Is HCD
	Ionization Source: Is Nano spray

4. The Sequest HT node is used to search the MS² spectra from the connected Spectrum Selector against a protein database. In the Parameters pane, choose the following:

Input Data	*Protein Database:* Appropriate .fasta file
	Enzyme Name: Trypsin
Tolerances	*Precursor Mass Tolerance:* 25 ppm
	Fragment Mass Tolerance: 0.03 Da
Dynamic Modifications	*C-Terminal Mod:* Methyl (+14.016 Da) (*see* **Note 20**)
	Dynamic Mod: Oxidation (+15.995 Da) (M)
	Dynamic Mod: Deamidation (+0.984 Da) (N, Q)
	Dynamic Mod: Methyl (+14.016 Da) (D, E, H, R, S, T)
Static Modifications	*Peptide N-Terminus:* DiLeu (+145.128 Da)
	Static Mod: DiLeu (+145.128 Da) (K)
	Static Mod: Carbamidomethyl (+57.021 Da) (C)

5. The Percolator node, connected to the Sequest HT node, is used to validate identified peptide spectrum matches and calculate statistically meaningful q-values for each PSM via a machine learning algorithm and target decoy comparison. It then filters the data to a fixed false discovery rate (FDR). The default values in the Parameters pane—0.01 for Target FDR (Strict) and 0.05 Target FDR (Relaxed)—are recommended.

6. Name the report and click the run button. The resulting report provides an abundance of information about the identified proteins, peptides, and peptide spectrum matches. Quantitative ratios are also displayed but are not corrected for isotopic interferences. In the Peptides tab, peptide identification details can be viewed by double clicking on a peptide entry. Available information includes a summary of peptide details (sequence, modifications, charge, m/z with mass error, MH+ mass, retention time, and XCorr value), a table listing b- and y-ions with matched ions highlighted, and the MS² spectrum with b- and y-ion series annotations.

Each DiLeu primary reporter ion peak is accompanied by low-intensity isotopic peaks that are greater or lesser in mass by one neutron. For each type of stable isotope (^{13}C, ^{15}N, ^{2}H) incorporated into the reporter, a discrete −1 isotopic "impurity" peak exists. For example, the 118b reporter (m/z 118.14067), which contains ^{13}C, ^{15}N, and ^{2}H, has three −1 isotopic peaks at m/z 117.13494 (^{2}H → ^{1}H), 117.13786 (^{13}C → ^{12}C), and 117.14363 (^{15}N → ^{14}N). The −1(^{13}C) and −1(^{15}N) isotopic peaks share the exact mass of the neighboring 117b and 117c channels and interfere, while the −1(^{2}H) isotopic peak has a unique mass that falls between channels 117a and 117b. To compensate for interferences and improve quantitative accuracy, correction factor equations that add each channel's isotopic peak intensities to the raw reporter ion signal while subtracting interfering isotopic contributions from neighboring reporter ions can be applied. The interference correction feature of Proteome Discoverer only allows the user to specify +/− isotope intensity percentages for a single +1 and single −1 isotope peak. This suffices for data acquired at low resolving power where the −1(^{13}C), −1(^{15}N), and −1(^{2}H) isotopic peaks go unresolved from each other and are observed as a single peak. However, the −1(^{13}C) and −1(^{15}N) isotopic peaks are baseline-resolved as two discrete peaks at a resolving power of 30 k (@ m/z 400) and interfere with two different neighboring channels. At a resolving power of 60 k, the −1(^{2}H) isotopic peak is also resolved, but instead of interfering with surrounding channels, they fall between them, spaced at a distance of ~3 mDa. In order to adequately handle interference correction of mass defect-based reporter ions, the software must allow the user to specify isotope intensity percentages for multiple −1 isotopic peaks and choose the particular channel with which each isotopic peak interferes.

As a workaround, interference correction of 12-plex DiLeu reporter ion signals is performed manually using the method introduced by Shadforth et al. [19]. Using 116a as an example, the signal Sn observed in the MS^2 spectrum is described by the formula:

$$S_{116a} = y_{116a}I_{116a} + z_{115a}I_{115a} + x_{117a}I_{117a}$$

where yn, zn, and xn are the percentages of the true reporter ion intensities, In, from the primary reporter ion, $y_{116a}I_{116a}$, and the interfering +1 and −1 isotopic peaks from neighboring channels, $z_{115a}I_{115a}$ and $x_{117a}I_{117a}$, that contribute to Sn.

Isotopic interference correction requires that the isotopic impurities for each channel be known. These values should be determined for each new batch of synthesized DiLeu tags by direct infusion MS analysis (*see* **Note 21**).

1. In Proteome Discoverer, open a report and view the Proteins tab.

2. Right-click on the column header bar and un-check Show Peptide Groups to display peptide spectrum matches. Expand a protein entry with the [+] to show the PSMs for that protein and confirm that intensities for each reporter channel are displayed. If not, modify the Quantification Method to show reporter ion intensities by checking Show the Raw Quan Values under the Ratio Calculation tab.

3. Right-click on the column header bar and choose Export to Excel Workbook. In the export window, check Layer 2: PSMs. The output file will contain grouped PSMs in an expandable layer for each protein entry.

4. Use interference correction equations to apply corrections to the raw reporter ion intensities in Excel as described in refs. [18] and [19]. Calculate quantitative ratios for each PSM for a particular protein, then calculate either the median or average of these values to get the corrected quantitative ratio for that protein.

4 Notes

1. Consider first following the DiLeu synthesis and purification procedure (as described in Subheadings 3.1.3 and 3.2) using non-isotopic reagents to prepare a "light" DiLeu 114 tag and become familiar with each step. This way, mistakes can be made without risking loss of the more expensive isotopic reagents. It may also be beneficial to proceed with activation and labeling of a simple protein digest with this DiLeu 114 tag followed by MS analysis for a more complete understanding of the overall method. Using the DiLeu 114 tag is also a useful and inexpensive way to test questionable samples or alternative methods.

2. Hydrochloric acid is corrosive. Avoid using caps that contain materials that may degrade by coming into contact with concentrated HCl acid or fumes.

3. Pressure may build inside the reaction vessel while under heat. If using a snap cap vial, seal and wrap securely with parafilm. If using a screw cap vial, seal using a cap with a pre-slit septum insert which will vent excess pressure. Alternatively, the vessel can be carefully vented manually periodically. Do not allow the reaction vessel to remain open, as the solution can evaporate and compromise the reaction.

4. Hydrochloric acid is corrosive. To avoid damage to MS source components during direct infusion, dilute aliquots by at least

1000-fold. Also suggested is to co-evaporate HCl from each aliquot with several water/EtOH washes and dry cycles in the SpeedVac.

5. Observation of ^{18}O-exchanged leucine ions at +2 m/z and +4 m/z indicate one and two ^{18}O incorporations, respectively, on the carboxylic moiety. The reaction is complete when non-exchanged leucine ion peak intensity is <1 % compared to the peak intensities of ^{18}O-exchanged leucine. Because $H_2^{18}O$ is in great molar excess compared to leucine, ^{18}O incorporation is ≥99 % complete despite the purity of $H_2^{18}O$ (~97 %).

6. Add a small amount of MeOH first to dissolve the reaction mixture gel, then add the DCM to around 15:1 DCM:MeOH. Heat may be used to encourage solubilization if necessary, but beware of fumes due to the low boiling point of DCM (~40 °C).

7. Observation of an abundant dimethyl leucine ion (m/z 164 for DiLeu 117–118; m/z 164 and 166 for DiLeu 115–116, due to ^{18}O exchange) and the labile reporter ion fragment (m/z 115–118), resulting from in-source fragmentation, confirms isolation of the dimethyl leucine product.

8. For samples of up to 100 μg protein digest, use 100 μL C18 SPE pipette tips. For samples greater than 100 μg protein digest, use Sep-Pak C18 SPE cartridges with a 1 mL syringe. For desalting many such samples at once, use Sep-Pak Vac 1 cc C18 SPE cartridges with a vacuum manifold.

9. The activation reaction is performed with DiLeu in molar excess to these reagents in order to minimize unreacted DMTMM/NMM that could cause labeling side reactions. In order to reduce the risk of the activation solution taking on moisture or DMTMM undergoing self-degradation, prepare this solution fresh and use it immediately upon preparation.

10. Storage of activated DiLeu is not recommended because trace amounts of water produced during the activation reaction will rapidly hydrolyze the amine-reactive group. Prepare the protein digest samples in advance so that they are ready as soon as the DiLeu activation is complete and begin the labeling procedure immediately.

11. If a dry solvent system is not available, acquire bottled anhydrous DMF from Sigma. Flame a glass dram vial over a Bunsen burner to evaporate any moisture and let the vial cool. Using a syringe with a small bore needle (≤18 GA), puncture the SureSeal septum, invert the bottle, and withdraw the solvent. Dispense the solvent into the dram vial and seal with rubber septum. Use this solvent immediately to minimize moisture intake.

12. A DiLeu:protein ratio of around 10:1 w/w or greater is typically sufficient for complete labeling. Each sample should contain approximately the same amount of protein digest, or at least be labeled at the same DiLeu:protein ratio. If adjusting the volume of DiLeu solution, ensure that the labeling reaction solution has an organic solvent concentration of around 70 % upon addition of activated DiLeu solution by also adjusting the volume of ACN or 0.5 M TEAB pH 8.5 accordingly.

13. For samples of up to 1 mg labeled protein digest, use SCX SpinTips (Protea Biosciences) or SCX TopTips (Glygen Corp.) with a benchtop microcentrifuge. Alternatively, for samples of 100 μg to 2 mg labeled protein digest, use a PolySULFOETHYL A SCX column with an HPLC system to fractionate the sample. Unreacted DiLeu and activation byproducts elute during early SCX fractions.

14. Unreacted DiLeu and activation byproducts are partially removed during wash steps and partially elute during early SCX fractions.

15. Other Orbitrap or Q-TOF platforms capable of a resolving power of at least 30 k (at m/z 400) are also appropriate for 12-plex DiLeu quantification. At an MS/MS resolving power of 60 k, several isotopic interference peaks falling between channels are resolved, allowing more accurate quantification following isotopic interference correction.

16. The original protein digest sample has incurred losses from sample handling and cleanup and is now spread across eight fractions—dissolve in volumes appropriate for one to three injections depending on anticipate signal and LC–MS replicate needs.

17. The optimal HCD NCE value for fragmentation of DiLeu-labeled peptides, based on PSM identification rate, is slightly lower than that for unlabeled peptides [20]. This is in contrast to TMT- and 8-plex iTRAQ-labeled peptides, which require elevated HCD NCE values.

18. Significantly greater quantitative precision and accuracy can be achieved on hybrid LTQ-Orbitrap systems using CID MS^2 for peptide sequence identification followed by HCD MS^3 acquisition for reporter ion quantification. Conventional isobaric labeling experiments obtain quantitative information from the reporter ions produced at the MS^2 stage. For complex proteomics samples, ubiquitous co-isolation of interfering isobaric background ions during precursor selection distorts reporter ion ratios [21–24]. By performing an MS^3 isolation and HCD fragmentation event on the most

intense labeled fragment ion of the target precursor from a CID MS² scan, accurate reporter ion signals can be generated [25]. Unfortunately, isolating only one fragment results in much lower reporter ion signal intensities, and the number of quantified peptides is drastically reduced. The Orbitrap Fusion hybrid mass spectrometer, however, can employ automated synchronous precursor selection during MS³ acquisition to isolate and fragment multiple MS² fragment ions simultaneously, improving reporter ion signal intensities and significantly increasing the number of quantified peptides [26]. If using an Orbitrap Fusion mass spectrometer, the use of CID MS² followed by HCD MS³ acquisition with synchronous precursor selection for quantification is recommended.

19. If HCD MS³ acquisition was used during LC–MS analysis on a hybrid LTQ-Orbitrap, specify MS³ as the MS order to filter only MS³ spectra into the Report Ions Quantifier node.

20. Methylation of C-termini and several other residues can occasionally occur during DiLeu labeling due to leftover byproducts from activation. Deamidation of asparagine and glutamine residues can occur due to the basic pH conditions necessary for enzymatic digestion [27]. Including these as dynamic modifications in the search parameters can increase the number of identified peptides but does not significantly impact the number of identified proteins.

21. Directly infuse each inactivated tag individually at a resolving power of 60 k. Fragment the DiLeu precursor ion using HCD with an isolation width of 2.5 Da, applying sufficient NCE to fragment the precursor completely. Acquire scans over 2 min. Average all scans and convert the intensities of the +1/−1 isotopic peaks and the primary reporter peak to percentages of their summed intensity (Σisotopic peak intensity% + primary reporter peak intensity% = 100 %). Table 1 shows as an example the primary and isotopic peak percentages determined for a particular batch of 12-plex DiLeu. Table 2 shows how the +1/−1 isotopic peaks of each channel (in columns) interfere with neighboring primary reporter ion peaks. Summing the row values gives the measured reporter ion signal (primary + interferences) for that channel.

Table 1
Primary and isotopic peak percentages of total reporter ion signal[a]

	115a	115b	116a	116b	116c	117a	117b	117c	118a	118b	118c	118d
−1(¹H)					13.64 % 115.13400		5.56 % 116.13104	5.47 % 116.13736		6.37 % 117.13439	5.04 % 117.14071	0.41 % 117.14656
−1(¹²C)		0.33 % 114.12773	0.78 % 115.12476	0.85 % 115.13108		0.74 % 116.12812		0.28 % 116.14028	1.69 % 117.14028	0.57 % 117.13147	0.92 % 117.14363	
−1(¹⁴N)	0.50 % 114.12773		0.61 % 115.13108			0.79 % 116.13444	0.60 % 116.14028		0.66 % 117.13780	0.50 % 117.14363		
0	92.36 % 115.12476	93.92 % 115.13108	93.73 % 116.12812	94.29 % 116.13444	81.25 % 116.14028	94.03 % 117.13147	87.86 % 117.13731	89.07 % 117.14363	94.12 % 118.13483	87.68 % 118.14363	89.90 % 118.14699	94.05 % 118.15283
+1(¹³C)	7.14 % 116.12812	5.75 % 116.13444	4.88 % 117.13147	4.85 % 117.13780	5.11 % 117.14363	4.44 % 118.13483	5.98 % 118.14067	5.21 % 118.14699	3.53 % 119.13819	4.87 % 119.14403	4.13 % 119.15035	5.54 % 119.15619

[a]The m/z values for each peak are displayed under the intensity percentage

Table 2
Isotopic peak interferences to neighboring primary reporter ion signals[a]

Primary signal + interferences	Primary and isotopic fractional signals											
	115a	115b	116a	116b	116c	117a	117b	117c	118a	118b	118c	118d
115a	92.36 %		0.78 %									
115b		93.92 %	0.61 %	0.85 %								
116a	7.14 %		93.73 %			0.74 %						
116b		5.75 %		94.29 %		0.79 %						
116c					81.25 %		0.60 %	0.28 %				
117a			4.88 %			94.03 %			1.69 %			
117b				4.85 %			87.86 %		0.66 %	0.57 %		
117c					5.11 %			89.07 %		0.50 %	0.92 %	
118a						4.44 %			94.12 %			
118b							5.98 %			87.68 %		
118c								5.21 %			89.90 %	
118d												94.05 %

[a]The fractional intensities of +1 and −1 isotopic peaks of each channel (in columns) interfere with primary reporter ion intensities (in rows) and contribute to the measured reporter ion signal for that channel (sum of row values)

Acknowledgments

The authors acknowledge support for this work by the National Institutes of Health grant (1R01DK071801). The Q-Exactive Orbitrap was purchased through the support of an NIH-shared instrument grant (NIH-NCRR S10RR029531). L.L. acknowledges an H.I. Romnes Faculty Research Fellowship.

References

1. Ong S-E, Blagoev B, Kratchmarova I et al (2002) Stable isotope labeling by amino acids in cell culture, SILAC, as a simple and accurate approach to expression proteomics. Mol Cell Proteomics 1:376–386. doi:10.1074/mcp.M200025-MCP200

2. Molina H, Yang Y, Ruch T et al (2009) Temporal profiling of the adipocyte proteome during differentiation using a five-plex SILAC based strategy. J Proteome Res 8:48–58. doi:10.1021/pr800650r

3. Hsu J-L, Huang S-Y, Chow N-H, Chen S-H (2003) Stable-isotope dimethyl labeling for quantitative proteomics. Anal Chem 75:6843–6852. doi:10.1021/ac0348625

4. Ji C, Guo N, Li L (2005) Differential dimethyl labeling of N-termini of peptides after guanidination for proteome analysis. J Proteome Res 4:2099–2108. doi:10.1021/pr050215d

5. Boersema PJ, Raijmakers R, Lemeer S et al (2009) Multiplex peptide stable isotope dimethyl labeling for quantitative proteomics. Nat Protoc 4:484–494. doi:10.1038/nprot.2009.21

6. Wu Y, Wang F, Liu Z et al (2014) Five-plex isotope dimethyl labeling for quantitative proteomics. Chem Commun 50:1708. doi:10.1039/c3cc47998f

7. Ross PL, Huang YN, Marchese JN et al (2004) Multiplexed protein quantitation in Saccharomyces cerevisiae using amine-reactive isobaric tagging reagents. Mol Cell Proteomics 3:1154–1169. doi:10.1074/mcp.M400129-MCP200

8. Choe L, D'Ascenzo M, Relkin NR et al (2007) 8-plex quantitation of changes in cerebrospinal fluid protein expression in subjects undergoing intravenous immunoglobulin treatment for Alzheimer's disease. Proteomics 7:3651–3660. doi:10.1002/pmic.200700316

9. Thompson A, Schäfer J, Kuhn K et al (2003) Tandem mass tags: a novel quantification strategy for comparative analysis of complex protein mixtures by MS/MS. Anal Chem 75:1895–1904. doi:10.1021/ac0262560

10. Dayon L, Hainard A, Licker V et al (2008) Relative quantification of proteins in human cerebrospinal fluids by MS/MS using 6-plex isobaric tags. Anal Chem 80:2921–2931. doi:10.1021/ac702422x

11. Werner T, Becher I, Sweetman G et al (2012) High-resolution enabled TMT 8-plexing. Anal Chem 84:7188–7194. doi:10.1021/ac301553x

12. McAlister GC, Huttlin EL, Haas W et al (2012) Increasing the multiplexing capacity of TMTs using reporter ion isotopologues with isobaric masses. Anal Chem 84:7469–7478. doi:10.1021/ac301572t

13. Viner R, Blank M, Bomgarden R, Rogers R (2013) Increasing the multiplexing of protein quantitation from 6- to 10-plex with reporter ion isotopologues. Thermo Scientific Technical Note 1–7. http://apps.thermoscientific.com/media/cmd/ASMS-TNGRoadshow/TNG/resouces/PN_ASMS13_W617_RViner.pdf

14. Xiang F, Ye H, Chen R et al (2010) N, N-dimethyl leucines as novel isobaric tandem mass tags for quantitative proteomics and peptidomics. Anal Chem 82:2817–2825. doi:10.1021/ac902778d

15. Hsu J-L, Huang S-Y, Shiea J-T et al (2005) Beyond quantitative proteomics: signal enhancement of the a1 ion as a mass tag for peptide sequencing using dimethyl labeling. J Proteome Res 4:101–108. doi:10.1021/pr049837

16. Fu Q, Li L (2005) De novo sequencing of neuropeptides using reductive isotopic methylation and investigation of ESI QTOF MS/MS fragmentation pattern of neuropeptides with N-terminal dimethylation. Anal Chem 77:7783–7795. doi:10.1021/ac051324e

17. Hui L, Xiang F, Zhang Y, Li L (2012) Mass spectrometric elucidation of the neuropeptidome of a crustacean neuroendocrine organ. Peptides 36:230–239. doi:10.1016/j.peptides.2012.05.007

18. Frost DC, Greer T, Li L (2015) High-resolution enabled 12-plex DiLeu isobaric tags

for quantitative proteomics. Anal Chem 87:1646–1654. doi:10.1021/ac503276z

19. Shadforth IP, Dunkley TPJ, Lilley KS, Bessant C (2005) i-Tracker: for quantitative proteomics using iTRAQ. BMC Genomics 6:145. doi:10.1186/1471-2164-6-145

20. Greer T, Lietz CB, Xiang F, Li L (2014) Novel isotopic N N-dimethyl leucine (iDiLeu) reagents enable absolute quantification of peptides and proteins using a standard curve approach. J Am Soc Mass Spectrom. doi:10.1007/s13361-014-1012-y

21. Bantscheff M, Boesche M, Eberhard D et al (2008) Robust and sensitive iTRAQ quantification on an LTQ Orbitrap mass spectrometer. Mol Cell Proteomics 7:1702–1713. doi:10.1074/mcp.M800029-MCP200

22. Ow SY, Salim M, Noirel J et al (2009) iTRAQ underestimation in simple and complex mixtures: "the good, the bad and the ugly". J Proteome Res 8:5347–5355. doi:10.1021/pr900634c

23. Karp NA, Huber W, Sadowski PG et al (2010) Addressing accuracy and precision issues in iTRAQ quantitation. Mol Cell Proteomics 9:1885–1897. doi:10.1074/mcp. M900628-MCP200

24. Altelaar AFM, Frese CK, Preisinger C et al (2013) Benchmarking stable isotope labeling based quantitative proteomics. J Proteomics 88:14–26. doi:10.1016/j.jprot.2012. 10.009

25. Ting L, Rad R, Gygi SP, Haas W (2011) MS3 eliminates ratio distortion in isobaric multiplexed quantitative proteomics. Nat Meth 8:937–940. doi:10.1038/nmeth.1714

26. McAlister GC, Nusinow DP, Jedrychowski MP et al (2014) MultiNotch MS3 enables accurate, sensitive, and multiplexed detection of differential expression across cancer cell line proteomes. Anal Chem 86:7150–7158. doi:10.1021/ac502040v

27. Krokhin OV, Antonovici M, Ens W et al. Deamidation of -Asn-Gly- sequences during sample preparation for proteomics: consequences for MALDI and HPLC-MALDI analysis. Anal Chem 78:6645–6650. doi:10.1021/ac061017o

Chapter 11

Isotopic *N,N*-Dimethyl Leucine (iDiLeu) for Absolute Quantification of Peptides Using a Standard Curve Approach

Tyler Greer and Lingjun Li

Abstract

Quantitative proteomics studies require an absolute quantification step to accurately measure changes in protein concentration. Absolute quantification using liquid chromatography–mass spectrometry (LC–MS) traditionally combines triple quadrupole instrumentation with stable isotope-labeled standards to measure protein concentrations via their enzymatically produced peptides. Chemical modification of peptides using labels like mass differential tags for relative and absolute quantification (mTRAQ) provides another route to determine protein quantities. This chapter describes a cost-effective and high-throughput chemical labeling method that utilizes five amine-reactive, isotopic *N,N*-dimethyl leucine (iDiLeu) reagents. These tags enable generation of four-point calibration curves in one LC–MS run to determine protein concentrations from labeled peptides. In particular, we provide a detailed workflow for protein quantification using the iDiLeu reagent that includes important considerations like labeling conditions and isotopic interference correction.

Key words Quantification, iDiLeu, DiLeu, Mass difference labeling, Proteomics, Calibration curve

1 Introduction

Relative quantification studies using liquid chromatography–mass spectrometry (LC–MS) can screen hundreds to thousands of proteins to determine their potential to predict biological pathologies. Biomarker candidates must then be investigated further with a targeted assay to validate and quantify changes in protein expression. Immunoassays provide researchers with a highly specific and sensitive method to quantify single proteins, but their high cost and development time significantly restrict the number of candidates investigated [1, 2].

Mass spectrometrists have developed LC–MS protein assays in an effort to maximize throughput and reduce the cost of validation studies. Absolute quantification using stable isotope-labeled peptide standards reigns as the most popular technique [3, 4]. Stable isotope-labeled peptides are spiked into a protein digest and

Salvatore Sechi (ed.), *Quantitative Proteomics by Mass Spectrometry*, Methods in Molecular Biology, vol. 1410, DOI 10.1007/978-1-4939-3524-6_11, © Springer Science+Business Media New York 2016

are used as internal standards during LC-multiple reaction monitoring (MRM) of isolated fragments to construct a calibration curve or quantify peptide abundance. The high specificity, sensitivity, and accuracy of LC-MRM assays coupled with their ability to sample multiple peptides in one run ensure this technique's continued popularity [5–7], but the great expense of synthesizing heavy peptides necessitated development of alternate quantification methods.

Chemical modification of peptides with amine-reactive labels is a cost-effective route to produce internal standard peptides in quantitative LC–MS experiments. Recognizing this fact, AB SCIEX produced a triplex set of amine-specific mass difference tags for relative and absolute quantification (mTRAQ) [8–10]. The mTRAQ reagent is especially useful for studies that quantify a large number of proteins since only synthesis of the less-expensive light peptides is required to generate internal standards.

We have advanced the capabilities of mass difference labels in targeted LC–MS quantification by developing an isotopic labeling reagent with five quantitative channels. This increase in multiplexing capacity was achieved by synthesizing isotopic N,N-dimethyl leucine (iDiLeu) reagents, as shown in Fig. 1, that are variants of our lab's original 4-plex DiLeu label [11]. Since the initial DiLeu

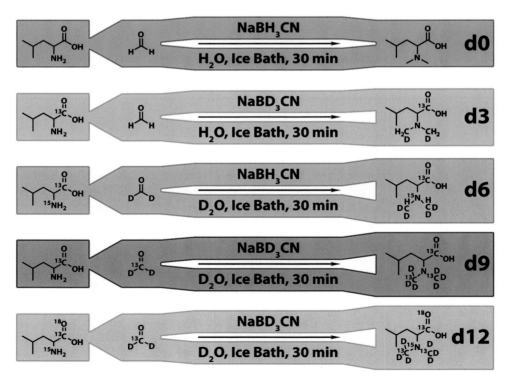

Fig. 1 iDiLeu synthetic scheme. Incorporated deuterium atoms do not affect the quantitative accuracy of the iDiLeu reagent [16]

publication, the tags have enjoyed expanded use [12–14] and modification [15], but iDiLeu is unique among DiLeu reagents in that quantitative values obtained from labeled peptides are not skewed by precursor co-isolation and co-fragmentation [16]. Each label adds a distinct mass shift to peptides so that accurate MS[1] quantification of extracted ion chromatograms (XICs) or distinct sequence-specific product ion transitions can be used to find absolute quantities of peptide analytes. While mass difference labels like iDiLeu have limited utility in early-stage protein screening studies of complex digests [17], they excel in targeted assays of discovered candidate biomarkers because of their superior accuracy. The iDiLeu reagent improves the throughput of previous absolute quantification methods by enabling construction of a calibration curve and peptide analyte quantification in a single LC–MS run. The general workflow to this new technique is shown in Fig. 2. In our previous study, an iDiLeu-labeled Allatostatin I (AST-I) peptide was quantified within 8 % error from ~1–1000 fmol using this "one-run" approach after being spiked into a mouse urine peptide mixture [16]. The iDiLeu method also benefits from the low cost of label synthesis, which was calculated to be less than $5 to label 100 µg of peptide standard or digest. The subsequent sections of this document provide a detailed protocol for iDiLeu label synthesis, peptide labeling, isotopic interference correction, and analyte quantification using XICs of iDiLeu-labeled peptides. Notes are also provided for portions of the workflow that may require additional consideration to optimize iDiLeu's performance.

2   Materials

2.1 Reagents and Equipment

1. Isotopic L-leucines: L-leucine, L-leucine-1-^{13}C, ^{15}N, and L-leucine-1-^{13}C (ISOTEC, Miamisburg, OH).

2. Isotopic formaldehydes: CH_2O (Sigma-Aldrich, St. Louis, MO), CD_2O and $^{13}CD_2O$ (ISOTEC).

3. Isotopic sodium cyanoborohydrides: $NaBH_3CN$ (Sigma-Aldrich) and $NaBD_3CN$ (ISOTEC).

4. 97 % ^{18}O water ($H_2^{18}O$) and deuterium water (D_2O, ISOTEC).

5. Hydrogen chloride gas (HCl, Sigma-Aldrich).

6. Iodoacetamide (IAM, Sigma-Aldrich).

7. Tris-hydrochloride (Tris–HCl, Sigma-Aldrich).

8. Reagent-grade formic acid (FA, Sigma-Aldrich).

9. Triethylammonium bicarbonate (TEAB, Sigma-Aldrich).

10. N,N-dimethylformamide (DMF, Sigma-Aldrich).

11. 4-(4,6-Dimethoxy-1,3,5-triazin-2-yl)-4-methylmorpholinium tetrafluoroborate (DMTMM, Sigma-Aldrich).

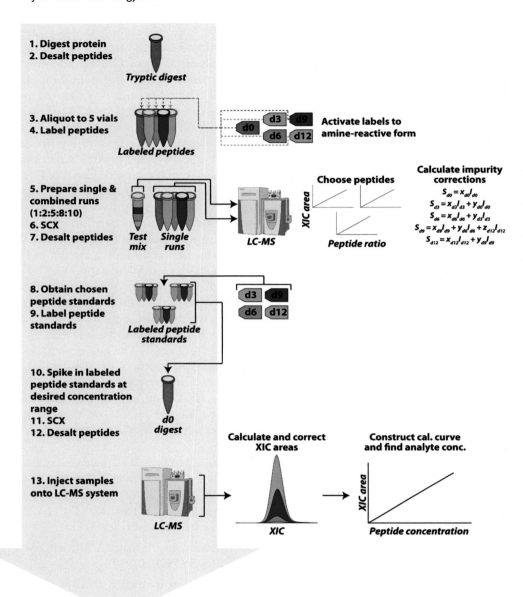

Fig. 2 Workflow of the iDiLeu quantification method

12. *N*-methylmorpholine (NMM, Sigma-Aldrich).

13. Ninhydrin (Sigma-Aldrich).

14. Sequencing grade trypsin (Promega, Madison, WI).

15. Urea (Fisher Scientific, Pittsburgh, PA).

16. ACS grade methanol (MeOH, Fisher Scientific).

17. ACS grade dichloromethane (DCM/CH_2Cl_2, Fisher Scientific).

18. ACS grade and Optima LC–MS grade acetonitrile (ACN/ C_2H_3N, Fisher Scientific).

19. Optima LC–MS grade water (Fisher Scientific).

20. Optima LC–MS grade formic acid (FA, Fisher Scientific).

21. StratoSpheres PL-HCO$_3$ MP resin (Agilent Technologies, Santa Clara, CA).

22. Hydroxylamine, 50 % aq. Soln. (H_3NO, Alfa Aesar, Ward Hill, MA).

23. Flash chromatography column (Ace Glass, Vineland, NJ).

24. Silica gel, 40–63 μm particle size (Silicycle, Quebec City, Quebec, Canada).

25. Büchi RE 111 Rotavapor (Flawil, Switzerland).

26. SepPak C_{18} SPE cartridges (Waters, Milford, MA).

27. C_{18} OMIX pipette tips (Agilent Technologies).

28. Savant SC 110 SpeedVac concentrator (Thermo Scientific, Waltham, MA).

29. Strong cation exchange (SCX) spin tips and buffers (Protea Biosciences, Morgantown, WV, USA).

30. Waters nanoAcquity UPLC system (Waters).

31. Symmetry C_{18} nanoAcquity trap column (180 μm × 20 mm, 5 μm, Waters).

32. 1.7 μm BEH C_{18} column (75 μm × 100 mm, 130 Å, Waters).

33. Q-Exactive Orbitrap (Thermo Scientific).

34. Xcalibur software (version 2.2, Thermo Scientific).

35. Proteome Discoverer software (version 1.4.0288, Thermo Scientific).

36. Mathcad software (version 14, Parametric Technology Corporation, Needham, MA).

3 Methods

3.1 ^{18}O Exchange of L-Leucine-1-^{13}C, ^{15}N

The *d*12 iDiLeu label requires an ^{18}O exchange before dimethylation.

1. Dissolve L-leucine-1-^{13}C, ^{15}N in 1 N HCl $H_2^{18}O$ solution (pH 1) and stir at 65 °C for 4 h (*see* **Note 1**).

2. Evaporate HCl using a Rotavapor and scavenge remaining acid with StratoSpheres PL-HCO$_3$ MP resin (*see* **Note 2**).

3.2 Synthesis of Isotopic N,N-Dimethyl Leucine (iDiLeu)

Isotopic L-leucines can be dimethylated by matching the L-leucine with the appropriate reagents as shown by Fig. 1.

1. Suspend L-leucine in H_2O or D_2O with a 2.5 molar excess of NaBH$_3$CN or NaBD$_3$CN (*see* **Note 3**).

2. Place the reaction vial in an ice water bath, add a 2.5 molar excess of 37 % w/w CH_2O, 20 % w/w CD_2O, or 20 % w/w $^{13}CD_2O$, seal the vial, and stir the reaction mixture for 30 min (*see* **Note 3**).

3. Verify complete dimethylation of the primary amine using a ninhydrin stain (*see* **Notes 4** and **5**).

4. Purify dimethyl leucine by collecting fractions in glass test tubes using flash column chromatography (DCM/MeOH) and dry with a Rotavapor (*see* **Notes 6–8**).

3.3 iDiLeu Activation to Amine-Reactive Form

1. Dissolve 1 mg of each dimethyl leucine in 50 μL of anhydrous DMF and combine with a 0.9× molar ratio of DMTMM and NMM (*see* **Note 9**).

2. Vortex the reaction mixture for 1 h and spin down the excess unactivated iDiLeu (*see* **Note 10**).

3.4 Protein Digestion

iDiLeu and DiLeu reagents have labeled tryptic peptides from *Saccharomyces cerevisiae* and human K562 cell lysates, human and mouse urine, and a variety of peptide and protein standards. This section provides a digest we commonly employ in our iDiLeu and DiLeu methods.

1. Reconstitute protein with ≥100 μL of 8 M urea, 50 mM Tris–HCl, pH 8 (*see* **Note 11**).

2. Reduce cysteine residues by adding 5 mM DTT and incubating at 37 °C for 1 h (*see* **Note 12**).

3. Alkylate free thiols with 15 mM IAM for 15 min in the dark at room temperature.

4. Quench the alkylation reaction using 5 mM DTT.

5. Dilute sample with enough 50 mM Tris–HCl, pH 8, to reduce the urea concentration to <1 M.

6. Add trypsin in a 1:50 enzyme:protein ratio.

7. Digest proteins for 16 h at 37 °C.

8. Quench digest by adding 10 % TFA to reduce the pH to <3.

9. Desalt peptides according to the manufacturer's protocol using either Waters SepPak C_{18} SPE cartridges or Agilent C_{18} OMIX tips (*see* **Note 13**).

10. Dry down peptides with a SpeedVac.

11. Reconstitute peptides in 0.5 M TEAB prior to labeling.

3.5 Peptide Labeling

While the iDiLeu quantification strategy is relatively simple, Fig. 2 provides a workflow of the technique so that repetitive steps like peptide labeling, SCX, and desalting need not be described multiple times.

1. Pipette a 10× *w/w* excess of activated iDiLeu supernatant to the peptide samples dissolved in 0.5 M TEAB.

2. Add anhydrous DMF to the reaction mixture so that the organic:aqueous ratio is ~75 %.

3. Vortex the reaction mixtures at room temperature for 2 h.

4. Quench the labeling reaction with the addition of 0.25 % *v/v* hydroxylamine (*see* **Note 14**).

5. After combining iDiLeu-labeled peptide channels as discussed in **Note 14**, dry down the samples and reconstitute the peptides in Protea Biosciences SCX resuspension buffer.

6. Remove iDiLeu reaction byproducts by using Protea Biosciences SCX spin tips according to the manufacturer's protocol.

7. Dry down the cleaned samples using a SpeedVac and desalt labeled peptides using either Waters SepPak C_{18} SPE cartridges or Agilent C_{18} OMIX tips.

3.6 Liquid Chromatography Mass Spectrometry (LC–MS)

This section provides LC, data-dependent acquisition (DDA), and MS-only settings from our previous iDiLeu report [15]. It is recommended to use the DDA method to confidently identify peptides and find their retention times while the MS-only method should be used for XIC iDiLeu quantification experiments.

1. Dissolve iDiLeu-labeled peptides in Optima LC–MS grade 0.1 % FA (*v/v*).

2. Perform reversed-phase separations using a Waters nanoAcquity UPLC system containing a Symmetry C_{18} nanoAcquity trap column (180 μm × 20 mm, 5 μm) and a 1.7 μm BEH C_{18} column (75 μm × 100 mm, 130 Å). Full LC parameters are provided in **Note 15**.

3. Ionize peptides into a Thermo Q-Exactive Orbitrap using a Nanospray Flex ion source and record spectra in DDA or MS-only mode. Method details from previous experiments are given in **Notes 16** and **17**.

3.7 Data Analysis

Data analysis software presented in this section can be substituted with equivalent programs.

1. Identify proteins with Thermo Proteome Discoverer 1.4. Previous experiments matched peptides to .fasta databases obtained from www.uniprot.org using the Sequest HT algorithm. Search parameters are provided in **Note 18**.

2. Process raw quantitative data using Thermo Xcalibur 2.2 to generate XIC peak areas from the Genesis peak detection algorithm.

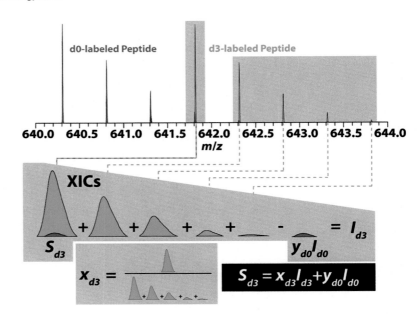

Fig. 3 Example of isotopic impurity corrections. The meaning of each variable is displayed in relation to the XIC of a typical iDiLeu-labeled peptide

3. Determine isotopic interference correction factors for chosen peptide surrogates using PTC Mathcad 14 and an adaptation of the previously reported i-Tracker method [18]. While the impurity correction equations for iDiLeu were previously described [16], Fig. 3 shows the physical descriptions of each variable for an example of peptide interference between *d*0- and *d*3-labeled peptides. Equation details are provided in **Note 19**.

4. Generate calibration curves from quantitative data using Microsoft Excel.

4   Notes

1. Make HCl $H_2^{18}O$ solution by gently bubbling HCl gas into $H_2^{18}O$ until a bright yellow color is observed.

2. Weigh a 1:1 *w/w* ratio of StratoSpheres PL-HCO$_3$ MP resin to L-leucine-1-^{13}C and place in a vial. Add 20 mL of EtOH to the resin and shake for 10 min. Pipette EtOH off of the resin and repeat this wash step two more times. Dissolve L-leucine-1-^{13}C in 20 mL of EtOH and transfer to the vial containing scavenger resin. Shake for 10 min and pipette the cloudy mixture off of the resin to a round-bottom flask. Repeat this step two more times to obtain maximum yield.

3. Sodium cyanoborohydride and formaldehyde are toxic by inhalation, in contact with skin, or if swallowed, and may cause

cancer and heritable genetic damage. These chemicals and reactions must be handled in a fume hood.

4. Make ninhydrin solution by dissolving 1.5 g of ninhydrin in 100 mL of *n*-butanol and adding 3 mL of AcOH.

5. Pipette 2 μL of dimethylation reaction mixture onto a thin-layer chromatography (TLC) plate. Dip the plate into the ninhydrin solution and gently heat using a heat gun. The absence of a pink spot confirms complete dimethylation of L-leucine.

6. Flash column mobile phase compositions: (1) 150 mL DCM:0 mL MeOH; (2) 150 mL DCM:10 mL MeOH; (3) 400 mL DCM:40 mL MeOH; (4) 160 mL DCM:20 mL MeOH; (5) 250 mL DCM:50 mL MeOH; and (6) 200 mL DCM:200 mL MeOH.

7. iDiLeu typically elutes at the beginning of the 1:1 DCM:MeOH ratio. Verify fractions by a $KMnO_4$ stain.

8. Make $KMnO_4$ solution by dissolving 1.5 g of $KMnO_4$, 10 g of K_2CO_3, and 1.25 mL of 10 % NaOH in 200 mL of water.

9. It is essential that dimethyl leucine is the excess reagent so that no excess DMTMM is available to react with peptides and form reaction side products.

10. Labeling efficiency is greatest when it occurs immediately after activation.

11. It may be necessary to perform a protein precipitation or buffer exchange step prior to reconstitution. Furthermore, a BCA assay or equivalent method to measure total protein concentration can be helpful when determining the volume of buffer to add and the amount of iDiLeu label needed.

12. The DTT concentration reflects its amount in solution after being added.

13. If the user does not wish to risk peptide loss with a desalting step, a digest protocol has been used in a previous DiLeu article that circumvents this step [11].

14. After quenching the labeling reaction, iDiLeu-labeled peptides can be: (1) Combined in ratios of 1:2:5:8:10 to choose those with the most linear response; (2) Aliquoted into separate vials to determine the isotopic corrections factors; or (3) Combined with labeled peptide standards (synthesized or purchased) to quantify the peptides chosen from (1). Refer to Fig. 2 to determine the appropriate action. Always combine the labeled peptide channels before SCX purification.

15. Mobile phase A: 0.1 % FA in Optima LC–MS grade water; Mobile phase B: 0.1 % FA in Optima LC–MS grade ACN. Previous experiments loaded peptides onto the trap column in 1 % B at a flow rate of 5 μL/min for 5 min. Peptide separation occurred on the analytical column at a flow rate of 300 nL/min with a 30 min gradient from 3 to 35 % B.

16. MS-only parameters: Previous XIC quantification experiments utilized profile mode and full MS scans from m/z 380 to 2000 at a resolution of 140 K. The automatic gain control (AGC) was 1×10^6, and the maximum ion injection time (IT) into the C-trap was 100 ms.

17. DDA parameters: Previous peptide identification experiments acquired MS spectra in profile mode from m/z 380–2000 at a resolution of 70 K. The automatic gain control (AGC) was 1×10^6, and the maximum ion injection time (IT) into the C-trap was 50 ms. The top 10 precursor ions were selected to be fragmented with higher-energy collisional dissociation (HCD) at a normalized collision energy (NCE) of 28. Tandem mass spectrometry (MS^2) parameters were: (1) Resolution = 17.5 K; (2) AGC = 1×10^5; (3) Maximum IT = 150 ms; (4) Fixed first mass = m/z 110; and (5) Underfill ratio = 0.1 %.

18. Search parameters: Enzyme = trypsin; maximum missed cleavages = 2; precursor mass tolerance = 50 ppm; fragment mass tolerance = 0.02 Da; static modification = cysteine carbamido-methyl (+57.0215 Da); and variable modification = methionine oxidation (+15.9949 Da). Static modifications representing iDiLeu modification of the N-terminus and lysine are: $d0$ = +141.1154 Da; $d3$ = +144.1313 Da; $d6$ = +147.1409 Da; $d9$ = +150.1631 Da; and $d12$ = +153.1644 Da.

19. iDiLeu-labeled peptide channels may significantly interfere with one another depending on the peptide's charge state and number of sites labeled. Our lab previously employed the i-Tracker method to correct isobaric label reporter ion intensities [11, 13], and we extended this technique to precursor quantification [16]. Figure 3 shows an example of the equations used to construct isotopic interference corrections. Briefly, XIC areas from the isotopic distribution of a peptide should be added to the area of that peptide's monoisotopic peak. The interfering signal from a different iDiLeu channel must be subtracted from this sum. Correction equations can be constructed for each labeled peptide by using the following system of equations:

$$S_{d0} = x_{d0}I_{d0}$$

$$S_{d3} = x_{d3}I_{d3} + y_{d0}I_{d0}$$

$$S_{d6} = x_{d6}I_{d6} + y_{d3}I_{d3}$$

$$S_{d9} = x_{d9}I_{d9} + y_{d6}I_{d6} + z_{d12}I_{d12}$$

$$S_{d12} = x_{d12}I_{d12} + y_{d9}I_{d9}$$

S = uncorrected area of a monoisotopic XIC.

I = sum of corrected monoisotopic and isotopic envelope XIC areas.

x =percentage of I from the monoisotopic XIC contribution.

y =percentage of I that interferes with a heavier mass-labeled peptide.

z =percentage of I that interferes with a lighter mass-labeled peptide.

This system of equations can be symbolically solved for using Mathcad and transferred to Microsoft Excel for efficient data processing.

Acknowledgments

The authors acknowledge support for this work by the National Institutes of Health grant (1R01DK071801). The Q-Exactive Orbitrap was purchased through the support of an NIH shared instrument grant (NIH-NCRR S10RR029531). L.L acknowledges an H.I. Romnes Faculty Research Fellowship.

References

1. Makawita S, Diamandis EP (2010) The bottleneck in the cancer biomarker pipeline and protein quantification through mass spectrometry-based approaches: current strategies for candidate verification. Clin Chem 56:212–222. doi:10.1373/clinchem.2009.127019

2. Addona TA, Shi X, Keshishian H et al (2011) A pipeline that integrates the discovery and verification of plasma protein biomarkers reveals candidate markers for cardiovascular disease. Nat Biotechnol 29:635–643. doi:10.1038/nbt.1899

3. Barr JR, Maggio VL, Patterson DG et al (1996) Isotope dilution – mass spectrometric quantification of specific proteins: model application with apolipoprotein A-I. Clin Chem 42:1676–1682

4. Gerber SA, Rush J, Stemman O et al (2003) Absolute quantification of proteins and phosphoproteins from cell lysates by tandem MS. Proc Natl Acad Sci U S A 100:6940–6945. doi:10.1073/pnas.0832254100

5. Zhao Y, Brasier AR (2013) Applications of selected reaction monitoring (SRM)-mass spectrometry (MS) for quantitative measurement of signaling pathways. Methods 61:313–322. doi:10.1016/j.ymeth.2013.02.001

6. Warnken U, Schleich K, Schnölzer M, Lavrik I (2013) Quantification of high-molecular weight protein platforms by AQUA mass spectrometry as exemplified for the CD95 death-inducing signaling complex (DISC). Cells 2:476–495. doi:10.3390/cells2030476

7. Sturm R, Sheynkman G, Booth C et al (2012) Absolute quantification of prion protein (90-231) using stable isotope-labeled chymotryptic peptide standards in a LC-MRM AQUA workflow. J Am Soc Mass Spectrom 23:1522–1533. doi:10.1007/s13361-012-0411-1

8. DeSouza L, Taylor A, Li W (2008) Multiple reaction monitoring of mTRAQ-labeled peptides enables absolute quantification of endogenous levels of a potential cancer marker in cancerous and normal. J Proteome Res 7:3525–3534

9. Zhou L, Wei R, Zhao P et al (2013) Proteomic analysis revealed the altered tear protein profile in a rabbit model of Sjögren's syndrome-associated dry eye. Proteomics 13:2469–2481. doi:10.1002/pmic.201200230

10. Zhang S, Wen B, Zhou B et al (2013) Quantitative analysis of the human AKR family members in cancer cell lines using the mTRAQ/MRM approach. J Proteome Res 12:2022–2033. doi:10.1021/pr301153z

11. Xiang F, Ye H, Chen R et al (2010) N,N-dimethyl leucines as novel isobaric tandem mass tags for quantitative proteomics and peptidomics. Anal Chem 82:2817–2825. doi:10.1021/ac902778d

12. Hui L, Xiang F, Zhang Y, Li L (2012) Mass spectrometric elucidation of the neuropeptidome of a crustacean neuroendocrine organ. Peptides 36:230–239. doi:10.1016/j.peptides.2012.05.007

13. Sturm RM, Lietz CB, Li L (2014) Improved isobaric tandem mass tag quantification by ion mobility mass spectrometry. Rapid Commun Mass Spectrom 28:1051–1060. doi:10.1002/rcm.6875

14. Hao L, Zhong X, Greer T et al (2015) Relative quantification of amine-containing metabolites using isobaric N,N-dimethyl leucine (DiLeu) reagents via LC-ESI-MS/MS and CE-ESI-MS/MS. Analyst 140:467–475. doi:10.1039/C4AN01582G

15. Frost DC, Greer T, Li L (2015) High-resolution enabled 12-plex DiLeu isobaric tags for quantitative proteomics. Anal Chem 87:1646–1654. doi:10.1021/ac503276z

16. Greer T, Lietz CB, Xiang F, Li L (2015) Novel isotopic N,N-dimethyl leucine (iDiLeu) reagents enable absolute quantification of peptides and proteins using a standard curve approach. J Am Soc Mass Spectrom 26:107–119. doi:10.1007/s13361-014-1012-y

17. Mertins P, Udeshi ND, Clauser KR et al (2012) iTRAQ labeling is superior to mTRAQ for quantitative global proteomics and phospho-proteomics. Mol Cell Proteomics. doi:10.1074/mcp.M111.014423

18. Shadforth IP, Dunkley TPJ, Lilley KS, Bessant C (2005) i-Tracker: for quantitative proteomics using iTRAQ. BMC Genomics 6:145. doi:10.1186/1471-2164-6-145

Chapter 12

Selecting Optimal Peptides for Targeted Proteomic Experiments in Human Plasma Using In Vitro Synthesized Proteins as Analytical Standards

James G. Bollinger, Andrew B. Stergachis, Richard S. Johnson, Jarrett D. Egertson, and Michael J. MacCoss

Abstract

In targeted proteomics, the development of robust methodologies is dependent upon the selection of a set of optimal peptides for each protein-of-interest. Unfortunately, predicting which peptides and respective product ion transitions provide the greatest signal-to-noise ratio in a particular assay matrix is complicated. Using in vitro synthesized proteins as analytical standards, we report here an empirically driven method for the selection of said peptides in a human plasma assay matrix.

Key words Targeted proteomics, Selected reaction monitoring, In vitro translation, Human plasma proteome, Proteotypic peptides

1 Introduction

Mass spectrometry has emerged as the dominant technology for the characterization of proteins in biological matrices due, in part, to its unequivocal combination of speed, selectivity, and sensitivity. Most classical mass spectrometry-based proteomic workflows have taken a shotgun approach in which the protein fraction is initially digested with a protease prior to analysis. The resulting peptides are then separated by nano-flow liquid chromatography, ionized, transferred to a mass spectrometer, and subjected to tandem mass spectrometry via data-dependent acquisition (DDA). In a DDA experiment, the mass information from periodic full-scan analyses of intact peptides is used to trigger subsequent tandem mass spectrometry (MS/MS) analyses of the most abundant precursor ions for sequence identification. The resulting peptide masses and fragment ions are then searched against protein sequence databases and ultimately used as a proxy for protein identification and/or relative abundance. This general discovery-based approach has

Salvatore Sechi (ed.), *Quantitative Proteomics by Mass Spectrometry*, Methods in Molecular Biology, vol. 1410, DOI 10.1007/978-1-4939-3524-6_12, © Springer Science+Business Media New York 2016

become extremely powerful for determining the protein content of moderately complex biological mixtures. However, the ability to accurately compare different samples is complicated by the semi-random sampling process of DDA. Some proteins of specific interest can go undetected in one or more compared samples. Furthermore, the immense dynamic range of relative protein concentration in clinically derived specimens usually necessitates laborious pre-analysis fractionation and chromatography protocols. These ultimately hinder the throughput of these methods and make them impractical for comprehensive studies with multiple biological and/or technical replicates.

Due to limitations of current discovery-based proteomic approaches, some laboratories have begun the development and application of technologies for the targeted analysis of proteins within complex mixtures. Numerous derivations of targeted mass spectrometry using the specific acquisition of tandem mass spectra of peptides predicted in silico have been reported. More recently, these methods have been based on the use of selected reaction monitoring (SRM) on triple quadrupole mass spectrometers [1–3]. These methods have high specificity within complex mixtures and can be performed in a fraction of the instrument time relative to discovery-based methods. In complex biological matrices, the chemical background of co-eluting analytes can often prohibit detection of a precursor ion in a DDA experiment. However, if the precursor ion m/z is known, a triple quadrupole mass spectrometer can be used to minimize the chemical interference using two orthogonal stages of mass analysis to selectively monitor a unique peptide. The combined specificity of chromatographic retention time, precursor ion mass, and product ion mass can enable the selective detection of a peptide within a complex matrix.

Targeted mass spectrometry measurements themselves are not necessarily quantitative. For an assay to be quantitative, the analyte response needs to be assessed using protein or peptide standards of known abundance. These assays can provide absolute quantitative measurements if they are thoroughly vetted like any classical quantitative mass spectrometry measurement by assessing the measurement linearity, variance, accuracy, limit of detection (LOD), and limit of quantitation (LOQ) [4]. The inter-laboratory consistency of these measurements can be robust between laboratories and across instrument platforms [5]. Furthermore, the high duty cycle of modern triple quadrupole instrumentation enables multiplexing of targeted SRM assays to measure multiple peptides for an array of proteins in any given experiment [3, 4, 6].

While the power of SRM assays is undeniable, the development of robust methodology for the selection of peptides to use as a proxy of a translated gene product is not straight forward.

Due to differences in their inherent physiochemical properties, equimolar peptides of different amino acid sequences can have drastically different responses in a mass spectrometer. A "proteotypic" peptide for targeted proteomics is defined here as one that is (1) unique to a given gene product; (2) lacking in high frequency, non-synonymous, single nucleotide polymorphisms; (3) devoid of known posttranslational modification sites; (4) has physiochemical properties amenable to a robust detection in the mass spectrometer; and (5) has salient features that generate characteristic MS/MS fragmentation patterns via collision-induced dissociation (CID). The selection of a set of best peptides for each protein-of-interest is a crucial step to the development of targeted SRM assays because considerable amounts of time and resources are often spent to produce quantitative standards such as synthetic peptides [1, 2], recombinant proteins from concatenated peptide sequences [3, 7], or developing immunoaffinity reagents for the enrichment of low abundance tryptic peptides [6]. Traditional approaches for selecting candidate peptides for SRM assays have relied on the mining of DDA spectral libraries [8, 9], the use of prediction algorithms trained on previous DDA experimental results [10, 11], or the costly synthesis of all in silico-predicted tryptic peptides [12]. The first two approaches are predicated on the assumption that the peptides most frequently identified in DDA experiments will produce peptides with the optimal signal-to-noise ratios for a targeted proteomic experiment. Unfortunately, this is not the case [13]. There are numerous reasons why a peptide may not be selected for MS/MS during a DDA experiment. Therefore, a peptide that is not observed in this type of experiment should not be excluded in a targeted experiment. Conversely, a peptide that is routinely sampled in a DDA style experiment might not necessarily be a suitable peptide for an SRM experiment. For these reasons and more, an important step in the development of SRM assays is the use of an analytical protein standard to assess empirically which peptides provide a good proxy of the target protein.

Here we demonstrate a general cost-effective strategy (*see* Fig. 1) for the systematic selection of best peptides for use in an SRM assay for a protein-of-interest in a human plasma matrix. We apply peptide selection criteria based on the sensitivity and specificity of a peptide's SRM signal while also making considerations for matrix effects, chromatographic properties, digestion kinetics, and post-digestion stability. Our strategy makes use of analytical standards expressed in vitro as C-terminal *Schistosoma Japonicum* Glutathione S-Transferase (*SJ-GST*) fusion proteins. Although discussed in the context of the human plasma proteome, this strategy is generalizable to other proteins and other biological matrices.

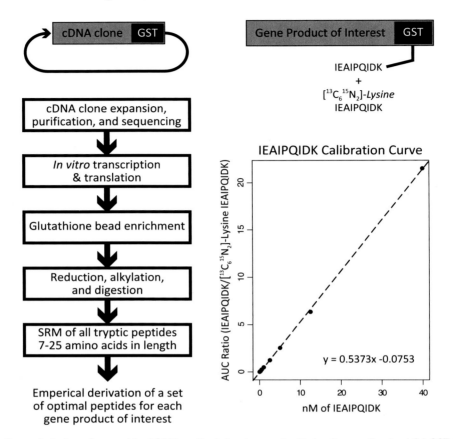

Fig. 1 General strategy for empirical SRM method development with in vitro synthesized SJ-GST fusion proteins as analytical standards

2 Materials

2.1 Preparation of C-Terminal GST Fusion Proteins to Use as Analytical Standards via In Vitro Protein Expression

1. Full-length cDNA clones for human proteins from the pANT7_cGST clone collection (Arizona State University Biodesign Institute Plasmid Repository, https://dnasu.org/DNASU/Home.do) (*see* **Note 1**).

2. Ampicillin-Supplemented LB Culture Medium: 10 g/L Tryptone (Becton, Dickinson, and Company, Sparks, MD), 5 g/L Yeast Extract (Becton, Dickinson, and Company, Sparks, MD), and 10 g/L Sodium Chloride (Fisher Scientific, Fairlawn, NJ) supplemented with 100 μg/mL ampicillin (Invitrogen, Carlsbad, CA).

3. Molecular biology grade USP sterile purified water (Corning, Manassas, VA).

4. Sterile 14 mL polypropylene round-bottom tubes with snap-on lids (BD Biosciences, Durham, NC).

5. Allegra X-12R Centrifuge with SX4750 rotor (Beckman-Coulter Inc., Brea, CA).

6. QIAprep Spin Miniprep Kit (Qiagen, Valencia, CA).

7. Bench-top centrifuge model 5417R (Eppendorf, Hauppauge, NY).

8. ND-1000 Spectrophotometer (Nanodrop, Wilmington, DE)

9. 1-Step Human Coupled in vitro protein synthesis kit (Pierce Biotechnology, Rockford, IL).

10. RiboLock RNAse inhibitor (Pierce Biotechnology, Rockford, IL).

11. Glutathione sepharose 4B beads (GE Healthcare Biosciences, Pittsburgh, PA).

12. Dulbecco's phosphate-buffered saline without $MgCl_2$ and $CaCl_2$ (Gibco, Grand Island, NY).

13. Sepharose Bead Wash Solution #1 Dulbecco's phosphate-buffered saline without $MgCl_2$ and $CaCl_2$ (Gibco, Grand Island, NY) supplemented with 863 mM sodium chloride (Fisher Scientific).

14. Sepharose Bead Wash Solution #2: 50 mM NH_4HCO_3, pH ~7.8 (Fisher Scientific, Fair Lawn, NJ) dissolved in 18Ω water.

15. Sepharose bead reconstitution buffer: 0.1 % PPS Silent Surfactant (Expedeon, San Diego, CA)/50 mM Ammonium Bicarbonate (pH-7.8) supplemented with 5 nM FasTrack crude "heavy" $[^{13}C_6^{15}N_2]$ (L)-lysine–labeled LLLEYLEEK and IEAIPQIDK peptides (Life Technologies, Grand Island, NY).

2.2 SDS-PAGE/ Western Blot Analysis of GST Fusion Proteins

1. Sample Loading Buffer: NuPage LDS 4× sample buffer (Life Technologies, Grand Island, NY).

2. SDS-PAGE Running Buffer: 20× NuPage MES SDS Running Buffer (Life Technologies, Grand Island, NY) diluted to 1× with deionized water.

3. Novex NuPage 4–12 % bis–tris mini gel, 10 well, 1.5 mm width (Life Technologies, Grand Island, NY).

4. Novex Sharp Pre-stained Protein Standards (Life Technologies, Grand Island, NY).

5. Invitrogen XCell SureLock mini-cell SDS-PAGE gel box (Life Technologies, Grand Island, NY).

6. PowerPac Basic Electrophoresis Power Supply (BioRad, Hercules, CA).

7. SilverQuest Staining kit (Life Technologies, Grand Island, NY).

8. XCell II Blot Module (Life Technologies, Grand Island, NY).

9. Blot Transfer Buffer: 20× NuPage transfer buffer (Life Technologies, Grand Island, NY) and Methanol (Fisher

Scientific, Pittsburgh, PA) diluted to 1× and 10 % (v/v), respectively, with deionized water.

10. PVDF filter paper sandwich (Life Technologies, Grand Island, NY).

11. Lab Rotator (Barnstead/Lab-Line).

12. Blot Blocking Buffer: Non-fat milk powder (Safeway Inc., Phoenix, AZ) and Tween-20 (Fisher Scientific, Pittsburgh, PA) diluted to 5 % (w/v) and 0.1 % (v/v), respectively, with Dulbecco's phosphate-buffered saline without $MgCl_2$ and $CaCl_2$ (Gibco, Grand Island, NY).

13. Blot Washing Buffer: Tween-20 (Fisher Scientific, Pittsburgh, PA) diluted to 5 % (w/v) and 0.1 % (v/v), respectively, with Dulbecco's phosphate-buffered saline without $MgCl_2$ and $CaCl_2$ (Gibco, Grand Island, NY).

14. Blot Primary Antibody Solution: Anti-GST Antibody (GE Healthcare Biosciences, Pittsburgh, PA) and Tween-20 (Fisher Scientific, Pittsburgh, PA) diluted 1000-fold and to 0.1 % (v/v), respectively, with Dulbecco's phosphate-buffered saline without $MgCl_2$ and $CaCl_2$ (Gibco, Grand Island, NY).

15. Blot Secondary Antibody Solution: Horseradish Peroxidase Anti-goat IgG Antibody (Pierce) and Tween-20 (Fisher Scientific, Fair Lawn, NJ) diluted 10,000-fold and to 0.1 % (v/v), respectively, with Dulbecco's phosphate-buffered saline without $MgCl_2$ and $CaCl_2$ (Gibco, Grand Island, NY).

16. ECL Prime Western Blotting Kit (GE Healthcare Bio-Sciences, Pittsburgh, PA).

17. Autoradiography cassette (Fisher Scientific, Pittsburgh, PA).

18. BioMax Light Chemiluminescence Film (Kodak, Rochester, NY).

19. X-OMAT 2000A Film Developer (Kodak, Rochester, NY).

2.3 Sample Digestion

1. Sepharose bead dilution buffer: 0.1 % PPS Silent Surfactant/50 mM NH_4HCO_3 dissolved in 18Ω water.

2. Commercially Sourced Human Plasma (Lampire Biological Laboratories, Pipersville, PA).

3. Plasma Dilution Buffer 1: 50 mM NH_4HCO_3 dissolved in 18Ω water.

4. Plasma Dilution Buffer 2: 0.2 % PPS Silent Surfactant/50 mM NH_4HCO_3 dissolved in 18Ω water.

5. BCA Protein Assay Kit (Pierce Biotechnology, Rockford, IL).

6. 500 mM dithiothreitol (Sigma Aldrich, St. Louis, MO) dissolved in 18Ω water.

7. 500 mM iodoacetamide (Sigma Aldrich, St. Louis, MO) dissolved in 18Ω water.

8. Sequencing grade modified porcine trypsin (Promega, Madison, WI) reconstituted at 0.5 µg/µL with Plasma Dilution Buffer 1.

9. 5 N solution of Hydrochloric Acid (Fisher Scientific, Pittsburgh, PA) in 18Ω water.

10. Bench-top centrifuge model 5417R (Eppendorf, Hauppauge, NY).

2.4 Nano-Flow Liquid Chromatography (nanoLC) Electrospray Ionization Tandem Mass Spectrometry of Digested Analytical Standards

1. Bench-top centrifuge model 5417R (Eppendorf, Hauppauge, NY).

2. In-house fritted trap columns: 200 µL of KASIL 1 potassium silicate (PQ Corporation, Malvern, PA) is mixed with 50 µL of formamide (Sigma Aldrich, St. Louis, MO), vortexed briefly, and centrifuged for 1 min at 10 K RPM in a bench-top centrifuge. Several 20 cm × 150 µm poly-amide-coated fused silica capillaries (Polymicro Technologies, Phoenix, AZ) are submerged in the resulting supernatant for 2–3 s and then cured overnight at 80 °C in a laboratory oven.

3. A homemade pressure bomb interfaced with a high-pressure helium gas cylinder as described in refs. [14, 15].

4. Trap Column: 5 cm × 150 µm poly-amide-coated fused silica capillary (Molex, Lisle, IL) fritted on one end with ~0.5 cm of polymerized potassium silicate and packed at 750 PSI with Jupiter Proteo 90Å C12 4 µ reversed-phase beads (Phenomenex, Ventura, CA).

5. Analytical Column: 20 cm × 75 µm poly-amide-coated fused silica capillary pulled to 10 µm emitter tip with a Sutter P-2000 laser puller (Sutter Instruments, Novato, CA) and packed at 750 PSI with ReproSil-Pur 120Å C18-AQ 3 µ reversed-phase beads (Dr. Maisch GmbH, Germany).

6. Polypropylene auto-sampler vials with snap-on lids (National Scientific, Rockwood, TN).

7. Easy-nLC 1000 Liquid Chromatography System (Thermo Fisher Scientific, San Jose, CA).

8. TSQ Quantiva Triple Quadrupole Mass Spectrometer (Thermo Fisher Scientific, San Jose, CA).

2.5 Quantification of In Vitro Expressed GST Fusion Proteins

1. Concentrated stock (1 µM each in 5 % Acetonitrile/0.1 % formic acid) of FasTrack crude "heavy" $[^{13}C_6{}^{15}N_2]$ (L)-lysine–labeled LLLEYLEEK and IEAIPQIDK peptides (Life Technologies, Grand Island, NY).

2. Concentrated stock (5 µM each in 5 % Acetonitrile/0.1 % formic acid) of AQUA unlabeled LLLEYLEEK and IEAIPQIDK peptides (Life Technologies, Grand Island, NY).

3. 1 nmol BSA Protein Digest Standard (Life Technologies, Grand Island, NY) reconstituted with 5 % Acetonitrile/0.1 % formic acid to 10 pmol/μL.

2.6 Software for Method Editing and Analysis of Quantitative Proteomics Data

1. Skyline: *see* http://skyline.maccosslab.org.
2. Panorama: *see* https://panoramaweb.org.

3   Methods

3.1 Preparation of Analytical Standards as C-Terminal GST Fusion Proteins via In Vitro Protein Expression

1. Each bacterial cDNA clone is grown overnight in 5 mL of Ampicillin-Supplemented LB Culture Medium. Bacterial cultures are performed in a floor shaker set to 200 RPM/37 °C.

2. Plasmid DNA is purified according to the manufacture's mini-prep protocol with the slight modification of an additional "PE buffer" wash step to help facilitate removal of any residual RNAse.

3. The concentration of plasmid stocks is estimated via the A_{260}/A_{280} ratio on a UV/Vis Spectrophotometer.

4. Plasmid stocks are Sanger sequenced using an M13 priming site upstream of the pANT7_cGST vector's T7 promoter. Plasmid sequencing is performed for the purpose of confirming the identity of the cDNA insert and assessing plasmid purity.

5. Purified plasmid DNA is used directly in the in vitro protein synthesis kit according to manufacturer protocol with a few minor modifications. Briefly, about 1 μg of plasmid DNA is used per 25 μL in vitro reaction mix supplemented with 12 Units of RNAse inhibitor. Protein synthesis reactions are carried out for 3.5 h in a floor shaker set to 200 RPM/30 °C.

6. Completed protein synthesis reactions are combined with a 125 μL aliquot of a 3 % slurry of glutathione sepharose 4B beads previously washed and equilibrated with DPBS.

7. Bead/protein mixture is rocked end-over-end for 16–18 h at 4 °C.

8. Bead/protein mixture is centrifuged at $500 \times g/4$ °C for 5 min in a bench-top centrifuge.

9. Supernatant is removed and saved for SDS-PAGE/Western Blot analysis to ensure efficient recombinant protein capture.

10. Sepharose bead pellets are washed twice with Sepharose Bead Wash Solution #1 and twice with Sepharose Bead Wash Solution #2.

11. Washed sepharose bead pellets are reconstituted with 50 μL of Sepharose Bead Reconstitution Buffer containing the heavy isotope-labeled peptides LLLEYLEEK and IEAIPQIDK from *SJ-GST*.

3.2 SDS-PAGE/ Western Blot Analysis of GST Fusion Proteins

1. A 5 μL aliquot of the undigested bead/protein mixture is combined with 1.7 μL aliquot of SDS-PAGE Sample Loading Buffer.

2. Mixtures are incubated for 5 min at 95 °C to facilitate protein denaturation.

3. Denatured protein extracts are resolved for 60 min on a pre-cast Novex NuPage 4–12 % bis–tris mini gel using an XCell SureLock mini-cell gel box interfaced with a PowerPac Basic Electrophoresis Power Supply set to 150 V.

4. Confirmation of protein expression can be performed by subjecting gels to either silver staining or immunoblotting against a polyclonal anti-GST antibody.

5. Silver staining, when applicable, is performed according to manufacturer's instructions.

6. Immunoblotting, when applicable, is performed by transferring SDS-PAGE-resolved proteins onto a PVDF membrane for 1 h at 30 V using the XCell II Blot Module according to manufacturer's instructions.

7. PVDF blots are rinsed briefly with deionized water and incubated in Blot Blocking Buffer while shaking for 60 min at room temperature or overnight at 4 °C.

8. PVDF blots are washed 2× with Blot Washing Buffer while shaking for 5 min each at room temperature.

9. PVDF blots are incubated in Blot Primary Antibody Solution with shaking for 60 min at room temperature.

10. PVDF blots are washed 3× with shaking for 5 min each at room temperature with Blot Washing Buffer.

11. PVDF blots are incubated in Blot Secondary Antibody Solution with shaking for 60 min at room temperature.

12. PVDF blots are washed 3× with shaking for 5 min each at room temperature with Blot Washing Buffer.

13. PVDF blots are visualized using the ECL Prime Western Blotting Kit with the BioMax Light Chemiluminescence Film according to manufacturer's instructions.

3.3 Sample Digestion

1. For each analytical standard, a 25 μL aliquot of enriched, bead-bound GST fusion proteins is diluted back out to 50 μL with Sepharose Bead Dilution Buffer.

2. For plasma samples, a 5 μL aliquot of Plasma is diluted out to 500 μL with Plasma Dilution Buffer 1.

3. Protein concentration of the diluted plasma is estimated via a Bovine Serum Albumin calibrated BCA assay according to manufacturer's protocol.

4. A 25 µL aliquot of diluted plasma is combined with 25 µL of Plasma Dilution Buffer 2.

5. Diluted bead-bound GST fusion proteins and twice diluted plasma samples are incubated for 5 min at 95 °C to facilitate protein denaturation.

6. Denatured proteins are reduced with the addition of 500 mM dithiothreitol to a concentration of 5 mM and incubation at 60 °C for 30 min.

7. Reduced samples are alkylated via the addition of 500 mM iodoacetamide to a final concentration of 15 mM and incubation at room temperature (22–25 °C) for 30 min in the dark.

8. Alkylation reactions are quenched via the addition of an additional aliquot of 500 mM dithiothreitol to bring the final concentration to 15 mM.

9. Each reduced and alkylated bead-bound protein mixture is digested with 1 µg of sequencing grade modified porcine trypsin for 2 h at 37 °C with mixing at $2.4 \times g$.

10. Digestion progress is quenched by the addition of 2.5 µL of 5 M HCl.

11. Acidified digests are incubated for 1 h at room temperature to facilitate hydrolysis of the PPS surfactant.

12. Digested standards are centrifuged at $13,000 \times g/4$ °C for 5 min in a bench-top centrifuge to pellet the sepharose bead fraction.

13. Supernatants are transferred to polypropylene auto-sampler vials with snap-on lids and stored a 4 °C while queued for injection.

3.4 Nano-Flow Liquid Chromatography Electrospray Ionization Tandem Mass Spectrometry of Digested Analytical Standards

1. A 3 µL aliquot of each digest is loaded from a 20 µL sample loop onto an in-house prepared trap column at a flow rate of 2 µL/min for 3 min.

2. Peptides are resolved on an Analytical Column using a 30 min linear gradient from 5 % acetonitrile in 0.1 % formic acid to 40 % acetonitrile in 0.1 % formic acid at a flow rate of 300 nL/min. The initial gradient was followed by a steeper 5 min gradient from 40 % acetonitrile in 0.1 % formic acid to 60 % acetonitrile in 0.1 % formic acid also at 300 nL/min. The column was then washed for 5 min with 95 % acetonitrile in 0.1 % formic acid at 500 nL/min. Prior to the next injection, the trapping column is re-equilibrated with 5 µL of 5 % acetonitrile in 0.1 % formic acid and the analytical column is re-equilibrated with 3 µL of

5 % acetonitrile in 0.1 % formic acid. All re-equilibration steps are performed at 250 bar.

3. Eluting peptides are ionized and emitted into a triple quadrupole mass spectrometer for tandem mass spectrometry analysis.

3.5 Empirical Selection of Optimal Peptides for Targeted Proteomic Workflows

1. For each GST fusion protein-of-interest, tryptic peptides and their respective fragment ions are chosen using Skyline, an open source document editor for building targeted proteomic methods and analyzing the ensuing mass spectrometry data.

2. Prior to beginning any experiments, all peptide and transition settings are configured in Skyline to match experimental design.
   Settings>Peptide Settings
   Settings>Transition Settings

3. For the current protocol, we monitor monoisotopic masses for all fully tryptic peptides from 7 to 25 amino acids in length in their (+2) charge state with all cysteines monitored as carbamidomethylated residues. For tandem MS analysis, we monitor singly charged y_3 to yn_{-1} fragment ions (*see* **Note 2**).

4. Amino acid sequences for each protein-of-interest are imported as FASTA files and digested in silico.
   File>Import>FASTA

5. In the initial round of MS/MS analysis, all peptides and respective MS/MS transitions that fulfill the criterion detailed above are considered.

6. Transition lists are exported from Skyline as instrument/vendor specific .csv files (*see* **Note 3**).

7. Exported transition lists are used to generate instrument/vendor specific SRM methods.

8. For nanoLC-SRM analysis, each analytical standard is injected separately. Data is acquired using a dwell time of 2 ms with both mass-filtering quadrupoles set to 0.7 FWHM resolution. Fragmentation is performed at 1.5 m Torr using optimized instrument-specific calculated peptide collision energies [16].

9. Results are imported into Skyline.
   File>Import>Results (Add single-injection replicates in files)

10. Chromatographic data for each peptide are manually inspected. Peptides not observed in these initial experiments are annotated and omitted from further consideration. For every other precursor, the peak area for each co-eluting transition is integrated and the relative distribution of y-ion intensities is noted for each peptide. An example of a relative y-ion distribution for an individual peptide from recombinant SERPINF2 (also commonly referred to as Alpha-2-Antiplasmin, Accession # P08697) is shown in Fig. 2c. Integrated peak areas from all

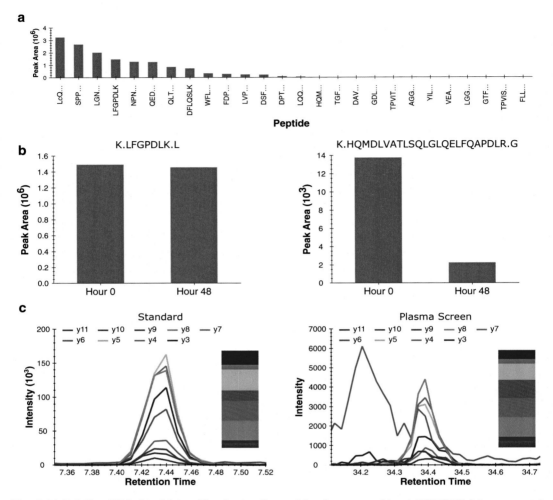

Fig. 2 (**a**) Relative SRM signal intensities for tryptic peptides from recombinant SERPINF2 (also commonly referred to as Alpha-2-antiplasmin, Accession # P08697) as an N-terminal *SJ-GST* fusion protein. (**b**) Comparison of two peptides from SERPINF2 *SJ-GST* fusion protein before and after 48 h auto-sampler incubation. (**c**) Co-eluting fragment ions for a peptide from SERPINF2 *SJ-GST* fusion protein (*left panel*) and co-eluting fragment ions for that same peptide from native SERPINF2 in human plasma (*right panel*). The relative contribution of each co-eluting fragment ion is displayed as a bar graph next to each respective chromatogram

monitored y-ion fragments for a given peptide are summed. The sum of each peptide's MS/MS intensities is ranked against the summed intensities of all other peptides derived from the same parent protein. Figure 2a represents an example of a relative peptide ranking for SERPINF2.

11. In a second round of nanoLC–MS/MS analysis, peptide stability is assessed and retention time calibration is performed. Each analytical standard digest is spiked with iRT calibration standards (*see* **Note 4**) and incubated at 4 °C in the auto-sampler for 48 h prior to re-injection. Integrated peak areas are compared for

each peptide at the 0 h time point (initial nanoLC–MS/MS) and at the 48 h time point (post auto-sampler incubation) to assess peptide degradation/modification.

12. Results are imported back into the original Skyline document and are again manually inspected. Relative retention times are calculated for each remaining peptide (*see* **Note 5**) and SRM signal intensities are compared to those from the initial round of LC–MS/MS analysis. Peptides with sub-optimal stability profiles (*see* **Note 6**) are annotated as such and omitted from further consideration.

13. Edited Skyline files are then uploaded to Panorama for the purpose of creating a chromatogram library (*see* **Note 7**). Chromatogram libraries provide a way to store results from previous curated targeted proteomic experiments by capturing peptide physiochemical properties such as relative product ion distribution, chromatographic peak shape, and relative retention time information.

14. In a third round of MS/MS analysis, a pooled plasma digest is spiked with iRT calibration standards and screened for each peptide that survived the initial two stages of selection. The combined specificity of peptide retention time and MS/MS fragmentation pattern provides a relatively straightforward way of confirming peptide detectability and selecting SRM transitions with the optimal signal-to-noise directly in the human plasma assay matrix.

3.6 Quantification of In Vitro Expressed GST Fusion Proteins

1. Absolute quantification of *SJ-GST* fusion proteins is facilitated by spiking the $[^{13}C_6{}^{15}N_2]$-lysine–labeled LLLEYLEEK and IEAIPQIDK peptides from *SJ-GST* into each into each in vitro protein synthesis reaction.

2. The unlabeled to labeled integrated peak area ratio is measured via nanoLC-SRM.

3. Observed peak area ratios are converted to absolute concentrations using an external calibration curve comprised of varying amounts of unlabeled LLLEYLEEK and IEAIPQIDK peptides spiked into a constant amount of the corresponding heavy isotope-labeled peptides (*see* **Note 8**).

4 Notes

1. The in vitro translation kit utilized in this protocol is also compatible with full-length cDNA clones in the pT7CFE-CHis expression vector.

2. Peptides longer that 12–15 amino acid residues tend to have singly charged fragments with *m/z*'s that exceed the functional

mass range of a triple quadrupole mass spectrometer. Thus, we find it useful to set the mass range in skyline to match instrument capabilities.

Settings>Transition Settings>Instrument Tab

3. Methods are designed to include both the light and heavy isotope-labeled *SJ-GST* peptides such that no more than 500 transitions are monitored in a single run. In the event that multiple injections are required for full protein coverage, the SJ-GST peptides are used to normalize signals across injections.

4. Multiple vendors now offer sets of peptides that can be used as retention time standards. It is also possible to use individual protein digests as a source of iRT peptides provided that there are at least ten stable reference peptides that span most of one's retention time range. For the current protocol, all peptides have been calibrated relative to the Biognosys iRT peptide standards [17].

5. Link to Skyline tutorial for iRT retention time calibration and prediction—https://skyline.gs.washington.edu/labkey/wiki/home/software/Skyline/page.view?name=tutorial_irt.

6. For the current protocol, we eliminate peptides that dropped in intensity more than 15 % from the first injection to the second injection following incubation in the 4 °C auto-sampler.

7. Link to Panorama tutorial for creating chromatogram libraries—https://panoramaweb.org/labkey/wiki/home/page.view?name=chromatogram_libraries

8. Calibration points used are dictated by the sensitivity of the mass spectrometer. For the current protocol, calibration points are measured at 0.1, 0.25, 0.5, 1, 2.5, 5, 12.5, and 40 nM of the unlabeled *SJ-GST* peptides spiked into a solution containing 5 nM of the heavy isotope-labeled *SJ-GST* peptides and 5 μM Bovine serum albumin digest in 5 % acetonitrile and 0.1 % formic acid.

Acknowledgments

This work was supported in part by National Institutes of Health grants P41 GM103533, R01 GM107142, and R01 GM107806.

References

1. Barnidge DR, Goodmanson MK, Klee GG et al (2004) Absolute quantification of the model biomarker prostate-specific antigen in serum by LC-MS/MS using protein cleavage and isotope dilution mass spectrometry. J Proteome Res 3:644–652

2. Gerber SA, Rush J, Stemman O et al (2003) Absolute quantification of proteins and phosphoproteins from cell lysates by tandem MS. Proc Natl Acad Sci U S A 100:6940–6945

3. Anderson LN, Hunter CL (2006) Quantitative mass spectrometric multiple reaction monitor-

ing assays for major plasma proteins. Mol Cell Proteomics 5(4):573–588

4. Rodriguez H, Tezak Z, Mesri M et al (2010) Analytical validation of protein-based multiplex assays: a workshop report by the NCI-FDA interagency oncology task force on molecular diagnostics. Clin Chem 56:237–243

5. Addona TA, Abbatiello SE, Schilling B et al (2009) Multi-site assessment of the precision and reproducibility of multiple reaction monitoring-based measurements of proteins in plasma. Nat Biotechnol 27:633–641

6. Anderson NL, Anderson NG, Haines LR et al (2004) Mass spectrometric quantitation of peptides and proteins using stable isotope standards and capture by anti-peptide antibodies (SISCAPA). J Proteome Res 3:235–244

7. Beynon RJ, Doherty MK, Pratt JM et al (2005) Multiplexed absolute quantification in proteomics using artificial QCAT proteins of concatenated signature peptides. Nat Methods 2:587–589

8. Picotti P, Lam H, Campbell D et al (2008) A database of mass spectrometric assays for the yeast proteome. Nat Methods 5:913–914

9. Prakash A, Tomazela DM, Frewen B et al (2009) Expediting the development of targeted SRM assays: using data from shotgun proteomics to automate method development. J Proteome Res 8:2733–2739

10. Mallick P, Schirle M, Chen SS et al (2007) Computational prediction of proteotypic peptides for quantitative proteomics. Nat Biotechnol 25:125–131

11. Fusaro VA, Mani DR, Mesirov JP et al (2009) Prediction of high-responding peptides for targeted protein assays by mass spectrometry. Nat Biotechnol 27:190–198

12. Picotti P, Rinner O, Stallmach R et al (2009) High-throughput generation of selected reaction-monitoring assays for proteins and proteomes. Nat Methods 7:43–46

13. Stergachis AB, MacLean B, Lee K et al (2011) Rapid empirical discovery of optimal peptides for targeted proteomics. Nat Methods 8:1041–1046

14. Von Haller P (2013) Packing capillary columns and pre-columns (traps), University of Washington Proteomics Resource document http://proteomicsresource.washington.edu/docs/protocols05/Packing_Capillary_Columns.pdf

15. Yates JR III, McCormack AL, Link AL et al (1996) Future prospects for the analysis of complex biological systems using, microcolumn liquid chromatography-electrospray tandem mass spectrometry. Analyst 121:65R–76R

16. MacLean B, Tomazela DM, Abbatiello SE et al (2010) Effect of collision energy optimization on the measurement of peptides by selected reaction monitoring (SRM) mass spectrometry. Anal Chem 82:10116–10124

17. Escher C, Reiter L, MacLean B et al (2012) Using iRT, a normalized retention time for more targeted measurement of peptides. Proteomics 12(8):1111–1121

Chapter 13

Using the CPTAC Assay Portal to Identify and Implement Highly Characterized Targeted Proteomics Assays

Jeffrey R. Whiteaker, Goran N. Halusa, Andrew N. Hoofnagle, Vagisha Sharma, Brendan MacLean, Ping Yan, John A. Wrobel, Jacob Kennedy, D.R. Mani, Lisa J. Zimmerman, Matthew R. Meyer, Mehdi Mesri, Emily Boja, Steven A. Carr, Daniel W. Chan, Xian Chen, Jing Chen, Sherri R. Davies, Matthew J.C. Ellis, David Fenyö, Tara Hiltke, Karen A. Ketchum, Chris Kinsinger, Eric Kuhn, Daniel C. Liebler, Tao Liu, Michael Loss, Michael J. MacCoss, Wei-Jun Qian, Robert Rivers, Karin D. Rodland, Kelly V. Ruggles, Mitchell G. Scott, Richard D. Smith, Stefani Thomas, R. Reid Townsend, Gordon Whiteley, Chaochao Wu, Hui Zhang, Zhen Zhang, Henry Rodriguez, and Amanda G. Paulovich

Abstract

The Clinical Proteomic Tumor Analysis Consortium (CPTAC) of the National Cancer Institute (NCI) has launched an Assay Portal (http://assays.cancer.gov) to serve as an open-source repository of well-characterized targeted proteomic assays. The portal is designed to curate and disseminate highly characterized, targeted mass spectrometry (MS)-based assays by providing detailed assay performance characterization data, standard operating procedures, and access to reagents. Assay content is accessed via the portal through queries to find assays targeting proteins associated with specific cellular pathways, protein complexes, or specific chromosomal regions. The position of the peptide analytes for which there are available assays are mapped relative to other features of interest in the protein, such as sequence domains, isoforms, single nucleotide polymorphisms, and posttranslational modifications. The overarching goals are to enable robust quantification of all human proteins and to standardize the quantification of targeted MS-based assays to ultimately enable harmonization of results over time and across laboratories.

Key words Multiple reaction monitoring, Selected reaction monitoring, MRM, SRM, PRM, Quantitative proteomics, Targeted mass spectrometry, Quantitative assay database, Harmonization, Standardization

Salvatore Sechi (ed.), *Quantitative Proteomics by Mass Spectrometry*, Methods in Molecular Biology, vol. 1410, DOI 10.1007/978-1-4939-3524-6_13, © Springer Science+Business Media New York 2016

1   Introduction

The CPTAC Assay Portal, developed in conjunction with the US National Cancer Institute (NCI) (http://assays.cancer.gov/), serves as a public repository of well-characterized, quantitative MS-based, targeted proteomic assays [1]. The goal of the CPTAC Assay Portal is to disseminate assays to the scientific community at-large, including standard operating protocols, reagents, and assay characterization data associated with targeted mass spectrometry-based assays. A primary aim of the portal is to facilitate the widespread adoption of targeted MS-based assays by bringing together clinicians or biologists and analytical chemists, enabling investigators to find assays to proteins relevant to their areas of interest, evaluate the performance of the assays, obtain information and materials pertinent to implementing assays in their own laboratories, and share characterization data from existing and newly-developed assays with the public.

There are several public databases containing lists or libraries of peptide analytes and transitions (e.g., SRMAtlas [2], PASSEL [3], GPMDB/MRM [4], QuAD [5], cancer peptide library [6]). The CPTAC Assay Portal distinguishes itself in that it contains characterization data to provide researchers with performance data for assays in real-world applications and matrices and provides standard operating protocols (SOPs) for download. In the context of the "Tiers" of targeted protein assays that were recently described [7], the experiments described in the portal are intended to provide preliminary validation data for assays to be used in Tier 2 applications.

This chapter details the structure and functions of the CPTAC Assay Portal, giving users instructions and guidelines for getting the most out of the portal. First, the overall structure is presented, followed by methods for utilizing the various features built into the portal. Finally, ongoing and future developments will be discussed.

2   Overview of Portal Pages and Data Structures

The overall structure of the portal is divided into four components (Fig. 1):

1. Database of qualified assays.

2. Repository of characterization data and processing scripts.

3. Links to external information and resources.

4. Web-based interaction tool for exploring, visualization, and features.

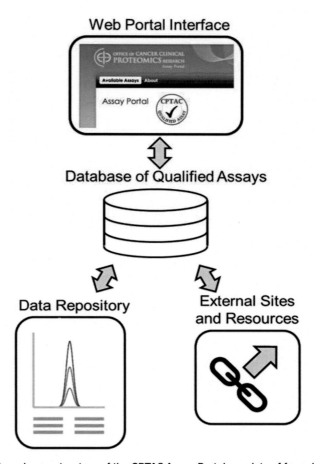

Fig. 1 The primary structure of the CPTAC Assay Portal consists of four elements: a database of qualified assays, a repository of targeted mass spectrometry data, links to external sites and resources, and a user interface for accessing records

2.1 Assay Database

The database of qualified assays contains all information pertaining to characterized assays. Users interact with the database through the portal pages and the links provided therein. Upon addition to the database, each assay is assigned a unique identifying number (e.g., CPTAC-ID#). This number is used to reference specific assays within the portal and outside of the portal. For example, researchers using an assay from the portal in a publication are asked to reference the CPTAC-ID# in the Methods section of their manuscript. Information stored in the database is collected from three primary sources, (1) a web-based metadata collection form, (2) a repository of characterization data (described below), and (3) links with external bioinformatics sites (also described below). The web-based metadata collection form is completed by contributing laboratories when uploading new assays. The form captures details that are displayed on the portal (e.g., instrument type, matrix type, method parameters, publications, etc.). In addition to capturing these experimental details, the metadata form allows users to upload detailed standard operating procedures (SOPs) into the database.

2.2 Repository of Characterization Data

Targeted mass spectrometry data are analyzed and manipulated via Skyline [8], an open-source tool for targeted proteomics experiments. Characterization data are stored in a vendor-neutral public repository called Panorama [9]. Panorama is an open-source repository server application for targeted proteomics that houses the assay data in the form of Skyline documents. Data in Panorama are organized using a directory structure. Each submitting laboratory has a folder. Within a laboratory, separate folders are made for different assay types, assays characterized in different matrices, or assays developed on different instruments. Within each assay type folder, subfolders are divided into assay characterization experiments. The subfolders for the assay characterization experiments contain the data files for a given assay. Note: these subfolders can contain data from multiple assays, given that the assays fit the characterized matrix, instrument, and assay type (i.e., the subfolder can contain multiple Skyline document uploads, but a peptide cannot be duplicated). The directory structure is designed as follows:

>Submitting Laboratory (ExampleU_PILab)

>Matrix/Instrument/AssayType   (CellLysate_Instrument_ directMRM)

   >ResponseCurves

   >ValidationSamples

   >Selectivity

   >Stability

   >EndogenousAnalyte

   >ChromatogramLibrary

Customized data processing scripts implemented as Panorama plug-ins analyze the characterization data and produce the graphics and data tables that are displayed on the portal. The portal assay database interacts with Panorama to gather the images and information needed for display on the portal.

2.3 External Links

The database also uses links to external sites to obtain information related to assays. Upon uploading assays, users are required to specify the target protein. This information is used by the database to collect information from several external bioinformatics websites. The portal uses bioDBnet [10] to collect biological information, as well as Uniprot, PhosphositePlus [11], KEGG, BioGRID, and GeneCards. The protein is also mapped to known pathways using KEGG and known protein–protein interactions through BioGRID. Finally, the chromosomal location is collected from GeneCards.

2.4 Web-Based User Interface

Information contained in the assay database is accessed through the portal user interface. The interface contains a main page which allows users to query and filter the available assays to identify interesting or desired targets. From there, users navigate to individual assay pages which describe in more detail the parameters associated with each assay, the validation data showing the performance of the assay, and downloadable content including raw data and SOPs. The following sections describe the features of the interface in more detail along with the methodology for using the interface.

3 Assay Portal Features

There are four main views associated with the assay portal: the database access page, the protein information panel, the assay details and parameters panel, and the assay resources and comments section.

3.1 The Database Access Page

The database access page is the main page for browsing assays. It is designed to be friendly to a wide array of researchers, allowing users to search the database for relevant assays based on biological interests. Query features are built into the portal to allow biologists to query the database for available assays according to a set of criteria (e.g., pathway, protein complex, chromosomal location). The table view displays currently available entries in the portal database, and is updated according to the filters applied through the queries. Figure 2 is a modified screenshot of the database access page. Labels in Fig. 2 correspond to the following feature descriptions.

1. The "Search" bar (Fig. 2a), located above the table to the left, can be used to search for a specific gene symbol or peptide sequence, as well as searching for fields contained in the table.

2. To search for assays to proteins within a specific pathway, use the KEGG pathways search box (Fig. 2b). KEGG pathways are grouped by category and listed as a drop-down menu. Selecting a pathway will limit the display table to those proteins/peptides in the assay database mapping to the selection.

3. To search by chromosomal location, select the chromosome number and input the start and/or stop coordinates (Fig. 2c), as the number of base pairs from the pter or qter (terminus of the short arm or long arm, respectively).

4. To search for assays in the database to proteins that interact with a specific protein, enter the gene symbol of the desired protein in the interaction form on the left of the table (Fig. 2d). Protein–protein interactions are collected from BioGRID database.

5. Pull-down menus on the left of the table are also provided for filtering the data by species and assay type (Fig. 2e). Assay type refers to the combination of sample preparation (e.g., enrichment

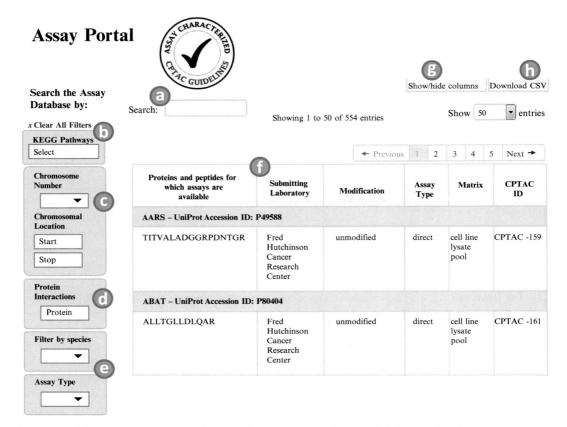

Fig. 2 The database access page is the main landing page for the portal. Labels in the figure correspond to features described in the text

required, fraction required, or direct LC–MS) and data type used in characterizing the assays (e.g., MRM, multiple reaction monitoring; PRM, parallel reaction monitoring). For example, direct-MRM refers to targeted MRM-MS assays with no enrichment prior to analysis.

6. A table of available assays is displayed according to protein target (Fig. 2f). The displayed table will change as filters or searchers are applied. Selecting an assay from the table for browsing in more detail is performed by clicking a peptide sequence in the table.

7. The table view can be reconfigured by selecting "Show/hide columns" in the upper right of the display table (Fig. 2g). Place a check next to fields you would like to display.

8. At any time, a table can be downloaded as a CSV file by selecting the "Download CSV" button (Fig. 2h).

3.2 The Protein Information Panel

Once an assay is selected, the portal displays a page with details about the target gene and the selected peptide assay. The top portion of the page displays protein-level information along with links to external sites. The availability of other peptide assays mapping to the selected protein is indicated in the protein sequence image and map. Figure 3 shows the protein information panel and protein map. Labels in Fig. 3 refer to the following features.

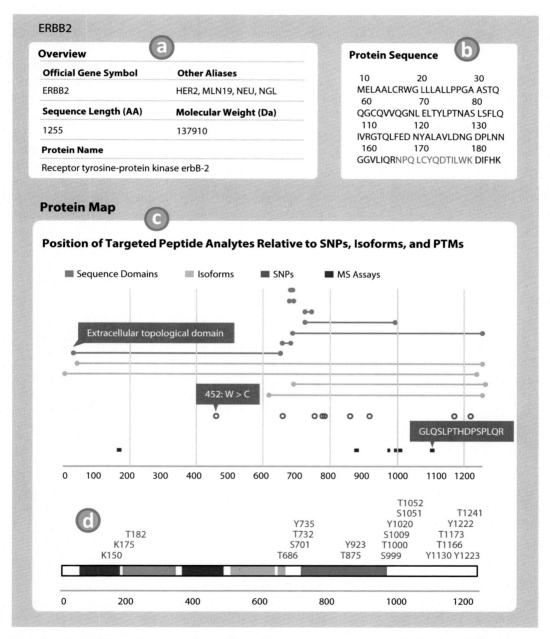

Fig. 3 The protein information panel shows information about the targeted protein. The Protein Map displays the location of available peptide assays relative to other prominent features in the protein of interest. The x-axis indicates amino acid location. *Green* bars indicate sequence domains, *yellow* isoforms, *red dots* single nucleotide polymorphisms, and *blue lines* are available peptide assays. Hovering over features displays more detail (shown in *gray boxes*). Posttranslational modifications are mapped to the protein sequence by the PhosphositePlus database. Labels in the figure correspond to features described in the text

1. The top section of the assay details page (Fig. 3a)shows the gene symbol, aliases, protein length, molecular mass, protein sequence, and protein description. Information is collected from external sites with links to the described protein provided within the panel.

2. The Protein Sequence window (Fig. 3b)is provided to visualize the location of available assays (highlighted in red) in relation to the entire protein sequence. For long proteins, hovering over the window expands the display to reveal the entire sequence. Clicking the highlighted peptide sequences displays the details pertaining to available assays.

3. The Protein Map (Fig. 3c) is a visual representation of the location of available assays in relation to other protein features. The top portion of the map shows sequence domains, isoforms, and SNPs (from Uniprot) in relation to targeted assays (peptide location is mapped as a blue line). Hovering over the nodes (i.e., circles) of features in the map will display further details, whereas clicking on the nodes of the features will link to more information.

4. Additional PTMs, mapped by the PhosphositePlus (phosphosite.org) database (Fig. 3d), are displayed below the protein map. Additional information for individual site modifications can be obtained by clicking the modification label.

3.3 The Assay Details and Parameters Panel

The bottom portion of the assay page allows users to browse details associated with the assay method and characterization data. The targeted assay information is reported under the "Assay Details" section and analytical parameters are reported in the "Assay Parameters" section. The Assay Details and Assay Parameters are depicted in Fig. 4 with labels referring to the following features.

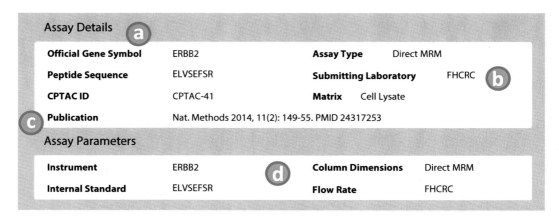

Fig. 4 Select details pertaining to the peptide assay and conditions for characterization are displayed for each assay. Labels in the figure correspond to features described in the text

1. Details pertaining to the peptide sequence are displayed under "Assay Details." (Fig. 4a) The CPTAD ID#, a unique identifier in the assay database, is used for referencing assays in the database. The CPTAC ID# should be used when referring to assays from outside sources (for example, when reporting on assays in the portal in publications). Additional fields in the details panel related to the peptide include modifications in the peptide sequence, the location of any modifications, the peptide molecular mass, and the relative start and stop location of the peptide within its native protein.

2. Conditions under which the assay was characterized by the submitting laboratory are also displayed (Fig. 4b). Assay type briefly describes the sample preparation protocol and data collection technique used in characterizing the assay. The matrix describes the background sample used to characterize the assay.

3. Publications associated with the assay are displayed (Fig. 4c).

4. Details pertaining to the instrumental parameters for assay characterization are displayed in the "Assay Parameters" panel (Fig. 4d). The specific mass spectrometry and liquid chromatography system along with column conditions are displayed. The type of peptide or protein standard (including purity and isotopic label type) used in the characterization experiments is also displayed.

A summary of characterization data from the submitting laboratory is displayed in the bottom portion of the page. This is intended to provide users with performance data related to the selected assays in the reported matrix. Results will help potential downstream users of assays feel more confident that investing time, money, and energy into adopting and deploying the assays will be beneficial. An "Assay Characterization Guidance Document" is posted on the portal (https://assays.cancer.gov/guidance-document/), with assay validation requirements and instructions for conducting the experiments. There are five experiments outlined (Response curve, Mini-validation of repeatability, Selectivity, Stability, and Reproducible detection of endogenous analyte). Chromatograms, response curves, and repeatability experiments are required for all assays uploaded to the Portal. Additional data evaluating the specificity, stability, and reproducibility of endogenous analyte measurement are encouraged; these data (when available) are found in the data repository. Figure 5 shows the layout of the data display.

1. The first panel of characterization data shows example chromatograms for the characterized assays (Fig. 5a). Chromatograms for the light and heavy channels of the assays are chosen by the submitting laboratory as being most representative of the assay. Chromatographic traces are compiled in a Chromatogram Library in Panorama.

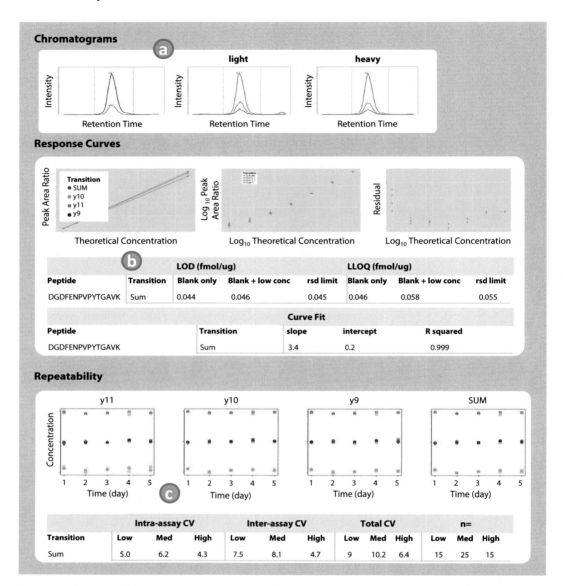

Fig. 5 Overview of the display for assay characterization data for available assays. Example chromatograms contain separate traces for light peptides, heavy peptides, and the combined trace. Response curves are plotted on a linear and log-scale, along with residuals for the curve fit. Curve fit parameters and performance figures of merit are displayed in a table below each plot on the portal. Repeatability of the assay is plotted for measuring three concentrations of analyte (denoted by *green*, *blue*, and *red points*) in triplicate over five separate days with a table reporting the intra-assay (within day), inter-assay (between day), and total CV

2. Response curves images are shown in the section below the chromatograms (Fig. 5b). A processing script in Panorama analyzes the data by performing a robust linear regression on the data points [12]. The display shows three plots, (1) the response curve plotted in linear space, (2) the curve plotted in

log10 space, (3) residuals from the curve fit for each point. The limit of detection (LOD) and lower limit of quantification (LLOQ) are determined three different ways. First, the LOD is determined from blank samples, using the average plus 3× the standard deviation of the blank peak area ratio (blank only). The LLOQ is the average plus 10× the standard deviation. Second, the LOD is calculated using the standard deviation of the peak area ratios observed for the blank samples and for the samples with the lowest concentration in the response curve (blank + low conc.). Again, the LLOQ uses 10× the standard deviation. Third, the LLOQ is determined based on values of the variability (i.e., RSD, relative standard deviation) measured in the curve (rsd limit) [12]. The LOD for the RSD limit method is LLOQ/3. Calculations for assays characterized using unpurified (crude) peptides as standards are performed in the same matter, substituting the estimated concentration of the peptide standards. Unpurified standards are denoted on the portal by asterisk and highlighting the axes labels in red. Note the peptide purity is also reported in the "Assay Details" section. Curve fit parameters are also displayed in the table.

3. The repeatability of assay measurement is displayed below the response curve data (Fig. 5c). Scripts in Panorama analyze the data automatically and produce the results for display on the portal. The repeatability image shows the peak area ratio (analyte:standard) measured in multiple replicates (minimum 3) for three different concentrations over 5 days. First, the intra-assay variability is calculated at each concentration as the CV of the three replicates on each of the 5 days. The CVs determined for each of the 5 days is averaged (this is the average intra-assay CV). Second, the inter-assay variability is calculated at each concentration by determining the CV of the first injection of each concentration across the 5 days, then the second injection, and then the third (if run in triplicate each day, continue this process for the fourth replicate and so on if more replicates are injected each day). These three (or more) CVs are averaged, and reported as the average inter-assay CV. The total CV is calculated as the square root of the sum of the (average intra-assay CV)^2 and the (average inter-assay CV)^2. The number of replicates at each concentration is reported in the column labeled "n = ." It is possible that the repeatability of the assays over multiple days is not as good as the response curves which are prepared on 1 day. Assays with repeatability data showing a total CV greater than 20 % in any concentration sample are highlighted in red.

Curve fit parameters and performance figures of merit are displayed in a table below each plot on the portal.

3.4 The Assay Resources and Comments Section

At the bottom of the Details page, there are links for more information about an assay, as well as a discussion board to allow users to share experiences and comments pertaining to individual assays.

1. Detailed standard operating protocols (SOPs) are associated with each assay. The documents are written by the submitting laboratory and available for download via links at the bottom of the assay details page. SOPs include descriptions of sample preparation, liquid chromatography, mass spectrometry, and the design of characterization experiments (e.g., run order and preparation of process replicates).

2. Links are also provided to the assay characterization data in Panorama. From Panorama, users can visualize any element of the data and download the associated Skyline documents, which can be used to generate transition lists and methods for laboratories desiring to implement the assays.

3. For assays using specialized reagents (like antibodies or peptide standards), links are provided to the source of the reagents (if available, for example at antibodies.cancer.gov).

4. Users can also share information, brief results, or experience with assays in a discussion board. This allows the community to exchange information about how specific assays behave in their laboratories or in previously uncharacterized matrices.

4 Ongoing and Future Development

Ongoing development is primarily focused on three areas: expanding search options, adding capabilities for processing characterization data, and increasing assay content. Incorporating additional search criteria (e.g., Gene Ontology and additional pathway databases) will further leverage biological information for identifying assays of interest. Scripts to collect, process, and display additional characterization experiments will allow further evaluation of the performance of assays. Finally, an interface for allowing any laboratory to contribute appropriately characterized assays to the portal is under development; an "Assay Characterization Guidance Document" is posted on the portal (https://assays.cancer.gov/about/), with assay validation requirements and instructions for designing characterization experiments. We highly encourage user feedback for suggestions in the development of improved design and usability of the portal. Feedback can be provided by clicking the "Contact Us" link from the About page (https://assays.cancer.gov/about/).

5   Availability and Requirements

The website is best viewed with the following browsers: Internet Explorer (Version 9 or higher), Firefox, Chrome, or Safari. To fully experience the site, we recommend using the latest version of any modern browser listed at BrowseHappy.com.

6   Resources for Help

The About page contains a list of resources that may be helpful for laboratories interested in targeted mass spectrometry-based assays. A FAQ page is also available to address common questions regarding the portal. Documents describing assay characterization guidelines are available for download from the About page. Finally, a guided tour of the portal features is available from the main landing page.

Acknowledgements

This work was funded by the Clinical Proteomic Tumor Analysis Consortium (CPTAC) of the US National Cancer Institute (U24CA160034, U24CA160036, U24CA160019, U24CA159988, and U24CA160035), R01 GM103551, and U01 CA164186.

References

1. Whiteaker JR, Halusa GN, Hoofnagle AN et al (2014) CPTAC Assay Portal: a repository of targeted proteomic assays. Nat Methods 11:703–704. doi:10.1038/nmeth.3002

2. Picotti P, Lam H, Campbell D et al (2008) A database of mass spectrometric assays for the yeast proteome. Nat Methods 5:913–914. doi:10.1038/nmeth1108-913

3. Farrah T, Deutsch EW, Kreisberg R et al (2012) PASSEL: the PeptideAtlas SRMexperiment library. Proteomics 12:1170–1175. doi:10.1002/pmic.201100515

4. Craig R, Beavis RC (2004) TANDEM: matching proteins with tandem mass spectra. Bioinformatics 20:1466–1467

5. Remily-Wood ER, Liu RZ, Xiang Y et al (2011) A database of reaction monitoring mass spectrometry assays for elucidating therapeutic response in cancer. Proteomics Clin Appl 5:383–396. doi:10.1002/prca.201000115

6. Yang X, Lazar IM (2009) MRM screening/biomarker discovery with linear ion trap MS: a library of human cancer-specific peptides. BMC Cancer 9:96. doi:10.1186/1471-2407-9-96

7. Carr SA, Abbatiello SE, Ackermann BL et al (2014) Targeted peptide measurements in biology and medicine: best practices for mass spectrometry-based assay development using a fit-for-purpose approach. Mol Cell Proteomics MCP 13:907–917. doi:10.1074/mcp.M113.036095

8. MacLean B, Tomazela DM, Shulman N et al (2010) Skyline: an open source document editor for creating and analyzing targeted proteomics experiments. Bioinformatics 26:966–968. doi:10.1093/bioinformatics/btq054

9. Sharma V, Eckels J, Taylor GK et al (2014) Panorama: a targeted proteomics knowledge base. J Proteome Res 13:4205–4210. doi:10.1021/pr5006636

10. Mudunuri U, Che A, Yi M, Stephens RM (2009) bioDBnet: the biological database network. Bioinformatics 25:555–556. doi:10.1093/bioinformatics/btn654

11. Hornbeck PV, Kornhauser JM, Tkachev S et al (2012) PhosphoSitePlus: a comprehensive resource for investigating the structure and function of experimentally determined post-translational modifications in man and mouse.

Nucleic Acids Res 40:D261–D270. doi:10.1093/nar/gkr1122

12. Mani DR, Abbatiello SE, Carr SA (2012) Statistical characterization of multiple-reaction monitoring mass spectrometry (MRM-MS) assays for quantitative proteomics. BMC Bioinformatics 13(Suppl 16):S9. doi:10.1186/1471-2105-13-S16-S9

Chapter 14

Large-Scale and Deep Quantitative Proteome Profiling Using Isobaric Labeling Coupled with Two-Dimensional LC–MS/MS

Marina A. Gritsenko, Zhe Xu, Tao Liu, and Richard D. Smith

Abstract

Comprehensive, quantitative information on abundances of proteins and their posttranslational modifications (PTMs) can potentially provide novel biological insights into diseases pathogenesis and therapeutic intervention. Herein, we introduce a quantitative strategy utilizing isobaric stable isotope-labeling techniques combined with two-dimensional liquid chromatography–tandem mass spectrometry (2D-LC–MS/MS) for large-scale, deep quantitative proteome profiling of biological samples or clinical specimens such as tumor tissues. The workflow includes isobaric labeling of tryptic peptides for multiplexed and accurate quantitative analysis, basic reversed-phase LC fractionation and concatenation for reduced sample complexity, and nano-LC coupled to high resolution and high mass accuracy MS analysis for high confidence identification and quantification of proteins. This proteomic analysis strategy has been successfully applied for in-depth quantitative proteomic analysis of tumor samples and can also be used for integrated proteome and PTM characterization, as well as comprehensive quantitative proteomic analysis across samples from large clinical cohorts.

Key words Quantitative proteomics, Isobaric labeling, iTRAQ, Two-dimensional liquid chromatography, Mass spectrometry

1 Introduction

Large-scale and deep characterization of the proteome and protein posttranslational modifications (PTMs) from clinical specimens holds great promise for better understanding of diseases pathogenesis and providing novel insights into therapeutic interventions [1, 2]. With modern mass spectrometry (MS) instrumentation, it is now feasible to routinely identify thousands of proteins from a biological sample [3]. The ability to utilize such proteomic datasets for improving the diagnosis, treatment, and prevention of diseases such as cancer, however, typically requires that in-depth proteomic analysis be carried out quantitatively across large clinical cohorts [4]. Therefore, a robust, efficient and high-throughput quantitative

Salvatore Sechi (ed.), *Quantitative Proteomics by Mass Spectrometry*, Methods in Molecular Biology, vol. 1410,
DOI 10.1007/978-1-4939-3524-6_14, © Springer Science+Business Media New York 2016

proteomics strategy is needed to address the challenges in typical clinical applications, such as high sample throughput, consistent quantitation for the entire sample cohort, extensive proteome coverage, and potential extension to characterization of PTMs.

In this chapter, we introduce a quantitative proteomics pipeline using multiplexed, isobaric labeling, and two dimensional reversed phase liquid chromatography coupled to tandem mass spectrometry (2D-LC–MS/MS), which is amenable to large-scale clinical proteomic applications. The workflow includes isobaric labeling of tryptically digested peptides using 4-plexed isobaric tags for relative and absolute quantitation (iTRAQ) [5], basic reverse-phase liquid chromatography (bRPLC) separation with fraction concatenation [6], and nanoelectrospray ionization LC–MS/MS analysis of the fractions. Peptides and proteins are identified through protein sequence database search, and quantified using the iTRAQ reporter ion intensities.

Compared to conventional label-free proteomic approaches, the 4-plex iTRAQ labeling method provides not only higher sample throughput, but also robust protein quantitation across large sample cohort when a "universal reference" strategy [7] is used, i.e., a common reference sample (typically a sample pooled from all samples involved in the comparison) is included in each 4-plex iTRAQ analysis to serve as a "bridge" for cohort-wide comparison. The bRPLC separation with fraction concatenation provides further enhanced proteome coverage and more streamlined sample processing (e.g., no need for sample clean up) compared to conventional strong cation exchange chromatography based fractionation [6]. This workflow is also amenable to integrated proteome and PTM (e.g., phosphorylation) characterization [8, 9], and use with other labeling and multiplexing approaches. For the scope of this chapter, we only describe details of the in-depth quantitative global proteome analysis of tissue sample as a demonstration of this quantitative proteome profiling strategy.

2   Materials

The materials required for this experimental workflow are listed by activity. Prepare all reagents in appropriate containers, preserving sterility when necessary. Ensure that all solvents are LC–MS grade, all chemicals are of high purity, and ultrapure water (prepared by purifying deionized water to attain a sensitivity of 18 MΩ cm at 25 °C) is used for preparation of solutions.

2.1   Materials and Equipment Needed During Each Step

1. BCA Protein Assay Kit (Thermo Fisher Scientific, Waltham, MA).

2. Infinite M200 plate reader (Tecan, Morrisville, NC).

3. Vortex Gene 2 (Scientific Industries, Bohemia, NY).

4. 5417R Refrigerated Microcentrifuge (Eppendorf, Westbury, NY).

5. Speed-Vac SC250 Express (Thermo Fisher Scientific, Waltham, MA).

6. Thermomixer R thermal mixer (Eppendorf, Westbury, NY).

2.2 Equipment Used for Partial Automation of Sample Processing (See Note 1)

1. epMotion 5075 Liquid Handler (Eppendorf, Westbury, NY).

2. GX-274 Aspec Automated SPE System with 406 Dual Syringe Pumps (Gilson, Middleton, WI).

2.3 Tissue Preparation, Protein Extraction, and Digestion

1. Kontes™ Pellet Pestle™ Cordless Motor and disposable pestles (Kimble Chase, Rochester, NY).

2. Sample Lysis/Denaturing buffer: 8 M urea, 50 mM NH_4HCO_3, pH ~8.0, cOmplete Protease Inhibitor Cocktail, EDTA-free (Roche, Indianapolis, IN).

3. Sequencing Grade Modified Porcine Trypsin (Promega, Madison, WI).

4. Branson Sonicator 1510 (Branson, Danbury, CT).

2.4 C18 SPE Clean-Up

1. 1 mL/100 mg SPE Discovery-C18 columns (SUPELCO, Bellefonte, PA).

2. Vacuum manifold with vacuum for SPE tubes (VisiPrep SUPELCO, Bellefonte, PA).

3. Washing buffer: 5 % acetonitrile (ACN) with 0.1 % trifluoroacetic acid (TFA).

4. Conditioning buffer: 0.1 % TFA.

5. Elution buffer: 80 % ACN.

2.5 iTRAQ Labeling

1. iTRAQ Reagent Multiplex Kit (Contains iTRAQ® reagents 114, 115, 116, 117, the appropriate buffers, and reagents for five 4-plex assays. Each individual reagent capable of labeling up to 100 µg of protein.) (AB SCIEX, Framingham, MA).

2. Dissolution Buffer: 500 mM triethylammonium Bicarbonate (TEAB); for dehydrating peptides samples.

2.6 bRPLC Fractionation and Concatenation

1. Agilent 1200 HPLC System equipped with a quaternary pump, degasser, diode array detector, peltier-cooled autosampler, and fraction collector (set at 4 °C) (Agilent, Santa Clara, CA).

2. XBridge C18 HPLC column, 250 mm×4.6 mm column containing 5-µm particles, and a 4.6 mm×20 mm guard column (Waters, Milford, MA).

3. Solvent A: 10 mM TEAB, pH 7.5.

4. Solvent B: 90 % ACN with 10 mM TEAB, pH 7.5.

5. 31-mm deep 96-well plates for collecting fractions.

6. Rehydrating solution: 50 % methanol (MeOH) with 0.05 % TFA.

2.7 LC–MS/MS
Analysis

1. nanoACQUITY UPLC system (Waters Corporation, Milford, MA).

2. LTQ Orbitrap Velos mass spectrometer (Thermo Fisher Scientific, Waltham, MA).

3. 3-μm Jupiter C18 bonded particles (Phenomenex, Torrence, CA).

4. 35 cm × 360 μm o.d. × 75 μm i.d. fused silica (Polymicro Technologies, Phoenix, AZ).

5. Mobile phase A: water with 0.1 % formic acid.

6. Mobile phase B: ACN with 0.1 % formic acid.

3 Methods

An overview of the quantitative proteomics strategy is illustrated in Fig. 1. Briefly, proteins from the tissue samples are first extracted, tryptically digested, and cleaned up. The resulted peptide samples are then labeled by the four different iTRAQ reagents separately, after which they are combined into one sample and separated using bRPLC with further fraction concatenation. The resulting fractionations (e.g., 24 fractions) are analyzed by capillary LC–MS/MS using the high resolution and high mass accuracy hybrid LTQ Orbitrap Velos mass spectrometer. Peptides and proteins are identified through protein sequence database search, and quantified using the iTRAQ reporter ion intensities. Typical performance of this quantitative proteomics strategy is simultaneous quantification of approximately 7000 proteins using the Orbitrap Velos instrument, or more than 10,000 proteins using the Q Exactive mass spectrometer [8].

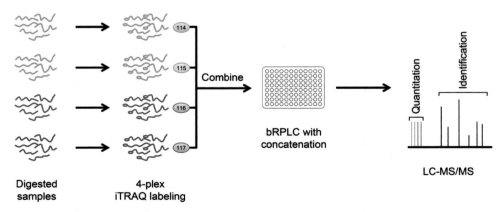

Fig. 1 Overview of the quantitative proteomics workflow. The samples were first converted into tryptic peptides and then each labeled by different iTRAQ reagents, combined, and fractionated (with concatenation) for quantitative proteome analysis using tandem mass spectrometry on the high resolution and high accuracy Orbitrap mass analyzer

3.1 Preparation of Tissue Sample for Protein Digestion

1. Place frozen tissue sample in 1.5-mL microcentrifuge tubes (*see* **Note 2**); spatula or tweezers may need to be used for tissue sample transfer (*see* **Note 3**).

2. Add appropriate amount of lysis buffer to the collection tube containing tissue sample (*see* **Note 4**).

3. Homogenize tissue sample in lysis solution by using Kontes™ Pellet Pestle™ Cordless Motor with disposable pestles (*see* **Note 5**). Keep microcentrifuge tube containing sample in lysis buffer on the ice or chill rack while performing homogenization.

4. In general, 15–30 s of the processing time is enough to satisfactorily homogenize ~150 mg of soft tissue sample (e.g., brain or liver) (*see* **Note 6**).

5. Keep collection tube containing homogenized tissue sample on ice or chill rack at all time (*see* **Note 7**).

3.2 Protein Extraction and Tryptic Digestion

1. Shake sample at 1200 rpm for 3 min at room temperature in Thermomixer and sonicate sample in sonication bath with ice for 3 min to homogenize sample further.

2. Centrifuge sample at $16,000 \times g$ for 10 min to clarify digest from residual tissue matter. Transfer supernatant into new microcentrifuge tube (or into 96-well plate) for the digestion.

3. Set aside small sample aliquots and run BCA protein assay to determine initial protein concentration of the sample before digestion (*see* **Note 8**).

4. In a mean time, to reduce proteins, add appropriate amount of 500 mM dithiothreitol (DTT) to reach 5 mM in sample and incubate sample at 37 °C for 1 h with 1200 rpm constant shaking in Thermomixer (*see* **Notes 8** and **9**).

5. To alkylate reduced proteins, add sufficient 500 mM iodoacetamide to reach 10 mM final concentration and incubate sample at room temperature in the dark with 1200 rpm constant shaking in Thermomixer for 1 h (*see* **Note 8**).

6. After sample is reduced and alkylated, dilute it eightfold with 50 mM NH_4HCO_3, pH ~8.0 containing 1 mM $CaCl_2$ (*see* **Notes 8** and **10**).

7. Add sequencing grade-modified trypsin to diluted sample at 1:50 (w/w) trypsin-to-protein ratio, vortex sample to mix, and incubate to digest for 3 h at 37 °C with 700 rpm constant shaking in Thermomixer (*see* **Note 8**).

8. After the 3-h incubation, stop the digestion by acidifying sample to 0.1 % TFA (~pH 3.0) with 10 % TFA. Centrifuge sample at $16,000 \times g$ for 20 min to clarify the digest. Transfer supernatant into a fresh vial without disturbing a debris pellet (*see* **Note 7**).

9. Proceed to manual (using vacuum manifold) or automated (using GILSON GX-274 Aspec Automated SPE System) SPE C18 to obtain purified peptides for iTRAQ isobaric tag labeling.

10. First, rinse and condition 1-mL SPE C18 column by slowly passing 3 mL of HPLC-grade methanol through the column follow by 2 mL of conditioning buffer.

11. Load digested sample onto preconditioned SPE C18 column and slowly pass it through.

12. Wash SPE C18 column with 4 mL of washing buffer.

13. Finally, elute purified peptides from SPE C18 column with 1 mL of elution buffer into clean low-retention tube.

14. Concentrate sample down to approximately 100 µL in Speed-Vac and perform BCA protein assay (*see* **Notes 7** and **8**).

3.3 iTRAQ Labeling

1. Postdigestion iTRAQ labeling is performed according to manufacturer's instructions (AB Sciex, Framingham, MA) [8].

2. To prepare samples for iTRAQ labeling, equal amount of four different desalted peptide samples was lyophilized in Speed-Vac.

3. Reconstitute the lyophilized samples in Dissolution Buffer (100 µg of peptides in 30.0 µL of Dissolution Buffer) by vigorously vortexing the samples (*see* **Note 11**).

4. Mix an appropriate amount of room temperature iTRAQ reagent (1 unit dissolved in 70 µL of ethanol) with the peptide sample (add 1 unit of iTRAQ reagent to 100 µg of peptide sample) according to the experiment labeling scheme (e.g., sample 1 to be labeled by reagent 114, sample 2 to be labeled by reagent 117; labeling scheme does not change the final quantitation results). For quantitative analysis of a large sample cohort using the "universal reference" strategy, the pooled reference sample is included in each iTRAQ experiment and labeled using the same iTRAQ reagent (e.g., 117) (*see* **Note 12**).

5. After 1 h incubation at room temperature, the labeling reaction is stopped by adding water (3× the volume) and 30 min incubation at room temperature.

6. Combine the content of each of the four iTRAQ-labeled samples (i.e., 114, 115, 116, and 117) into a fresh microcentrifuge tube and concentrate it in Speed-Vac to remove ethanol and reduce sample volume (*see* **Note 7**).

7. Perform SPE C18 to clean up the concentrated labeled peptide sample follow the same procedure and using the same type of the SPE C18 columns as described in Subheading 3.2 above (*see* **Note 7**).

3.4 Peptide Fractionation by bRPLC

1. iTRAQ-labeled sample is separated on a Waters reversed-phase XBridge C18 column (250 mm × 4.6 mm column containing 5-μm particles, and a 4.6 mm × 20 mm guard column) using Agilent 1200 HPLC System.

2. Reconstitute the sample in 900 μL of Solvent A and inject onto the column at a flow rate of 0.5 mL/min.

3. After sample loading, the C18 column is washed for 35 min with solvent A, before applying a 90-min LC gradient with solvent B. The LC gradient starts with a linear increase of solvent B to 10 % in 10 min, then a linear increase to 20 % B in 15 min, and 30 min to 30 % B, 15 min to 35 % B, 10 min to 45 % B, and another 10 min to 100 % solvent B. The flow rate is 0.5 mL/min.

4. A total of 96 fractions are collected into a 96-well plate throughout the LC gradient in equal time intervals (*see* **Note 7**).

5. These 96 fractions are concatenated into 24 fractions by combining 4 fractions that are 24 fractions apart (i.e., combining fractions #1, #25, #49, and #73; #2, #26, #50, and #74; and so on) (*see* **Note 8** and **13**).

6. Concentrate the resulting 24 fractions in Speed-Vac and perform BCA protein assay for each fraction to determine peptide concentration. Each fraction is analyzed using LC–MS/MS (*see* **Notes 7** and **8**).

3.5 LC–MS/MS Analysis of iTRAQ-Labeled Fractions

1. Peptide samples are analyzed using a Waters nanoACQUITY UPLC system coupled online to a LTQ Orbitrap Velos mass spectrometer outfitted with a custom electrospray ionization interface.

2. The capillary column is prepared in-house by slurry packing 3-μm Jupiter C18 bonded particles into 35-cm × 360 μm o.d. × 75 μm i.d. fused silica using a 1-cm sol-gel frit for media retention.

3. Electrospray emitters are custom built using 150 μm o.d. × 20 μm i.d. chemically etched fused silica.

4. Mobile phase flow rate is 300 nL/min and consisted of 0.1 % formic acid in water (A) and 0.1 % formic acid in ACN (B) with a gradient profile as follows (min:%B): 0:5, 2:7, 120:25, 125:68, 129:80, 130:5.

5. The LTQ Orbitrap Velos mass spectrometer is operated under the following conditions: the ion transfer tube temperature and spray voltage are 300 °C and 1.8 kV, respectively; Orbitrap spectra (automatic gain control (AGC): 3×10^6) are collected from 300 to 1800 m/z at a resolution of 30,000 followed by data-dependent higher energy collisional dissociation (HCD) MS/MS (centroid mode, at a resolution of 7500, collision energy 45 %, activation time 0.1 ms, AGC 5×10^4) of the 10

most abundant ions using an isolation width of 2.5 Da; charge state screening is enabled to reject unassigned and singly charged ions; A dynamic exclusion time of 30 s is used to discriminate against previously selected ions (within –0.55 to 2.55 Da).

3.6 Data Analysis

1. Peptides and proteins are first identified using MS-GF+ [10] which considers static mods of carbamidomethylation (+57.0215 Da) on Cys residues, 4-plex iTRAQ modification (+144.1021 Da) on the peptide N-terminus and Lys residues, and dynamic oxidation (+15.9949 Da) on Met residues when searching against a human protein sequence database (e.g., UniProt; trypsin and keratin contaminant sequences are often included as well). Other settings for MS-GF+ database search include ±10 ppm tolerance for parent ion mass error and, 0.5 m/z tolerance for fragment ion mass error, partially tryptic search, and target decoy database searching strategy [11, 12] for estimation of false discovery rate (FDR).

2. Spectral identification files from **step 1** above are converted to IDPicker3 index files (idpXML) and then used for protein assembly using IDPicker3 [13]. Peptide identification stringency is set at a maximum of 1 % peptide-to-spectrum matches (PSMs) FDR and a minimum of two unique peptides to identify a given protein within the full data set.

3. The intensities of all four iTRAQ reporter ions are extracted using MASIC software [14]. The PSMs which pass the confidence threshold as described above are linked to the extracted reporter ion intensities by scan number. The reporter ion intensities from different PSMs resulting in the same peptide identification (i.e., different scans in the same and different bRPLC fractions) are summed to represent arbitrary abundance measurement for that peptide; the reporter ion intensities are further summed across all the peptides derived from the same protein to represent arbitrary protein abundance measurement.

4. The relative protein abundances are then log2 transformed and accessed for the errors during tryptic peptide concentration measurement and pipetting steps (prior to combining the four samples labeled by different iTRAQ reagents), and sample-to-sample normalization coefficients (shifts in log2 scale) are calculated. For a given sample, the log2 normalization coefficient is derived as average log2 for a subset of proteins that are quantified across all samples.

5. Changes in protein abundance across the four different samples can be accessed by comparing the normalized and log2 transformed reporter ion intensity values.

6. The same data analysis workflow can also be applied for comprehensive quantitative proteomic analysis across a large sample cohort (*see* **Note 14**).

4 Notes

1. Incorporating the automated solutions in proteomics high throughput sample preparation needs to be tailored to the specific workflow and has to be determined by the number of samples, the size and type of these samples, experimental protocols applied, and instruments available in the laboratory [15]. It is problematic to establish a fully automated protocol in extensive and elaborate proteomics workflows due to all the variables in the multiple often disjointed sample preparation steps.

   One way to address this challenge is to identify and automate the most redundant and time consuming steps of the experimental workflow. For example, different liquid handling systems are the most commonly used automation systems in the high-throughput sample preparation labs. In our case, the types of liquid handling systems used are Eppendorf epMotion 5075 and GILSON GX-274 Aspec Automated SPE System. For example, making the sample serial dilutions and loading them onto 96-well plates for the BCA assays, can be performed in automated manner using Eppendorf epMotion throughout the process to ensure accuracy and reproducibility of the assay measurements.

2. 1.5-mL microcentrifuge tubes are most preferable at this step; smaller microcentrifuge tubes are not sufficiently large to handle the sample resulted from this step.

3. It is very important to keep tissue sample and tools (spatulas, tweezers) chilled on dry ice (or in liquid nitrogen) until beginning of the processing. Warming sample up or use warmed tools to handle sample may introduce unnecessary changes in biological properties.

4. During homogenization, do not exceed volume of 500 μL if using 1.5-mL microcentrifuge tubes to avoid lysis buffer splashing and subsequent sample loss.

5. QIAGEN TissueRuptor with disposable probes will work as well for low-throughput in-solution tissue homogenization as Kontes™ Pellet Pestle™ Cordless Motor. QIAGEN TissueLyser is well suited for disruption of the tissue samples for high-throughput, 96-well format and automated workflows.

6. Prior to protein extraction, hard and fibrous tissue (e.g., muscle, skin, and heart) could be cryo-powderized in liquid nitrogen and stored at −80 °C until further processing.

7. At this point of time sample may be snap frozen in liquid nitrogen and store in −80 °C until further processing.

8. This step can be automated using Eppendorf epMotion 5075 Liquid Handler.

9. 500 mM stock solutions of DTT and Iodoacetamide are to be made freshly for the digestion.

10. Before adding activated trypsin to the sample confirm and adjust sample solution to pH 7–9.

11. It is important to make sure sample solution pH to be ~8 after adding the Dissolution Buffer.

12. The pooled reference sample does not have to be always labeled using the same iTRAQ reagent (e.g., 117); however, doing so would eliminate any potential bias resulted from the different iTRAQ reagents (e.g., due to reagent quality issues).

13. Fist, fractions are dried all way down in Speed-Vac. Then, each fraction is reconstituted in 100 µL of 50 % MeOH, 0.05 % TFA for concatenation of 96 fractions into 24 samples. Fractions are concatenated into the same plate or into fresh vials, concentrated down to remove MeOH in the samples. Protein BCA assay may be performed on each concatenated sample if accurate peptide concentration is needed (e.g., for appropriate sample loading in the final LC–MS/MS analysis).

14. For comparing protein abundances across the entire sample cohort, the quantification relies on the common pooled reference sample that is labeled with a particular iTRAQ reagent (e.g., 117). Therefore to convert arbitrary reporter ion intensities to relative abundances that can be compared across the different 4-plex iTRAQ experiments, the 117 reporter ion cannot be a missing value. After discarding spectra with a missing 117 reporter ion, the sample channels (114, 115, and 116) for the remaining spectra are divided by the intensity of the reference channel (117). Relative protein or abundances from the individual iTRAQ experiments are then simply linked together and form a crosstab pivot table. Prior to further data manipulations the relative abundances are log2 transformed.

Acknowledgments

Portions of this work were supported by the grant U24CA160019, from the National Cancer Institute Clinical Proteomic Tumor Analysis Consortium (CPTAC) and National Institutes of Health grant P41GM103493. The experimental work described herein was performed in the Environmental Molecular Sciences Laboratory, a national scientific user facility sponsored by the DOE and located at Pacific Northwest National Laboratory, which is operated by Battelle Memorial Institute for the DOE under Contract DE-AC05-76RL0 1830.

References

1. Baker ES, Liu T, Petyuk VA et al (2012) Mass spectrometry for translational proteomics: progress and clinical implications. Genome Med 4:63

2. Rifai N, Gillette MA, Carr SA (2006) Protein biomarker discovery and validation: the long and uncertain path to clinical utility. Nat Biotechnol 24:971–983

3. Domon B, Aebersold R (2006) Mass spectrometry and protein analysis. Science 312: 212–217

4. Zhang B, Wang J, Wang X et al (2014) Proteogenomic characterization of human colon and rectal cancer. Nature 513 :382–387

5. Ross PL, Huang YN, Marchese JN et al (2004) Multiplexed protein quantitation in Saccharomyces cerevisiae using amine-reactive isobaric tagging reagents. Mol Cell Proteomics 3:1154–1169

6. Wang Y, Yang F, Gritsenko MA et al (2011) Reversed-phase chromatography with multiple fraction concatenation strategy for proteome profiling of human MCF10A cells. Proteomics 11:2019–2026

7. Qian WJ, Liu T, Petyuk VA et al (2009) Large-scale multiplexed quantitative discovery proteomics enabled by the use of an (18)O-labeled "universal" reference sample. J Proteome Res 8:290–299

8. Mertins P, Yang F, Liu T et al (2014) Ischemia in tumors induces early and sustained phosphorylation changes in stress kinase pathways but does not affect global protein levels. Mol Cell Proteomics 13:1690–1704

9. Mertins P, Qiao JW, Patel J et al (2013) Integrated proteomic analysis of post-translational modifications by serial enrichment. Nat Methods 10:634–637

10. Kim S, Gupta N, Pevzner PA (2008) Spectral probabilities and generating functions of tandem mass spectra: a strike against decoy databases. J Proteome Res 7:3354–3363

11. Elias JE, Gygi SP (2007) Target-decoy search strategy for increased confidence in large-scale protein identifications by mass spectrometry. Nat Methods 4:207–214

12. Qian WJ, Liu T, Monroe ME et al (2005) Probability-based evaluation of peptide and protein identifications from tandem mass spectrometry and SEQUEST analysis: the human proteome. J Proteome Res 4:53–62

13. Zhang B, Chambers MC, Tabb DL (2007) Proteomic parsimony through bipartite graph analysis improves accuracy and transparency. J Proteome Res 6:3549–3557

14. Monroe ME, Shaw JL, Daly DS et al (2008) MASIC: a software program for fast quantitation and flexible visualization of chromatographic profiles from detected LC-MS(/MS) features. Comput Biol Chem 32:215–217

15. Dayon L, Nunez Galindo A, Corthesy J et al (2014) Comprehensive and scalable highly automated MS-based proteomic workflow for clinical biomarker discovery in human plasma. J Proteome Res 13:3837–3845

Chapter 15

Multiple and Selective Reaction Monitoring Using Triple Quadrupole Mass Spectrometer: Preclinical Large Cohort Analysis

Qin Fu*, Zhaohui Chen*, Shenyan Zhang, Sarah J. Parker, Zongming Fu, Adrienne Tin, Xiaoqian Liu, and Jennifer E. Van Eyk

Abstract

Multiple reaction monitoring (MRM), sometimes referred to as selective reaction monitoring (SRM), is a mass spectrometry method that can target selective peptides for the detection and quantitation of a protein. Compared to traditional ELISA, MRM assays have a number of advantages including ease in multiplexing several proteins in the same assay and independence from the necessity for high-quality, expensive, and at times unreliable antibodies. Furthermore, MRM assays can be developed to quantify multiple proteoforms of a single protein allowing the quantification of allelic expression of a particular sequence polymorphism, protein isoform, as well as determining site occupancy of posttranslational modification(s). In this chapter, we describe our workflow for target peptide selection, assay optimization, and acquisition multiplexing. Our workflow is presented using the example of constrained MRM assays developed for the serum protein ApoL1 in its various proteoforms to highlight the specific technical considerations necessary for the difficult task of quantifying peptide targets based on highly specific amino acid sequences by MRM.

Key words Quantification, Multiple reaction monitoring, Selective reaction monitoring, Mass spectrometry, APO L1

1 Introduction

Multiple reaction monitoring (MRM), also known as selective reaction monitoring (SRM), employs targeted mass spectrometry (MS)-based data acquisition by monitoring prespecified peptide precursor ions along with a defined handful of their fragments for highly sensitive and specific quantification. Due to its high sensitivity, accuracy, and reproducibility MRM has become increasingly used for targeted peptide and protein quantification [1] and has been successfully used in several high-throughput clinical applications [2–4]. Compared to traditional ELISA, MRM assays have a

---

*Author contributed equally with all other contributors.

Salvatore Sechi (ed.), *Quantitative Proteomics by Mass Spectrometry*, Methods in Molecular Biology, vol. 1410, DOI 10.1007/978-1-4939-3524-6_15, © Springer Science+Business Media New York 2016

number of advantages. For instance, a key aspect of MRM assays is that each protein has multiple, semi-independent, layered observations which collectively indicate protein quantity (e.g., independent quantification of each peptide selected for a protein along with independent quantification of multiple transitions for each selected peptide, if so chosen). This differs from ELISA where there is a single dependent measurement based on the binding of antibodies to two sites physically located within the protein. Additionally, an implication of antibody-based assays is that interferences in, differences in, or modifications on the amino acid sequences and/or protein tertiary structures on these protein analytes will still be recognized by these antibodies. An ELISA may, therefore, miss critical information in patients who could have unknown posttranslational modifications (PTMs) and SNPs and thus MRM has the potential for analytical specificity and physical validity above that of an ELISA. Further, MRMs have the advantage of ease in multiplexing several proteins in the same assay. Finally, one can develop MRM assays to quantify multiple proteoforms of a single protein without developing separate reagents (e.g., antibodies) for each proteoform separately. This enables (1) the quantification of allelic expression of a particular single nucleotide polymorphism (SNP) or panel of known SNPs associated with a given disease (facilitating genotype-to-phenotype correlation inferences); (2) different protein isoforms, which can enable characterization of changes in protein isoform expression between biological states; and (3) determining and quantifying site occupancy of co- or PTMs. With MRM, all relevant proteoforms can theoretically be quantified simultaneously within the same assay multiplex. When peptide selection is constrained to a specific amino acid sequence because of its diagnostic value to discriminate between various proteoforms, we refer to them as "constrained MRM assays" [5].

MRM is a standard quantitative method for small molecules, but it is still being developed for larger molecules, like peptides and proteins, and adapted for large cohort analysis. For protein quantification, multiple peptide(s) composed of an amino acid sequence that is uniquely represented within only one protein in the genome (termed proteotypic peptides) are isolated and quantified within the mass spectrometer [6–8]. Triple quadrupole mass spectrometers are commonly used due to their high specificity, sensitivity, and scan speed although other instruments like Thermo Scientific™ QExactive™ can be used in a similar workflow [9]. Peptide identification is confirmed by monitoring multiple transitions of each target peptide, with each transition representing the mass-to-charge ratios (m/z) of a pair of precursor and breakdown product ions arising from that particular peptide occurring at well-defined retention time windows. Each step in MRM assay development involves considerable investment of time and effort to optimize. A generic workflow for developing MRM assays is provided in Fig. 1.

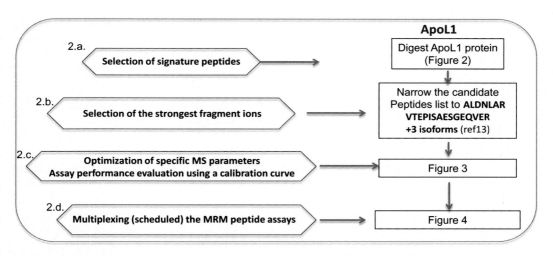

Fig. 1 Schematic workflow of an MRM assay (Apolipoprotein L1 (ApoL1) as a model protein). A generic workflow for developing MRM assays using Apolipoprotein L1 (ApoL1) as a model protein is displayed

First, the design of the assay itself requires careful selection of peptides proteotypic to a protein target of interest, or alternatively their "constrained" form specific to a PTM, allelic SNP, or protein isoform. Second, the detection of these target peptides must be robust across diverse patient populations, and the selected transition pairs used should be minimally impacted by interference from other ions and this must be confirmed experimentally or at least predicted from existing proteomic knowledge databases. Third, peptides ideally should not contain amino acid residues which are variably modified in an unpredictable way either during sample preparation (Cys, Met) or biologically (Asn, Gln) which can alter peptide detection properties by LC-MS. Fourth, in order to achieve optimal quantitative accuracy, stable isotopic labeled peptide standards, most commonly using heavy isotopic (N^{15} or C^{13}), should be synthesized for the same sequence as the assay target and used as "known" quantitative references and can be analyzed against a standard curve to calculate absolute quantity of assay targets. Finally, preparation of samples for analysis by MRM must be efficient and robust against technical variability in order to achieve acceptable percent coefficient of variance (%CV) performance, and this preparation scheme must scalable depending on the throughput of the application. For MRM to be used in extreme high-throughput applications with the goal of quantifying multiple proteins across hundreds to thousands of samples, ideally within a clinically applicable setting, requires the establishment of robust pipelines that meet all of these critical assay design requirements.

One important advantage of MS-based approaches for quantification is the ability to build constrained MRM assays that are able to quantify the biological diversity of proteins. It is often challenging to develop suitable high-quality antibodies for such a targeted ELISA to a particular SNP, isoform, or PTM and it is not possible to obtain both measurement of the modified protein version and total protein concentration in the same ELISA, requiring two analyses at increased cost and the use of twice the sample quantity. Constrained MRM assays overcome these issues. As described above, constrained MRM assays consist of an analytical toolset that allows quantitative differentiation between specific protein isoforms, alleles of proteins encoded by genes with biologically important SNP mutations, or a particular PTM. When multiplexed with a proteo-specific peptide present across all proteoforms, the expression of differential proteoforms can be quantified relative to total protein concentration to estimate the extent of each proteotypic form. Our laboratory has successfully built constrained MRM to differentiate protein isoforms such as TGF beta isoforms 1, 2, and 3 [5], disease-induced posttranslational modifications such as cardiac troponin I phosphorylation [10], and citrullination on glial fibrillary acidic protein (GFAP) [11] which we hypothesize may correlate to unique biological state to predict the progression of a chronic disease.

To generate peptides from protein samples, trypsin is generally the preferred protease for MS due to its cleavage specificity, robustness to work under a wide spread of buffer conditions (ionic strength and denaturants etc.), and because the cleavage produces a positive charge at the C-terminus of the peptide that can produce a diagnostic y ion aiding peptide spectrum interpretation. Due to the abundance of Arg and Lys residues in proteins, trypsin digestion typically will generate peptides in the 800–2000 Da mass range most easily detectable by existing MS instruments. For some applications, in particular with constrained MRM assays, it is necessary to either select a tryptic peptide with a weak response in the MS or use other enzymes, for example Lys-C or Glu-C, to capture the targeted amino acid sequence of interest. Enzyme selection should be used in concert with peptide sequence analysis and, if possible, the peptides should not contain Met or Cys residues, which have the potential for in vitro artificially induced oxidation or Asn or Gln which can undergo deamination and importantly any known natural sequence variations or PTMS (unless targeting these). Even so, biological variability that could be encountered in clinical samples cannot be fully predicted, and thus tracking (correlating) independent measurements from two or more peptides per protein on each patient sample can be helpful to ensure that there are no other (unknown) modifications in the population being studied.

The selection of fragment ions from a chosen peptide sequence monitored by MRM involves at least two nonexclusive approaches. The sensitivity of the MRM assay will be driven by the most sensitive peptide quantified, and thus careful selection of quantification parameters is critical. If the purified proteins are available, trypsin digestion of the pure protein can be scanned to identify transitions with the most robust MS responses [12]. Alternatively, potential transitions can be assigned based on discovery MS data from cell or tissue lysates using software such as Skyline to generate predicted MRM assays [13], predicted based upon the amino acid sequence using an instrument's vendor software, selected from one of several spectrum library websites, such as PeptideAtlas (https://db.systemsbiology.net/sbeams/cgi/PeptideAtlas), NIST (http://chemdata.nist.gov/dokuwiki/doku.php?id=peptidew:start), and GPM (ftp://ftp.thegpm.org/projects/xhunter/libs/). As a general rule of thumb, it is our experience that empirical data obtained on the instrument ultimately being used is the shortest route to creating a robust assay. For absolute quantification the inclusion of heavy isotopic labeled peptides (used as internal standards) for each peptide comprising the multiplex is required. Internal standards are important as they also help to correct for instrument drift, matrix suppression or enhancement of the peptide signal, or other stochastic factors induced during acquisition that will affect measured intensities. For absolute quantitation, a calibration curve must be generated by using a serial dilution of intact proteins and a fixed concentration of isotope-labeled peptides into a matrix as similar as possible to that of the samples being analyzed. When the intact protein is not available, the labeled peptides alone can be used. The lower limit of detection (LLOD) and lower limit of quantitation (LLOQ) are defined as the concentrations that yield peaks with signal to noise ratio of 3 and 10, respectively. For a sample to be quantifiable its value must be above the LLOQ.

In the following sections, we present our current method for MRM assay development and provide an example of a constrained MRM assay for ApoL1. The APOL1 MRM was developed and previously published by Zhou et al. [14]. Their APO L1 assay is based on two common peptides and its three isoforms represented by three peptides (*see* Fig. 2). To date, antibodies that discriminate between all three ApoL1 variant forms (wild type, G1 and G2) are not available. Thus, an MRM multiplex is advantageous as it is able to detect and quantify ApoL1 and its various proteoforms. We have transferred this assay to our laboratory on Sciex 5500 and 6500 QTRAP MS instruments.

2 Materials

2.1 Sample Preparation, Digestion, and MS Analysis

1. Water, HPLC grade (Thermo Fisher Scientific, San Jose, CA).
2. HPLC Buffer A: 0.1 % formic acid.
3. HPLC Buffer B: 95 % acetonitrile in 0.1 % formic acid.
4. Acetonitrile (ACN), HPLC grade (Thermo Fisher Scientific, San Jose, CA).
5. Formic acid, (Thermo Fisher Scientific, San Jose, CA).

MEGAALLRVS VLCIWMSALF LGVGVRAEEAGARVQQNVPS
GTDTGDPQSK PLGDWAAGTM DPESSIFIED AIKYFKEKVS
TQNLLLLLTD NEAWNGFVAA AELPRNEADE LRKALDNLAR◄━━━━━
QMIMKDKNWH DKGQQYRNWF LKEFPRLKSE LEDNIRRLRA
LADGVQKVHK GTTIANVVSG SLSISSGILT LVGMGLAPFT
EGGSLVLLEP GMELGITAAL TGITSSTMDY GKKWWTQAQA
HDLVIKSLDK LKEVREFLGE NISNFLSLAG NTYQLTRGIG
KDIRALRRAR ANLQSVPHAS ASRPRVTEPI SAESGEQVER◄━━━━━
VNEPSILEMS RGVKLTDVAP VSFFLVLDVV YLVYESKHLH
EGAKSETAEE LKKVAQELEE KLNILNNNYK ILQADQEL

Wt NIL.. NYK; G1 NNL..NYK; G2 NIL...NK ◄━━━

APOL1 (O14791)	3 C terminal peptides found	
ILQADQEL (c terminal)	VAQELEEK	VTEPISAESGEQVER
y5 (464.7-575.3)	y6 (472.7-775.4)	y10 (815.4-1091.5)
b7 (464.7-798.4)	b6 (472.7-670.3)	b5 (815.4-540.3)

Fig. 2 Selection of MRM signature peptides from a tryptic digest of ApoL1. The amino acid sequence of 398 aa in ApoL1 (molecular weight 43,975 Da) is displayed. There are 21 predicted tryptic peptides, but we only observe three C terminal peptides in a discovery mass spectrometer data set, shown in the Table (also in *orange color* in sequence). The five arrow-pointed peptides (two common peptides ALDNLAR [14], VTEPISAESGEQVER (unique to our observation) and three isoforms Wild type, G1 and G2 at amino acid region LNILNNNYK [14]) were chosen for the evaluation of the MRM assay

6. Sample vials (VWR International, Radnor, PA).

7. Oasis HLB Extraction Cartridge, 1 cc/30 mg WAT094225 (Waters, Milford, MA).

8. Sequencing Grade Modified Trypsin (Promega, Madison, WI).

9. Tris(2-carboxyethyl) phosphine hydrochloride (TCEP) (Pierce, Rockford, IL).

10. DTT (Dithiothreitol), No-Weigh Format (Pierce, Rockford, IL).

11. Iodoacetamide (Sigma, St Louis, MO).

12. Sodium deoxycholate (DOC) (Sigma, St Louis, MO).

13. 37 and 55 °C incubators.

14. Speed/Vac concentrator, Savant SPD 20/D (Thermo Fisher Scientific, San Jose, CA).

15. Vac Elut SPS 24 Manifold (Varian, Inc.).

16. Low retention 1.5-mL microfuge tubes (Thermo Fisher Scientific, San Jose, CA).

17. Beta-galactosidase (Sigma, St Louis, MO).

18. Xylene (Sigma-Aldrich, St. Louis, MO).

19. 2-D clean-up kit (Biorad, Hercules, California).

20. CB-X assay kit (G-Biosciences MO, USA).

21. Human Lipoprotein HDL from Intracel Resources (Frederick, MD).

22. Human Lipoprotein LDL from Intracel Resources (Frederick, MD).

23. Human Lipoprotein-Deficient Plasma (Frederick, MD).

24. Human ApoL1/apolipoprotein L1 Protein from Sino Biological Inc. (Beijing, China).

25. Human plasma from BioreclamationIVT (New York, NY).

26. NEP synthesized isotopic labeled peptides from New England Peptide (Gardner, MA).

27. Resuspension buffer: 20 % acetonitrile, 0.1 % formic acid.

28. LC-MS Instrumentation: A5500 or 6500 triple quadrupole LC-MS/MS system (ABSciex, Farmingham, MA) coupled with a Prominence UFLCXR high performance liquid chromatography system (Shimadzu Scientific Instruments, Columbia, MD) with an XBridge BEH30 C18 reverse-phase chromatography column (2.1 mm × 100 mm, 3.5 μm, Waters, Milford, MA).

2.2 Software

1. Skyline (version 3.1) [13], Skyline software is a windows application for building multiple reaction monitoring (MRM), quantitative methods, and analyzing the resulting mass spectrometer data (https://brendanxuw1.gs.washington.edu/labkey/wiki/home/software/Skyline/page.view?name=default).

2. Analyst, Version 1.6.2 (Sciex, Farmingham, MA) for data acquisitions.

3. MultiQuant, Version 3.0.1 (Sciex, Farmingham, MA) for quantitation and quantitative data analysis.

2.3 Peptide and Protein Standards

Heavy isotope-labeled (e.g., C^{13}, N^{15}) peptide standards were synthesized from New England Peptide (Gardner, MA). Beta-galactosidase protein standard can be purchased from various vendors.

3 Methods

There are a number of steps involved in assay development that we have outlined in our recent method papers [12, 15]. Five basic and essential steps described below (Subheadings 3.1–3.5) are needed to establish an MRM assay and these are: Selection of peptide(s) unique to the protein of interest (signature peptides); Selection of the strongest fragment ion (transitions); Optimization for specific MS parameter (e.g., collision energy); Assay performance evaluation; and Multiplexing the peptides in a single assay.

We classify our MRM assays in MRM and immunoMRM (iMRM, although other types of enrichment methods are feasible and can be coupled to MRM quantification) (*see* **Note 1**). In MRM, quantification consists of data that is normally obtained from two or more specific peptides, each comprised of an amino acid sequence that is unique to the target analyte, as well, up to five transitions selected for each peptide. Peak area is used for quantification.

3.1 Selection of Signature Peptides

The candidate peptides can be determined by searching the public databases (PeptideAtlas) or predicted using Skyline. It is possible that same protein digest would generate slightly different results with different instruments [16, 17]. In our hands, however, as long as the discovery and validation instruments use similar front-end sources (such as TripleTOF 6600 and QTRAP 6500), interchanging instruments between discovery and quantitative assay can save time and effort. In unconstrained assays, peptide sequences should be selected in the following order: (1) proteotypic to the protein being quantified; (2) high intensity and reliably detected by MS; (3) free from common modification sites; (4) if possible, peptides that are conserved across different species can be selected to increase the generalizability of the assay to other experimental systems; (5) peptide retention times can serve as a final discriminating factor in the design of highly multiplexed assays targeting multiple protein analytes. With constrained assays, peptide selection is dictated by the amino acid sequence of interest for the biological question. For multiplexing multiple assays, retention time of each peptide is an important determining factor in building multiplexed MRM assay.

3.2 Selection of the Strongest Fragment Ions

1. Skyline is an ideal tool for the selection of transitions for a given peptide and is outlined in Subheading 3.1. We recommend the visual inspection of the extracted ion chromatograms for up to ten of the fragments generated from a selected peptide analyte for (1) alignment of fragment retention time at the peak apex, (2) peak shape, and (3) peak height. Transitions whose retention time, shape, or intensity varies substantially from the other transitions in a peak group may indicate interference from other ions in the matrix and reflect poorly performing transitions for assay design. Again, with constrained assays, transitions diagnostic for a specific amino acid or PTM on an amino acid are required and must be selected for the assay regardless of overall performance.

2. As a general rule, identify and quantify 3–4 peptides and 2–5 well-performing product ions with the highest intensity (*see* **Note 2**).

3.3 Optimization of Specific MS Parameters

1. Predict Collision Energy (CE) via Skyline's MRM method build option.

2. Modify the MRM method to run five scans for each peptide with collision energies ranging from –4, –2, 0, +2, and +4 V from Skyline predicted CE to obtain data for CE optimization.

3. Inject 5 µL of the peptides (2000–5000 fmol on column) in triplicate using this modified MRM method.

4. Quantify the integrated peak areas for all transitions and CE settings using MultiQuant.

5. For each transition, select the CE producing the highest peak area with the lowest variability (e.g., the percentage of coefficient of variation (%CV) <20 %).

6. As a final check, all peak integrations for selected transitions monitored during the CE optimization step should be manually inspected to ensure the correctness in peak integration for quantification, appropriate peak shapes, and normal peak widths for the chromatography used. In addition, the retention time should be same between all transitions of the same peptides, and between the isotopically labeled heavy peptide standards and their native light peptides.

3.4 Assay Performance Evaluation Using a Calibration Curve

A standard curve is generated by comparing the ratios (light/heavy) for each transition to digested protein concentration. For each protein, LLOQ, linearity, recovery, and reproducibility need to be evaluated.

1. Selection of peptides and setup of dilution curve. Usually, a four- to fivefold serial dilution of the trypsin-digested control protein solution, prepared in 0.1 % formic acid, 20 % ACN.

2. Inject each dilution point in triplicate using the optimized MRM method.

3. Extract intensity-by-time ion chromatograms (XICs) for each transition from the assay using quantitative software such as MultiQuant (Sciex). Using the calculated intensity area under the curve (AUC) of each transition analyte, calculate the ratio of the "light" target analyte to its "heavy" internal standard, then plot these values against the known mass of each analyte to construct a linear standard curve.

4. To calculate analyte concentration, perform the same XIC analysis and AUC normalization to internal standards as described in **step 3** for the standard curve analytes. Calculate the %CV of measured, normalized intensities and expected concentration using the area ratio data utilizing the formula derived from fitting the linear curve in Multiquant. We usually calculate analyte concentration based on each transition of the

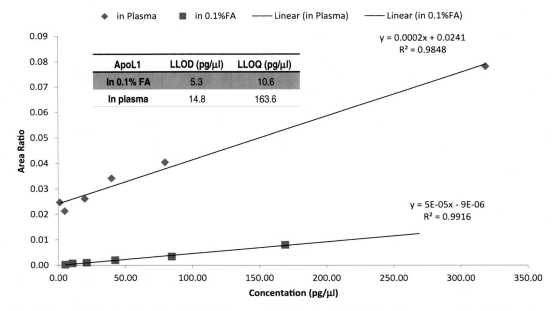

Fig. 3 Example of a significant matrix effect on quantification by MRM. Calibration curves of ApoL1 (based on the measurement of an unique peptide, ALDNAR) in 4 μL of (**a**) 0.1 % FA; (**b**) Human plasma with deficient lipoprotein is displayed. Recombinant ApoL1 was spiked into either 0.1 % FA or human plasma prior to tryptic digestion. Following online desalting by a divert valve, peptides were analyzed by LC-MS/MS. Calibration curves were generated from ratios of the ALDNAR peak area (digested from various concentration of ApoL1) over isotopic labeled ALDNAR^ peak area (fixed concentration; added before the LC-MS/MS). LLOD calculated on the signals >4* noise STDEV; LLOQ: >10*noise STDEV

same peptides and then calculate the mean value as the inferred intensity for each peptide.

5. Determine the lower limit of quantification LLOQ. LLOQ is the lowest concentration defined by a signal to noise ratio (S/N) above 5, accuracy between 80 and 120 % recovery (% of calculated concentration over expected concentration from dilution curves), and a %CV below 20 % [18]. *See* Fig. 3 for example of standard curves for the APO L1 peptide multiplex assay and lists of the LLOD and LLOQ.

3.5 Multiplexing (Scheduled) the MRM Peptide Assays

One of the advantages for MS-based MRM quantification is that it allows the monitoring of multiple peptides (hence proteins) in a single MS run. To set up multiplexed assays, determine the expected retention time of a given peptide analyte. Predict a retention time window within which a particular analyte is expected to elute. Peptides with different retention times can then be scheduled for analysis at different points across the chromatographic gradient. Multiplexing can be optimized by careful analysis of retention time precision, reproducibility, and peak width to identify how closely spaced any two analytes can be along the chromatogram and still be acquired in separate MRM acquisitions (*see* Fig. 4).

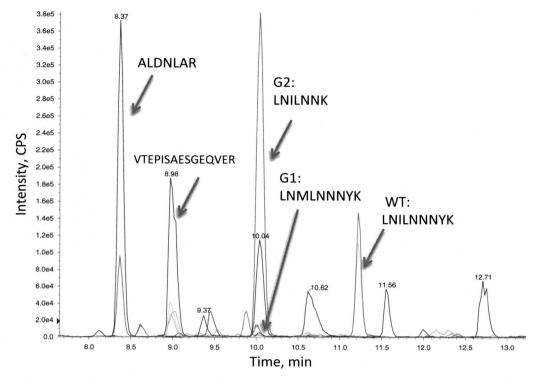

Fig. 4 Multiplexing (scheduled) MRMs for the ApoL1 MRM peptide in a single assay. The five ApoL1 peptides (two common peptides ALDNLAR [14] (retention time 8.37 min), VTEPISAESGEQVER (retention time 8.76 min) and three isoforms Wild type (wt), G1 and G2 at amino acid region LNILNNNYK [14]) were able to be measured in a 12 min liquid chromatographic run

3.6 Sample Preparation

Sample processing quality control (QC) is carried out for all samples and is based on quantification of beta-galactosidase (β-gal, a bacterial protein not expected to be present in human or other mammalian samples), which is added at a known quantity to each sample prior to sample processing.

3.6.1 Formalin Fixed Paraffin Embedded (FFPE) Tissues

1. To extract proteins from the FFPE tissue samples (e.g., aortas [19–21]), the tissues are first removed from paraffin blocks using a razor blade with all visible paraffin scratched off the tissues.

2. The isolated tissues are deparaffinized by incubation through two changes of xylene, and rehydrated through a series of graded alcohols and water washes for 5 min.

3. The wet tissues are then homogenized using Dunce homogenizer with 100 strokes in 0.2 % Rapigest 50 mM DTT, 50 mM Tris–HCl (pH 8.8).

4. The homogenized tissue mixtures are then incubated at 100 °C for 20 min in a heating block. The mixture is then incubated at 80 °C for 2 h with 600 rpm at a thermomixer (Fisher Scientific Waltham, MA), or incubated at 90 °C at 40,000 psi using a NEP 2320 Barocycler (Pressure Biosciences, South Easton, MA).

5. Centrifuge at $18,000 \times g$ for 20 min. The supernatants are then collected.

6. The protein concentrations are measured using CB-X assay kit (G-Biosciences MO, USA) (*see* **Note 3**).

7. Store at −80 °C until digested and MS analysis (outlined below).

3.6.2 Frozen or Fresh Tissue or Cell Culture

1. Tissue or cell pellets are homogenized and solubilized in 8 M urea, 2 M thiourea, 4 % CHAPs, and 1 % fresh DTT (SDS can be used but needs to be diluted out).

2. Centrifuge sample at 52684 g-force for 20 min at room temperature. Collect supernatant.

3. Protein concentration of the supernatant is assessed using CB-X assay kit (G-Biosciences MO, USA).

4. Proteins are precipitated using 2-D clean-up kit (Biorad, Hercules, CA) to remove detergents and then reconstituted in 6 M urea, 50 mM ammonium bicarbonate.

5. Dilute sample to 2 M urea in 50 mM ammonium bicarbonate.

6. Add stable isotope-labeled peptides as internal standard.

7. Proceed with digested and MS analysis (outlined below).

3.6.3 Serum or Plasma

1. Thaw frozen plasma slowly on ice or at 4 °C, gently vortex, and then centrifuge for 10 min at $10,000 \times g$ at room temperature to remove precipitates.

2. Add 30 μL of 50 mM ammonium bicarbonate to 4 μL plasma.

3. Add beta-galactosidase, wing peptide (extra amino acids added at the non-miscleavaged tryptic digested peptide) for quality control digestion, and 10 % sodium deoxycholate (1 % as final concentration) as denaturant (*see* **Note 4**).

4. Add stable isotope-labeled peptides as internal standard.

5. Proceed with digestion and MS analysis as outlined below.

3.6.4 Urine

1. Thaw frozen urine on ice or room temperature.

2. Gently vortex, and centrifuge for 10 min at $10,000 \times g$ at room temperature to remove precipitate.

3. Protein concentration is measured by pyrogallol red-molybdate (PRM) assay (Sigma Microprotein-PR Reagent) (*see* **Note 5**).

4. Add β-galactosidase and stable isotope-labeled peptides as internal standard.

5. Proceed with samples digestion and MS analysis as outlined below.

3.7 Sample Digestion and Desalting

To optimize trypsin digestion conditions and ensure the complete proteolysis, pooled matrix (serum, plasma, cell/tissue or urine) and a mixture of purified target proteins were digested with varying amounts of trypsin over various time periods and then analyzed with an MRM assay targeting as many peptides as possible. Figure 3 illustrates the effect of different matrices for APO L1 peptides.

1. Proteins are reduced with 5 mM TCEP (*see* **Note 6**) for 30–45 min at the following specified temperature and buffers. For plasma samples, 4 μL of plasma plus 30 μL of 50 mM ammonium bicarbonate, 1 % sodium deoxycholate to a total of 40 μL, reduction reaction was done at 60 °C. For urine samples, 20 μL urine with 6 μL of NH_4HCO_3 (1 M), 4 μL Rapigest (1 %, Waters) to a total of 40 μL, incubated at 60 °C for 1 h. For tissue samples, proteins from tissue are reconstituted in 6 M urea, 50 mM ammonium bicarbonate, reduction should be performed at 37 °C for 30 min.

2. Add freshly prepared 10 mM iodoacetamide (an alkylating agent) and incubate the reaction for 15–30 min at 37 °C in the dark.

3. Add trypsin and incubate the reaction at 37 °C. For each protein assay the quantity of trypsin and incubation time need to be optimized. Usually, ratio of trypsin to sample protein (μg trypsin to μg protein) can be 1:100, 1:50, 1:20 and time needed for the complete protein digestion ranges from 2 h to overnight.

4. Digested peptides were desalted on an HLB microplate in a vacuum manifold (Waters) or desalted online using divert valve.

5. HLB wells were preconditioned with 700 μL methanol and then equilibrated 3× with 700 μL of 0.1 % formic acid.

6. The digested peptides were diluted to 300 μL in 0.1 % formic acid, then added 300 μL H_3PO_4 (4 %) and loaded, dropwise, into HLB well.

7. The wells were washed 3× with 1 mL 0.1 % formic acid.

8. Peptides were slowly eluted with 400 μL of 80 % acetonitrile, 0.1 % formic acid.

9. The eluent was dried in a speed-vacuum.

10. Add 100 μL of resuspension buffer (the resuspension buffer: 20 % acetonitrile, 0.1 % formic acid).

11. Add desired heavy peptide as internal standard before mass spectrometry analysis.

3.8 Mass Spectrometry Analysis

In our lab, MRM assays were performed on a high-flow LC/MS system with a reverse-phase column (XBridge BEH30 C_{18} column, 2.1 mm × 100 mm, 3.5 μm) plumbed into an HPLC (Shimadzu Prominence) linked to a triple quadrupole mass spectrometer (QTRAP 6500 or QTRAP 5500, Sciex).

1. Digested peptides (5 μL) are injected in triplicate at a flow rate of 200 μL/min.

2. A linear A/B gradient is used where % buffer A increased from 13 to 40 % over 7 or 15 min.

3. Process data with Multiquant (Sciex) as described above.

3.9 Blocking and QC Analysis of the Large Cohorts

The order that samples are run on the MS instrument (termed blocking) can exert technical bias into sample quantification and must be randomized. The additional QC assurances critical for accurate MRM assessment of protein quantities are as follows:

1. All samples are blinded and randomized for MRM assays.

2. MRM assays are performed in duplicate or triplicates.

3. Standard curves, which often comprised of ten different concentrations for all peptides, are run at the beginning of the run series. The highest and lowest concentrations should bracket the range of the unknown samples.

4. At the end of every 10 % block of samples in a cohort, a quality control curve should be included. In our lab, a quality control curve is comprised of three different concentrations (high, medium, and low concentrations) of all peptides. The concentrations of the quality control samples should bracket the range of the samples.

5. The %CV (see **Note 7**) for sample processing is calculated from the amount of β-gal (or other reference protein) spiked in initially compared to the quantity estimated from peptides after processing and MS. Similar calculation can also be done for each target protein. Samples with a %CV of more than 20 % are repeated along with appropriate standard curves and QC curves.

6. Any sample with CV% over 20 % (calculated based on ratio of IS and native peptides) should be re-acquired with appropriate standard curves and controls.

4   Notes

1. In iMRM quantification, the antibodies or other affinity capture agents are used to capture the target protein or peptide prior to MRM assay when an analyte is at such a low concentration that it cannot be detected in unenriched samples. The protein

is isolated using a capture antibody or other capture agents, then reduced, alkylated, and digested using trypsin prior to quantified, as described above.

2. Usually y- and b-ions are common fragment ions produced in tandem MS. For specificity, we prefer to select fragments with m/z greater than the precursor m/z as these tend to be most specific for a peptide sequence of interest.

3. Another popular protein assay for plasma and serum is BCA protein assay (Pierce™ BCA Protein Assay Kit).

4. The common denaturants for serum and plasma are 8 M urea, 0.1 % rapigest (Waters), and 1 % sodium deoxycholate. Sodium deoxycholate could be removed at the end of digestion by adding equal volume of 1 % Formic acid, vortex and spun at $16,000 \times g$ for 15 min (sodium deoxycholate is acid precipitable).

5. For urine, we have tested more than six total protein quantification assays and we found that only pyrogallol red-molybdate (PRM) assay gives accurate results with satisfied sensitivity. Other protein assays such as Pierce BCA kit, in our hands, provided inaccurate results due to the urine matrix interference.

6. 5 mM DTT as reducing agent works equally well, except DTT needs to be freshly made.

7. The CV usually is calculated based on the ratio of IS and native peptides.

Acknowledgements

This work was supported by NHLBI Johns Hopkins Proteomic Innovation Center in Heart Failure—HHSN268201000032C (JVE)—and partially supported by the Chronic Kidney Disease Biomarker Consortium funded by NIDDK U01-U01DK085689. Special thanks to Drs. Josef Coresh, Lesley Inker, Chi-yuan Hsu, John Eckfeldt, Paul Kimmel, Dr. Vasan Ramachandran, and Harold I. Feldman.

References

1. Addona TA, Abbatiello SE, Schilling B et al (2009) Multi-site assessment of the precision and reproducibility of multiple reaction monitoring-based measurements of proteins in plasma. Nat Biotechnol 27(7):633–641. doi:10.1038/nbt.1546

2. Hoofnagle AN, Roth MY (2013) Clinical review: improving the measurement of serum thyroglobulin with mass spectrometry. J Clin Endocrinol Metab 98(4):1343–1352

3. Bystrom CE, Salameh W, Reitz R et al (2010) Plasma renin activity by LC-MS/MS: development of a prototypical clinical assay reveals a subpopulation of human plasma samples with substantial peptidase activity. Clin Chem 56(10):1561–1569

4. Chen Z, Caulfield MP, McPhaul MJ et al (2013) Quantitative insulin analysis using liquid chromatography-tandem mass spectrometry in

a high-throughput clinical laboratory. Clin Chem 59(9):1349–1356

5. Liu X, Jin Z, O'Brien R et al (2013) Constrained selected reaction monitoring: quantification of selected post-translational modifications and protein isoforms. Methods 61(3):304–312

6. Yuan M, Breitkopf SB, Yang X et al (2012) A positive/negative ion-switching, targeted mass spectrometry-based metabolomics platform for bodily fluids, cells, and fresh and fixed tissue. Nat Protoc 7(5):872–881

7. Anderson L, Hunter CL (2006) Quantitative mass spectrometric multiple reaction monitoring assays for major plasma proteins. Mol Cell Proteomics 5(4):573–588

8. Liebler DC, Zimmerman LJ (2013) Targeted quantitation of proteins by mass spectrometry. Biochemistry 52(22):3797–3806. doi:10.1021/bi400110b

9. Gallien S, Domon B (2015) Detection and quantification of proteins in clinical samples using high resolution mass spectrometry. Methods. doi:10.1016/j.ymeth.2015.03.015

10. Zhang P, Kirk JA, Ji W et al (2012) Multiple reaction monitoring to identify site-specific troponin I phosphorylated residues in the failing human heart. Circulation 126(15):1828–1837

11. Jin Z, Fu Z, Yang J et al (2013) Identification and characterization of citrulline-modified brain proteins by combining HCD and CID fragmentation. Proteomics 13(17):2682–2691

12. Grote E, Fu Q, Ji W et al (2013) Using pure protein to build a multiple reaction monitoring mass spectrometry assay for targeted detection and quantitation. Methods Mol Biol 1005:199–213

13. MacLean B, Tomazela DM, Shulman N et al (2010) Skyline: an open source document editor for creating and analyzing targeted proteomics experiments. Bioinformatics 26(7):966–968. doi:10.1093/bioinformatics/btq054

14. Zhou H, Hoek M, Yi P et al (2013) Rapid detection and quantification of apolipoprotein L1 genetic variants and total levels in plasma by ultra-performance liquid chromatography/tandem mass spectrometry. Rapid Commun Mass Spectrom 27(23):2639–2647

15. Fu Q, Schoenhoff FS, Savage WJ et al (2010) Multiplex assays for biomarker research and clinical application: translational science coming of age. Proteomics Clin Appl 4(3):271–284

16. Jones KA, Kim PD, Patel BB et al (2013) Immunodepletion plasma proteomics by triple-TOF 5600 and Orbitrap elite/LTQ-Orbitrap Velos/Q exactive mass spectrometers. J Proteome Res 12(10):4351–4365

17. Toprak UH, Gillet LC, Maiolica A et al (2014) Conserved peptide fragmentation as a benchmarking tool for mass spectrometers and a discriminating feature for targeted proteomics. Mol Cell Proteomics 13(8):2056–2071

18. Fu Q, Zhu J, Van Eyk JE (2010) Comparison of multiplex immunoassay platforms. Clin Chem 56(2):314–318

19. Fu Z, Yan K, Rosenberg A et al (2013) Improved protein extraction and protein identification from archival formalin-fixed paraffin-embedded human aortas. Proteomics Clin Appl 7(3–4):217–224

20. Nirmalan NJ, Harnden P, Selby PJ et al (2009) Development and validation of a novel protein extraction methodology for quantitation of protein expression in formalin-fixed paraffin-embedded tissues using western blotting. J Pathol 217(4):497–506

21. Nirmalan NJ, Hughes C, Peng J et al (2011) Initial development and validation of a novel extraction method for quantitative mining of the formalin-fixed, paraffin-embedded tissue proteome for biomarker investigations. J Proteome Res 10(2):896–906

Chapter 16

Methods for SWATH™: Data Independent Acquisition on TripleTOF Mass Spectrometers

Ronald J. Holewinski, Sarah J. Parker, Andrea D. Matlock, Vidya Venkatraman, and Jennifer E. Van Eyk

Abstract

Data independent acquisition (DIA also termed SWATH) is an emerging technology in the field of mass spectrometry based proteomics. Although the concept of DIA has been around for over a decade, the recent advancements, in particular the speed of acquisition, of mass analyzers have pushed the technique into the spotlight and allowed for high-quality DIA data to be routinely acquired by proteomics labs. In this chapter we will discuss the protocols used for DIA acquisition using the Sciex TripleTOF mass spectrometers and data analysis using the Sciex processing software.

Key words Data independent acquisition (DIA), SWATH, Quantitative proteomics, Mass spectrometry, Spectral ion library

1 Introduction

Data Independent Acquisition Mass Spectrometry (DIA-MS) is a long-standing technique [1, 2] that has garnered increased attention recently due to the development of new pipelines for extracting, identifying, and quantifying peptides using a targeted analysis approach [3, 4]. SWATH™ couples DIA-MS with direct searching of individual samples against an established, and often a more exhaustive, peptide MS spectral library [3, 5, 6]. SWATH™ is, therefore, a two-step process (Fig. 1), development of the MS spectral library, commonally on a pooled sample representing the breath of the experimental collection, using information dependent acquisition (IDA also termed data dependent acquistion (DDA)) (*see* **Note 1**) and then the subsequent analysis of each individual sample by DIA. Thus, a major advantage of SWATH™ is that it can maximize the peptides observed both within an individual sample and across all of the samples in an experimental set, thereby increasing proteome coverage, experimental efficiency,

Salvatore Sechi (ed.), *Quantitative Proteomics by Mass Spectrometry*, Methods in Molecular Biology, vol. 1410, DOI 10.1007/978-1-4939-3524-6_16, © Springer Science+Business Media New York 2016

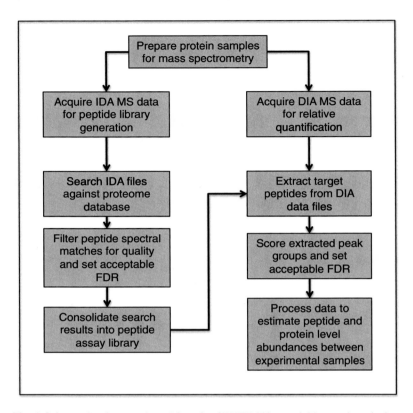

Fig. 1 Schematic of general workflow for SWATH-MS acquisition and analysis

reducing quantitative variability, and minimizing missing data across an experimental matrix. It is important to note that SWATH™ is an emerging approach and methods for estimating peptide identification confidence and false discovery rates as well as the ideal approach for estimating peptide and protein quantity from transition extracted ion chromatograms are continuing to evolve along with the sensitivity and capabilities of the instrumentation itself. As with any large-scale quantitative screening method, care should be taken to confirm and validate the biological differences and conclusions that are derived from a SWATH™ experiment.

In a SWATH™ experiment, proteins are digested and either directly infused or, more often, separated by liquid chromatography (LC) prior to analysis on a TripleTOF mass spectrometers (5600 or 6600, Sciex), a Q-Exactive mass spectrometer (Thermo Scientific), or any instrument with sufficiently high scan speed and a quadrupole mass filter. On the Triple TOF instruments, precursor peptide ion selection is performed by filtering precursors collectively through mass-to-charge windows, typically $4-10\,m/z$ wide, sequentially across the entire m/z range of interest rather than selectively isolating a single precursor mass/charge (m/z) per

MS/MS scan as performed in IDA-MS experiments. Due to the typically wider isolation windows used in DIA experiments, two or more co-eluting precursors are often fragmented collectively to produce an MS2 spectrum containing a convoluted mixture of fragment ions from multiple precursor ions.

One approach used to increase the ability to find and confidently identify peptides from these complex mixed spectra is to associate specific peptides with defined regions within the chromatographic elution profile. Currently, in order to accomplish this, retention time (RT) determination and alignments across samples are key aspects of searching IDA data. Exogenous supplied RT standards [6] or endogenous RT standards [7] that are composed of peptides consistently observed across large number of samples must be used for RT calibration in order to properly align individual ion chromatograms across the entire sample's elution profile.

Optimization of m/z window number and dwell time/ion accumulation time per window is performed so that the instrument cycles through the entire desired precursor m/z range (e.g., 400–1250 m/z). This is largely instrument and sample specific. For the 6600 triple TOF, you can go up to 2250 m/z but we typically analyze between 400 and 1250 m/z for tryptic digests. When analyzing middle down or any peptides larger than the average tryptic peptides, the full range can be used with the appropriate considerations to SWATH™ windows and cycle times. Ultimately, the key is to allow the instrument to cycle rapidly enough to capture multiple observations across the chromatographic elution profile for a given ion.

The data are subsequently searched against a sample-specific peptide library that allows a set number of transition ion chromatograms to be extracted for a peptide within the window of its predicted RT (determined by its observed or normalized RT from the peptide library). The peak groups are scored according to several factors intended to discriminate a "true" peptide target from nonspecific noise, and the distribution of these target scores is modeled against the distribution of scores attributed to decoy peak groups to determine a score cut off resulting in an acceptable false discovery rate. Relative peptide abundance is then inferred from the aggregate of the area under the curve for each transition extracted ion chromatograms (XICs), and various statistical approaches are used to roll transition intensity XICs into peptide intensity estimates, which can then be used to estimate the overall protein intensity. In this chapter, we present the typical workflow used currently by our group to prepare, acquire, and analyze proteomic data for a DIA-MS experiment of cell or tissue samples. For simplicity and pragmatism, we present the workflow as completed using SCIEX TripleTOF® instruments and data analysis platform exclusively, with mention of alternative approaches as appropriate.

1.1   Quality Assurance and Quality Control (QA/QC) Considerations

Robust quality assurance (QA) or quality control (QC) protocols are essential to monitor instrument performance and improve reproducibility and reliability of data. A QC standard run can be analyzed at fixed times such as the beginning and end of an experiment or day to assess variation in a variety of quality control metrics [8]. For the TripleTOF instruments, we conduct internal mass calibrations of mass accuracy and sensitivity for both MS1 and MS2 scans every 3–5 runs by monitoring at least eight peptides from 100 fmol digested beta-galactosidase standard (Sciex) and seven transition ions from the 729.3652 [M+2H]$^{2+}$ ion (Table 1). What also needs to be tracked is sample processing to ensure the quality of the peptide mixture being analyzed, which is not addressed at in this manuscript but is well established in targeted multiple and selective monitoring workflows. To do this one can include an exogenously protein, such as beta-galactosidase, into the sample prior to digestion. Beta-galactosidase selected peptides can be quantified (if N^{15} labeled peptides are added after digestion to the sample) or assessed in each sample (for more details *see* Chen et al., in this book).

Internal peptide retention time (RT) standards are currently an essential component of both peptide library generation and DIA-MS data analysis, and must be (1) detectable across all individual samples and (2) spread evenly across the chromatogram. Retention time of a given peptide from the library is used to set an

Table 1
Beta-galactosidase peptides used for autocalibration and quality control

Beta-galactosidase peptide sequence	[M+2H]$^{2+}$	Transition ions for 729.36	Fragment
YSQQQLMETSHR	503.2368		
RDWENPGVTQLNR	528.9341		
GDFQFNISR	542.2654		
IDPNAWVER	550.2802		
DVSLLHKPTTQISDFHVATR	567.0565		
VDEDQPFPAVPK	671.3379		
WENPGVTQLNR	714.8469		
APLDNDIGVSEATR	729.3652		
		175.1190	y1
		347.2037	y3
		563.2784	y5
		729.3652	b7
		832.4523	y8
		1061.5222	y10
		1289.6332	y12

extraction window for its peak group identification from the SWATH™/DIA-MS data file, and subsequently also used in scoring the confidence of a given peak group assignment to a peptide sequence from the library. If SWATH™/DIA-MS data files and peptide library files are collected absolutely sequentially with nearly identical chromatography, one might bypass the use of RT alignment standards. Much more commonly, differences in sample matrix, chromatographic setups, timing of instrument batch acquisitions, and many other factors will contribute to imperfect chromatographic alignment necessitating RT standards to normalize peptide assay library retention time to the SWATH™ acquisition file retention time. Used alone or in combination with retention time standards that are spiked into a sample, endogenous reference peptides can also be used for the calibration of retention times across samples [7]. These can be unique to a specific library (sample); however, there are common and conserved peptides that may be present in most, if not all, mammalian cells and tissues which can be used as a complement or replacement to synthetic, externally spiked RT reference peptides [7]. Note, that new methods to analyze DIA data sets are being developed and the need for RT standards may change, however, expectations are that RT alignment will remain part of the QC for assessment of LCMS runs. As well, QC tools are available to assess quality control metrics in a shotgun or targeted proteomic workflow that allows chromatographic performance and systemic error to be monitored [9]. Tracking RT standards across sample runs can also serve to assess instrument performance.

Finally, as larger numbers of individual samples are analyzed adopting other routine QC such as randomization or blocking of sampled to minimize sample analysis bias and regular collection of quality control samples spaced evenly and strategically throughout acquisition batches will be necessary components of SWATH™ experimental design.

1.2 Spectral Library Building—Data Generation

A spectral ion library is most often used for the targeted analysis of SWATH™/DIA-MS data, although other methods (as mentioned above) are being explored and developed [10, 11], and can be primarily cell or tissue and species specific or a broader library assembled from all relevant peptide observations from a given species [5]. Spectral ion libraries are most commonly built using traditional shotgun proteomics in IDA- MS mode. In some cases spectral ion libraries previously generated have been made available to the public from various labs [5, 12, 13]. Here we will discuss the creation of new spectral ion libraries from IDA analysis of proteolytic digestions. Additional detailed information regarding the generation of spectral ion libraries, including the management of protein redundancy and isoform specificity, can be found in Schubert et al. [5]. It is important to consider differences in peptide fragmentation patterns between instruments, and ideally use

IDA data acquired on the same instrument from which you will perform your SWATH™/DIA-MS acquisition [14].

Spectral ion libraries can be constructed in a number of ways. The first and most straightforward way to create an ion library is to analyze a proteolytic digestion in IDA mode of a pooled sample created from all of the individual samples that will be subsequently analyzed by DIA or of samples composing the extremes of the phenotype. This will give the most basic ion library comprising the peptides identified in a single IDA run that can then be used against the SWATH™ acquired version of itself and any other SWATH™/DIA-MS acquired sample of the same general proteome. In an attempt to expand the number of ions selected for fragmentation for library generation from a single IDA run of the pooled sample, multiple runs or technical replicates might help increase the proteome coverage provided to the sample library beyond what may be obtained from a single run and thus may help compensate for the error in sampling that is inherent to DIA methods. Alternatively, deeper and more inclusive ion libraries can be constructed post-digestion using off-line peptide fractionation and analysis of these fractions independently in IDA mode. The IDA runs are then combined to create a more complete and inclusive ion library for the given sample proteome. This should ultimately increase the power of DIA-based protein identifications by increasing the number of peptides used to quantitate highly abundant proteins while harnessing the sensitivity of MS2-based quantitation necessary for low abundance proteins and peptides. Some methods commonly used for peptide fractionation are basic-reverse phase HPLC (bRP-HPLC) [15], strong cation exchange (SCX), and strong anion exchange (SAX) [16] (*see* **Notes 2** and **3**). Our lab typically uses bRP-HPLC or a solid phase extraction SCX [17] method for peptide fractionation prior to MS analysis. For SWATH™ analysis of post-translational modifications, it is recommended to employ enrichment strategies (if applicable) either independently or in combination with the peptide fractionation techniques described and as typically performed in shotgun experiments.

The following protocol is for library generation using Sciex TripleTOF™ systems with an Eksigent® 415 nano LC and ekspert 400 autosampler, although alternative LC and autosamplers may be used with the TripleTOF systems.

2   Materials

1. Proteolytic peptide mixture, most often MS-grade trypsin (Promega).
2. 5600 or 6600 TripleTOF system.
3. Nano-LC and autosampler (e.g., Eksigent® 415 nano LC, ekspert™ 400 autosampler) and ekspert™ cHiPLC (optional).

4. Trap and analytical LC columns (Eksigent® P/N 804-00006 and 804-00001).

5. Proteolytic peptide mixture, most often MS-grade trypsin (Promega).

6. 5600 or 6600 TripleTOF system.

7. Retention time standards, either commercial peptides that are spiked in right before MS analysis (e.g., Biogynosis cat# KI-3002-2) or endogenous peptides present in all samples, can be used (Parker et al., in press) (*see* **Note 4**).

Software Needed (*See* **Note 5**)

1. Analyst TF 1.7.

2. PeakView 2.0 or higher.

3. Variable Window Calculator.

4. Protein Pilot 4.5 or higher.

5. SWATH™ microapp.

6. Microsoft Excel.

7. MarkerView (optional).

3   Methods

3.1   IDA Analysis of Proteolytic Digests for Spectral Ion Library Building

1. Create an IDA method in Analyst TF 1.7 with one survey scan and 20 candidate ion scans per cycle (*see* **Note 6**). Check the **Rolling Collision Energy** box.

2. For TOF MS (MS1)

   (a) Under the **MS Tab** set the accumulation time to 250 ms and the mass range from 400 to 1250 Da (Fig. 2, *see* **Note 7**). Set the method duration to match the length of your LC gradient method.

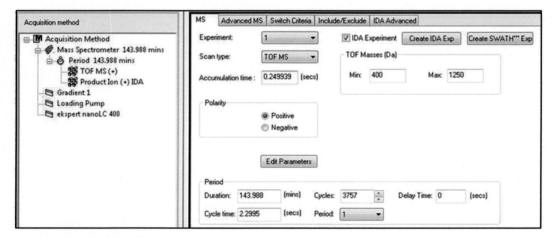

Fig. 2 Example of TOF MS parameters for TripleTOF MS instruments

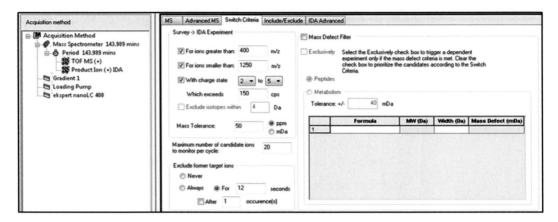

Fig. 3 Example of Switch Criteria parameters for TripleTOF MS instruments

(b) Under the **Switch Criteria** tab set the range to match what you selected under the above window, monitor charge states from 2 to 5 which exceed 150 counts, set the mass tolerance to 50 ppm, and set your exclusion criteria (Fig. 3, *see* **Note 8**).

(c) Under the **Include/Exclude** tab put in any masses you want to monitor or exclude in your analysis.

(d) Under the **IDA Advanced** tab make sure Rolling Collision Energy is checked and make any other necessary changes that would be pertinent to your experiment.

(e) Default settings do not need to be changed under the **Advanced MS** tab.

3. For Product Ion (MS2)

(a) Under the **MS Tab** set the accumulation time to 100 ms and the mass range from 100 to 1800 Da and check whether you want high resolution or high sensitivity (the high sensitivity function is most commonly selected for proteomics experiments).

(b) All other tabs should maintain the same parameters as for the TOF MS and do not need to be changed.

4. Load the sample appropriate Gradient, Loading Pump, and autosampler methods and save your Acquisition File.

5. Analyze your peptide samples.

3.2 SWATH-MS Data Acquisition (DIA-MS acquistion)

3.2.1 Creation of Variable Window SWATH™ Methods

Optimized SWATH™ methods can be constructed for specific samples using the Sciex Variable Window Calculator application. The steps for creating the customized SWATH™ variable windows for a specific sample are listed in the Variable Window Calculator under the **Instructions and Controls** tab. After following these directions select the number of variable windows (*see* **Note 9**) you want to analyze in your method and the mass range of the

SWATH™ analysis. For general proteomics experiments the window overlap is usually left at 1 Da and the collision energy spread (CES) is usually left at 5. The minimum window width should be set no lower than 4 due to the default parameters in the PeakView software. After the Variable Window calculator is finished creating the optimal windows for your analysis go to the **OUTPUT for Analyst** tab and copy columns A, B, and C into a new Excel file and save as a Text (Tab Delimited) file which can then be loaded into the SWATH™ method within Analyst TF 1.7.

3.2.2 Creation of a SWATH™ Method in Analyst TF 1.7

1. In Analyst TF 1.7 go to the **Build Acquisition Method** tab on the left-hand side of the window. Click on TOF MS and select Create SWATH™ Exp button then select the **Manual** tab within this window.

2. Under **SWATH™ Analysis Parameters** select the mass range of the analysis (typically 400–1250 Da for tryptic peptides). Under **Fragmentation Conditions** make sure Rolling Collision energy is checked (the CES set in the Variable Window Calculator will overwrite the CES value inputted on this screen). Under **SWATH™ Detection Parameters** select the mass range to monitor for the SWATH™ MS2 spectra (typically 100–1800 Da) and the accumulation time for each window (typically for 100 VW 30 ms is adequate) (*see* **Note 10**). Lastly, click the **Read SWATH™ Windows from Text File** box and load in your .txt file created in the Variable Window Calculator.

   The accumulation time for the MS1 can be set between 50 and 150 ms to give a quick survey scan for each cycle (*see* **Note 11**). Select the appropriate loading pump, gradient, and autosampler methods for the file (*see* **Note 12**). The gradient method chosen should be the same one that was used during the IDA analysis preformed to generate the proteome-specific spectral library.

3.3 SWATH™ Data Analysis Using PeakView 2.1 and SWATH™ Microapp 2.0

3.3.1 Introduction to SWATH™ Data Analysis Procedure

As with many methodologies, there are several options for processing SWATH™ data and analyzing results. Here, we present the protocol to process data through the SCIEX proprietary software. In our lab, we also regularly utilize two alternative pipelines, Skyline [18] and OpenSWATH [4]. Skyline is a free and open-source tool built in Windows computing environments for analysis of multiple MS data types, including DIA. OpenSWATH™ is a free and open-source built within the openMS data analysis tool space, and operates optimally in a Linux computing environment. A summary of the basic information pertaining to using these two alternate data analysis pathways is provided in Table 2 located at the end of this section. In this final section, we will provide a cursory summary specific to the approach used in our lab for the general implementation of the SCIEX software tools. We recommend referring to the SCIEX software user manuals for additional guidance.

Table 2
Selected alternative DIA-MS data analysis approaches

Parameters	Skyline[a]	OpenSWATH[b]
Input DIA File format	.WIFF	.mzML/.mzXML[c]
Peptide Ion Library	Built from DDA search result files (e.g., pep.xml, .group) or imported as a "transition list"	Built using TPP tools and custom Python scripts[d]
SWATH workflow	Internal to Skyline	OpenSwathWorkflow.exe
Output File Format	.csv transition report	.tsv transition report
Visualization	Internal to Skyline	TAPIR[e]
Peak Picking Algorithm	mProphet[f] adaptation	pyProphet[g]
Multi-Run Alignment	–	Feature Alignment[h]
Quantitative Statistics	Linked External Tool MSstats[i]	External Tools (e.g., MapDIA,[j] MSstats)

[a]MacLean, B. et al. Skyline: an open source document editor for creating and analyzing targeted proteomics experiments. Bioinformatics 26, 966–968 (2010)
[b]Röst HL et al. OpenSWATH™ enables automated, targeted analysis of data independent acquisition MS data. Nature Biotechnology 10;32(3):219–223 (2014)
[c]Conversion to mzML or mzXML can be done using the tool msconvert, available at: (http://proteowizard.source-forge.net/tools/msconvert.html). Do not select peak picking, files may expand 10× or more from raw file size
[d]Schubert OT et al., Building high-quality assay libraries for targeted analysis of SWATH™ MS data. Nature Protocols, 10(3):426–441 (2015). *Note*: Libraries generated using the pipeline described in the Schubert et al. paper can be formatted for use in the PeakView microapp, and substituted in the workflow above
[e]https://github.com/msproteomicstools/msproteomicstools/blob/master/gui/TAPIR.py
[f]http://www.mprophet.org/
[g]https://pypi.python.org/pypi/pyprophet
[h]Python script, available to download from https://github.com/msproteomicstools, found in folder msproteomicstools/analysis/alignment/feature_alignment.py
[i]http://www.msstats.org/
[j]http://mapdia.sourceforge.net/Main.html

3.3.2 Creation of Spectral Ion Library Using Protein Pilot Paragon Method

1. Prepare the protein reference database that you will use for matching DDA spectra to peptide sequences. For instance, FASTA documents for annotated proteomes can be downloaded from the Uniprot website: (http://www.uniprot.org/proteomes). Typically, we chose to use the curated, or reference proteomes, for a given organism of interest.

   (a) If external retention time standards were used in the experiment, such as the Biognosys iRT (*see* **Note 13**) peptides, copy their sequences and append to your FASTA file by opening it in a text editor. FASTA proteome databases should be saved in the appropriate folder within the Protein Pilot software files on your computer as per the software manual instructions.

2. In Protein Pilot, select the option for an LC MS search and prepare a database search method appropriate for your experiment, including all of the raw data files you would like to include to build the ion library.

3. Once the search is completed open the "FDR report" generated for the search and record the number of proteins identified at 1 % Global FDR to be used as input in the following section.

3.3.3 Importing Ion Libraries into the SWATH™ Microapp and Analyzing SWATH™ Data

1. Open PeakView and using the tabs at the top of the screen, navigate to Quantitation → SWATH™ Processing → Import Ion Library (Fig. 4).

2. Find the .group file produced from the Protein Pilot search and set the number of proteins to import to the 1 % Global FDR (*see* **Note 14**) recorded in the previous section from the FDR report generated by Protein Pilot. Typically peptides shared by more than one protein are not imported. Under **Select sample type**, chose the option appropriate for whether the samples were unlabeled (typical) or labeled with a chemical tag (i.e., iTRAQ, SILAC).

3. Select all of the SWATH™ files to be analyzed for a given experiment.

4. Set your processing settings. For protein quantitation analysis, examples of typical parameter settings are given in Fig. 5 (*see* **Note 15**).

5. After setting your processing settings click "Process" to analyze your SWATH™ data.

6. Once completed you can export the data for visualization in MarkerView by clicking **Quantitation → SWATH™ Processing → Export → Areas** or **Export → All** to get a complete list of all parameters for the analysis in Excel format (Fig. 6).

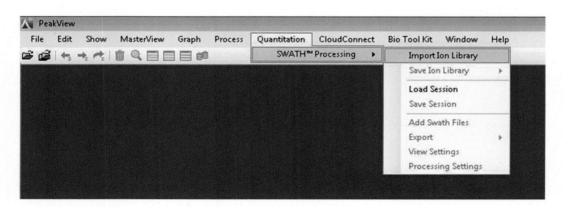

Fig. 4 Schematic for importing ion library into PeakView software

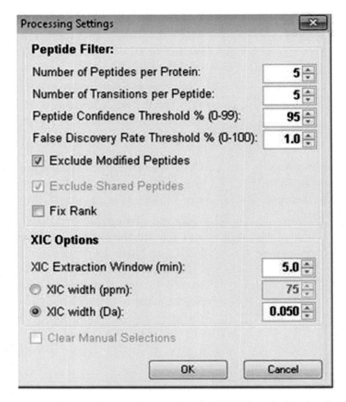

Fig. 5 Example of typical processing settings for SWATH analysis using PeakView software

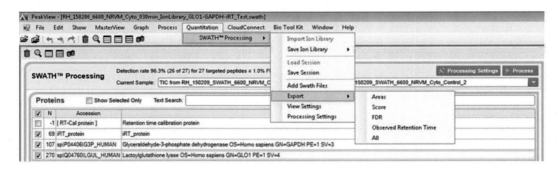

Fig. 6 Schematic for exporting SWATH results from PeakView software

4   Notes

1. The Sciex terminology Information Dependent Acquisition (IDA) is the same as Data Dependent Acquisition (DDA) and this is the terminology used in the Sciex software for shotgun proteomics experiments. In this paper we will be using the

IDA acronym to be consistent with the Sciex terminology and software.

2. bRP-HPLC fractionation may be preferred over SCX or SAX fractionation if downstream phosphopeptide enrichment or analysis of other negatively charged peptides is desired. This is due to a more equal distribution of phosphopeptides throughout basic RP fractions compared to SCX and SAX fractions, in which phosphopeptides are most dense in the early and late fractions, respectively.

3. The SCX method published by Dephoure and Gygi [17] was based on 10 mg of starting material and was used upstream of phosphopeptide enrichment. Our lab has used this method for both phosphoproteomic and general proteomic analysis and we have scaled back the protocol for 1 mg of starting material, in which we have cut the reagents used in the Dephoure and Gygi paper by 1/10th. If using less than 1 mg of starting material, scale back the reagents accordingly [13].

4. If large number of samples, include beta-galactosidase for sample preparation assessment and N^{15} labeled peptides to track (*see* Chen et al., this book).

5. Sciex software can be downloaded at http://www.absciex. com/downloads/software-downloads.

6. The number of survey scans desired for the analysis of concatenated or single run samples for library generation is a matter of user discretion but a typical IDA method on a TripleTOF system uses 20 candidate ions.

7. The 5600 TripleTOF system can go up to 1250 m/z and the 6600 TripleTOF can go up to 2250 m/z. However, we find that for tryptic digests there is little additional peptide data obtained above 1250 m/z. The larger mass range on the 6600 system is beneficial when doing large protein modifications such as glycoproteomics or when using alternative proteolytic methods that produce larger peptides (i.e., Lys-C, CNBr).

8. These values are meant to be used as a general guide in setting up an IDA method. Optimization for individual systems and sample types may be required for optimal results. For PTM and low abundant peptide analysis the accumulation times may be adjusted to allow for increased signal in both the MS1 and MS2 scans.

9. The number of variable windows chosen should be considered carefully as the more windows selected the shorter the dwell time will have to be for each window. For general purposes 100 VW and a 30 ms dwell time should be sufficient to yield good quantitation of peptides.

10. If accumulation times less than 30 ms are desired, it is recommended that they be tested prior to large-scale sample analysis

to ensure the accumulation time chosen will give adequate signal for quantitation.

11. If using the 5600 TripleTOF system, the minimum accumulation time for the MS1 should be set to 150 ms to ensure the MS1 quality is sufficient to perform the background calibrations during the run. The 6600 TripleTOF system does not use this background calibration so a shorter MS1 accumulation time (50 ms) may be used to get a quick survey scan.

12. The LC and autosampler methods will vary between labs and the gradient lengths will vary depending on the complexity of the samples. Typically, for complex mixtures a gradient of 5–35 % B over 90–120 min is suitable and for less complex samples (i.e., immunoprecipitations, purified proteins) shorter gradients between 30 and 60 min may be sufficient.

13. iRT FASTA sequence is available at www.biognosys.com, or type the following into your FASTA file:

    (a) >Biognosys iRT Kit Fusion
        A G G S S E P V T G L A D K V E A T F G V D E
        S A N K Y I L A G V E S N K D A V T P A D F S E W S K F L L Q
        F G A Q G S P L F K L G G N E T Q V R T P V I S G P Y Y E R T
        P V I T G A P Y Y E R G D L D A A S Y Y A P V R T G F I I D P G G
        V I R G T F I I D P A A I V R

14. FDR threshold can be set higher or lower depending on the user preference; the higher the FDR is set the more proteins will be incorporated into the library but the confidence of these proteins will not be as high as if a lower FDR threshold is used.

15. These parameters are meant as a guideline and can be adjusted based on user preferences. Refer to the Sciex PeakView software documentation and the literature regarding optimizing these settings for your particular experiment. Importantly, for PTM analysis, un-check the Exclude Modified Peptides box and increase the number of peptides per protein to a larger value (i.e., 100) to import all peptides identified at the confidence level selected or create a PTM enriched peptide library.

Acknowledgements

We would like to acknowledge funding to JVE from 1R01HL111362-01A, 1P01HL112730-01A1, 1U54NS091046-01, NHLBI-HV-10-05 [2], HHSN268201000032C and the Erika Glazer Endowed Chair for Women's Heart Health and funding to SP from the National Marfan Foundation Victor E McKusick Post-Doctoral Fellowship; as well as the technical and intellectual support of Brigitt Simons and Christie Hunter at Sciex in many helpful discussions.

References

1. Venable JD, Dong MQ, Wohlschlegel J, Dillin A, Yates JR (2004) Automated approach for quantitative analysis of complex peptide mixtures from tandem mass spectra. Nat Methods 1(1):39–45. doi:10.1038/nmeth705

2. Dong MQ, Venable JD, Au N, Xu T, Park SK, Cociorva D, Johnson JR, Dillin A, Yates JR 3rd (2007) Quantitative mass spectrometry identifies insulin signaling targets in C. elegans. Science 317(5838):660–663. doi:10.1126/science.1139952

3. Gillet LC, Navarro P, Tate S, Rost H, Selevsek N, Reiter L, Bonner R, Aebersold R (2012) Targeted data extraction of the MS/MS spectra generated by data-independent acquisition: a new concept for consistent and accurate proteome analysis. Mol Cell Proteomics 11(6):O111.016717. doi:10.1074/mcp.O111.016717

4. Rost HL, Rosenberger G, Navarro P, Gillet L, Miladinovic SM, Schubert OT, Wolski W, Collins BC, Malmstrom J, Malmstrom L, Aebersold R (2014) OpenSWATH enables automated, targeted analysis of data-independent acquisition MS data. Nat Biotechnol 32(3):219–223. doi:10.1038/nbt.2841

5. Schubert OT, Gillet LC, Collins BC, Navarro P, Rosenberger G, Wolski WE, Lam H, Amodei D, Mallick P, MacLean B, Aebersold R (2015) Building high-quality assay libraries for targeted analysis of SWATH MS data. Nat Protoc 10(3):426–441. doi:10.1038/nprot.2015.015

6. Wang J, Perez-Santiago J, Katz JE, Mallick P, Bandeira N (2010) Peptide identification from mixture tandem mass spectra. Mol Cell Proteomics 9(7):1476–1485. doi:10.1074/mcp.M000136-MCP201

7. Parker SJ, Rost H, Rosenberger G, Collins BC, Malmström L, Amodei D, Venkatraman V, Raedschelders K, Van Eyk JE, Aebersold R. Mol Cell Proteomics. 2015 Oct;14(10):2800–13. doi:10.1074/mcp.O114.042267

8. Bereman MS (2015) Tools for monitoring system suitability in LC MS/MS centric proteomic experiments. Proteomics 15(5–6):891–902. doi:10.1002/pmic.201400373

9. Bereman MS, Johnson R, Bollinger J, Boss Y, Shulman N, MacLean B, Hoofnagle AN, MacCoss MJ (2014) Implementation of statistical process control for proteomic experiments via LC MS/MS. J Am Soc Mass Spectrom 25(4):581–587. doi:10.1007/s13361-013-0824-5

10. Tsou CC, Avtonomov D, Larsen B, Tucholska M, Choi H, Gingras AC, Nesvizhskii AI (2015) DIA-Umpire: comprehensive computational framework for data-independent acquisition proteomics. Nat Methods 12(3):258–264. doi:10.1038/nmeth.3255

11. Ting S, Egertson J, MacLean B, Kim S, Payne S, Noble W, MacCoss MJ (2014) Pecan: Peptide Identification Directly from Data-Independent Acquisition (DIA) MS/MS Data. American Society for Mass Spectrometry, Baltimore, MD

12. Toprak UH, Gillet LC, Maiolica A, Navarro P, Leitner A, Aebersold R (2014) Conserved peptide fragmentation as a benchmarking tool for mass spectrometers and a discriminating feature for targeted proteomics. Mol Cell Proteomics 13(8):2056–2071. doi:10.1074/mcp.O113.036475

13. Kirk JA, Holewinski RJ, Kooij V, Agnetti G, Tunin RS, Witayavanitkul N, de Tombe PP, Gao WD, Van Eyk J, Kass DA (2014) Cardiac resynchronization sensitizes the sarcomere to calcium by reactivating GSK-3beta. J Clin Invest 124(1):129–138. doi:10.1172/JCI69253

14. Escher C, Reiter L, MacLean B, Ossola R, Herzog F, Chilton J, MacCoss MJ, Rinner O (2012) Using iRT, a normalized retention time for more targeted measurement of peptides. Proteomics 12(8):1111–1121. doi:10.1002/pmic.201100463

15. Wang Y, Yang F, Gritsenko MA, Wang Y, Clauss T, Liu T, Shen Y, Monroe ME, Lopez-Ferrer D, Reno T, Moore RJ, Klemke RL, Camp DG 2nd, Smith RD (2011) Reversed-phase chromatography with multiple fraction concatenation strategy for proteome profiling of human MCF10A cells. Proteomics 11(10):2019–2026. doi:10.1002/pmic.201000722

16. Han G, Ye M, Zhou H, Jiang X, Feng S, Jiang X, Tian R, Wan D, Zou H, Gu J (2008) Large-scale phosphoproteome analysis of human liver tissue by enrichment and fractionation of phosphopeptides with strong anion exchange chromatography. Proteomics 8(7):1346–1361. doi:10.1002/pmic.200700884

17. Dephoure N, Gygi SP (2011) A solid phase extraction-based platform for rapid phosphoproteomic analysis. Methods 54(4):379–386. doi:10.1016/j.ymeth.2011.03.008

18. MacLean B, Tomazela DM, Shulman N, Chambers M, Finney GL, Frewen B, Kern R, Tabb DL, Liebler DC, MacCoss MJ (2010) Skyline: an open source document editor for creating and analyzing targeted proteomics experiments. Bioinformatics 26(7):966–968. doi:10.1093/bioinformatics/btq054

Chapter 17

Measurement of Phosphorylated Peptides with Absolute Quantification

Raven J. Reddy, Timothy G. Curran, Yi Zhang, and Forest M. White

Abstract

Mass spectrometry, when coupled to on-line separation such as liquid chromatography or capillary electrophoresis, enables the identification and quantification of protein expression and post-translational modification changes under diverse conditions. To date most of the methods for mass spectrometry-based quantification have either provided relative quantification information (e.g., comparison to a selected condition) or utilized one-point calibration curves, or calibration curves in a different biological matrix. Although these quantitative methods have been used to generate insight into the differences between biological samples, additional biological insight could be gained by accurately measuring the absolute quantity of selected proteins and protein modifications. To address this challenge, we have developed the MARQUIS (Multiplex Absolute Regressed Quantification with Internal Standards) method, designed to provide absolute quantification for potentially hundreds of peptides across multiple samples in a single analysis, using a multi-point internal calibration curve derived from synthetic, isotopically distinct standard peptides.

Key words Absolute quantification, Phosphorylation, EGFR, IMAC, LC-MS/MS

1 Introduction

Characterizing active signaling pathways in disease states has led to the development of many therapeutics that have significantly impacted clinical outcomes [1]. Central to these studies is the ability to identify and quantify levels of protein expression and post-translational modification (PTM). The myriad proteomics technologies designed for this task can be divided into two categories: recognition (usually by antibodies or aptamers) and physical measurement. Immunoblotting has been the most common proteomics approach for decades, and has recently been scaled up to facilitate high-throughput measurement in the form of protein microarrays [2]. However, recognition-based techniques are fundamentally constrained by their requirement of a priori knowledge, which prohibits identification of new targets. This is especially limiting in the context of PTMs, which are not hard coded into the

Salvatore Sechi (ed.), *Quantitative Proteomics by Mass Spectrometry*, Methods in Molecular Biology, vol. 1410,
DOI 10.1007/978-1-4939-3524-6_17, © Springer Science+Business Media New York 2016

genome and must first be identified experimentally. Mass spectrometry measures mass to charge ratios of thousands of peptides in complex mixtures, which can be used to identify novel sequences and PTMs. Recent quantification advances, including chemical-stable isotope labeling, metabolic-stable isotope labeling, and label-free quantification, have permitted comparison of thousands of proteins between multiple samples in a single analysis [3, 4]. Combining these methods with immunoaffinity enrichment against specific phosphorylation modifications has yielded extensive coverage of the phosphoproteome [5].

Despite these technical advances facilitating relative quantification, a key deficiency lies in the inability to measure the absolute amount of peptide, protein, or protein PTM in a cell. Stoichiometric information enables additional comparisons often yielding greater biological insight. For example, two proteins that show a twofold relative change could be increasing from 100 to 200 copies or 1 to 2 million copies; two scenarios which may have very different functional consequences. Additionally, absolute quantification of multiple phosphorylation sites indicates which are more prevalent, suggesting more probable interactions mediated by proteins' docking domains.

Previously, these measurements were difficult to obtain. An accurate quantification of protein expression with antibodies requires the use of recombinant standard proteins at known concentrations. Quantification of protein PTMs with this strategy is much more challenging due to the difficulty of establishing the modification state of the recombinant protein. Mass spectrometry offers a simple solution with the inclusion of synthetic standard peptides. One of the most common methods for obtaining absolute quantification is isotope dilution (commercially available as AQUA), in which a known amount of heavy-labeled synthetic peptide is added during processing, and quantification is obtained by comparing the elution profiles of the corresponding endogenous and standard peptides [6]. However, the reliance of a single point for calibration compresses the dynamic range of measurement and amplifies stochastic measurement errors. To address these issues we have developed an alternative method, termed Multiplex Absolute Regressed Quantification of Internal Standards (MARQUIS) [7]. This technique uses isotopically distinct synthetic standard peptides in conjunction with isobaric chemical labeling to create a multi-point internal calibration curve for each target peptide, including peptides with PTMs. Doing so, it faithfully compensates for nonlinear response due to multiple factors, including dynamic range of the instrument, signal-to-noise at the detection limits, and isobaric tag ratio compression that occurs from contaminants during precursor isolation.

The protocol presented here applies MARQUIS to quantify phosphorylation dynamics of the epidermal growth factor receptor (EGFR) signaling network in response to growth factor stimulation. This complex test case illustrates absolute quantification of key signaling targets rapidly changing within a cell. EGFR is a receptor tyrosine kinase capable of transmitting information from the extracellular environment to intracellular decision-making machinery that ultimately elicits behavioral responses. Ligand binding to the extracellular domain of the receptor induces dimerization and activation of the cytoplasmic kinase domain initiating several phosphorylation cascades that affect broad transcriptional programs. The resulting phenotypic changes include proliferation, migration, and differentiation [8]. Though previous work has catalogued the relative phosphorylation dynamics in response to a variety of growth factors, without stoichiometric information these studies have been limited in their ability to ascribe the individual contributions of particular signaling pathways to phenotypic response [9, 10]. Absolute quantification of phosphorylation dynamics in the EGFR signaling network has generated novel insight and yielded testable hypothesis about the structure of this signaling network [7].

The methods presented here are broadly applicable to a variety of systems, including other PTMs such as acetylation, methylation, and ubiquitination, or protein expression. We present standard peptide synthesis guidelines that should be considered during experimental design and highlight processes that can be adapted to multiple different mass spectrometry pipelines. These techniques will provide a new dimension to the application of mass spectrometry to proteomic studies.

2 Materials

2.1 Cell Culture
(See Note 1)

1. MCF10A Human Mammary Epithelial Cells.
2. Complete Media: DMEM/F12 supplemented with 5 % Horse Serum, 20 ng/mL EGF, 0.5 µg/mL hydrocortisone, 100 ng/mL Cholera Toxin, 10 µg/mL Insulin, 1× Pen-Strep.
3. Starve Media: DMEM/F12 supplemented with 0.5 µg/mL hydrocortisone, 100 ng/mL Cholera Toxin, 1× Pen-Strep.
4. 1× Dulbecco's Phosphate-Buffered Saline (PBS).

2.2 Cell Stimulation and Lysis

1. 100 µg/mL EGF dissolved in MilliQ water.
2. Ice-cold PBS.
3. Urea Lysis Buffer: 8 M urea dissolved in MilliQ water, made immediately before use, kept on ice (see Note 2).
4. Cell scrapers.

2.3 Protein Reduction, Alkylation, Digestion

1. BCA Protein Assay.
2. Urea Lysis Buffer.
3. Synthetic standard peptides.
4. Ammonium Acetate Solution: 100 mM ammonium acetate (pH 8.9).
5. Dithiothreitol Solution: 1 M dithiothreitol in ammonium acetate solution.
6. Iodoacetamide Solution: 800 mM iodoacetamide in ammonium acetate solution.
7. Sequencing grade trypsin.

2.4 Peptide Desalting and Lyophilization

1. C-18 Sep-Pak cartridges.
2. Cleanup Equilibration Solution: 90 % acetonitrile in 0.1 % acetic acid.
3. Cleanup Elution Solution: 40 % acetonitrile in 0.1 % acetic acid.
4. Vacuum centrifuge.

2.5 TMT Labeling

1. Dried 800 μg TMT aliquots.
2. Dissolution Buffer: 0.5 M TEAB (triethylammonium bicarbonate, $N(Et)_3HCO_3$).
3. Anhydrous acetonitrile.
4. Ethanol (200 proof).
5. TMT Wash Solution: 40 % acetonitrile in 0.1 % acetic acid.
6. Vacuum centrifuge.

2.6 Precolumn Preparation

1. Fused silica capillary tubing OD 360 μm, ID 100 μm.
2. Silicon cutter.
3. Kasil.
4. YMC gel, ODS-A, 12 nm, S-10 μm, AA12S11 (10 μm beads).
5. Trypsin-digested angiotensin.

2.7 Immunoprecipitation

1. Protein G Plus agarose beads.
2. 4G10 anti-phosphotyrosine antibody, PT-66 anti-phosphotyrosine antibody, pY-100 anti-phosphotyrosine antibody.
3. TMT IP Buffer: 100 mM Tris–HCl, 1 % NP-40 (pH 7.4).
4. IP Buffer: 100 mM Tris–HCl, 0.3 % NP-40 (pH 7.4).
5. Tris Buffer: 500 mM Tris (pH 8.5).
6. IP Rinse Buffer: 100 mM Tris–HCl (pH 7.4).
7. IP Elution Buffer: 100 mM glycine (pH 2).

2.8   IMAC Enrichment

1. IMAC column: see Ref. 11 for details on preparing and testing IMAC columns.
2. EDTA: 100 mM EDTA (pH 8.9).
3. Iron Chloride: 100 mM iron (III) chloride.
4. Organic Rinse: 25 % acetonitrile, 1 % acetic acid, 100 mM NaCl.
5. IMAC Elution Buffer: 250 mM NaH_2PO_4 (pH 8.9).
6. HPLC Solvent A: 0.2 M acetic acid in ultrapure water.

2.9   LC-MS/MS Analysis

1. HPLC Solvent A.
2. HPLC Solvent B: 70 % acetonitrile, 0.2 M acetic acid in ultrapure water.
3. Thermo Scientific Easy-nLC 1000 in conjunction with a Thermo Scientific Q Exactive Orbitrap mass spectrometer.
4. MASCOT Distiller version 2.5 in conjunction with MASCOT Server version 2.4 (Matrix Science, Boston, MA).

2.10   Synthetic Peptide Design and Testing

1. Endogenous target peptide sequences.
2. Thermo Scientific Easy-nLC 1000 in conjunction with a Thermo Scientific Q Exactive Orbitrap mass spectrometer.
3. HPLC Solvent A.
4. HPLC Solvent B.

3   Methods

3.1   Cell Culture

1. Grow MCF10A cells in 10 cm dishes in 10 mL Complete Media at 37 °C and 5 % CO_2, splitting 1:4 when confluent.
2. Seed one plate per experimental condition in Complete Media for 48 h.
3. Aspirate Complete Media, rinse with 10 mL PBS, add 10 mL Starve Media.
4. Incubate cells at 37 °C for 24 h prior to stimulation.

3.2   Cell Stimulation and Lysis

1. Add growth factor directly to starve media to desired concentration, incubate at 37 °C for desired stimulation time.
2. With plate on ice, aspirate media, rinse cells with 10 mL ice-cold PBS to remove residual media, aspirate PBS.
3. Lyse cells by covering the dish with 1 mL Urea Lysis Buffer, using a cell scraper to remove adherent cells or remaining cell debris. Collect lysate in conical tube.
4. Vortex lysate, remove 10 µL aliquot for protein assay.
5. Store at −80 °C until further processing.

**3.3 Protein
Reduction, Alkylation,
Digestion**

1. Perform BCA Protein Assay with 10 µL lysate aliquot to determine protein concentration of each lysate (*see* **Note 3**).

2. Thaw lysate, add 400 µg protein from each sample into a fresh conical tube, and equalize sample volumes by adding Urea Lysis Buffer.

3. Add desired range of heavy-labeled synthetic peptides to samples (*see* **Note 4**).

4. Add 1:100 Dithiothreitol Solution to sample, to a final concentration of 10 mM dithiothreitol. Incubate at 56 °C for 1 h.

5. Add 1:14.5 Iodoacetamide Solution to sample, to a final concentration of 55 mM iodoacetamide. Incubate on rotor at room temperature for 1 h. Exposure of iodoacetamide to light should be limited by wrapping tubes in aluminum foil.

6. Dilute samples by adding 2.5× original lysate volume of ammonium acetate (e.g., 4 mL lysate would receive 10 mL ammonium acetate).

7. Add sequencing grade trypsin at a ratio of 1:50 (8 µg trypsin:400 µg lysate), allow digestion to proceed on rotor overnight (16 h) at room temperature.

8. Stop digestion by adding glacial acetic acid to 10 %.

9. Add any synthetic peptides containing missed cleavage sites at desired range.

10. Spin down samples to remove any debris that may clog Sep-Pak cartridge. Digested samples may be stored at –80 °C.

**3.4 Peptide
Desalting
and Lyophilization**

1. Acidify C-18 Sep-Pak cartridge (Waters WAT023501) with 10 mL 0.1 % acetic acid at a flow rate of 2 mL/min. Use a syringe pump for multiple samples, if available.

2. Equilibrate the cartridge with 10 mL Cleanup Equilibration Solution at a flow rate of 2 mL/min.

3. Wash the cartridge with 10 mL 0.1 % acetic acid at a flow rate of 2 mL/min.

4. Load the acidified peptide samples at a flow rate of 1 mL/min.

5. Wash the sample loaded cartridge with 10 mL 0.1 % acetic acid at a flow rate of 2 mL/min.

6. Elute the peptides into a clean conical tube with 10 mL Cleanup Elution Solution at 1 mL/min.

7. Reduce the total volume of each sample to less than 1 mL in a vacuum centrifuge.

8. Freeze the sample by immersing in liquid nitrogen for 10 min.

9. Lyophilize the sample overnight, or until all solvent has sublimated. Lyophilized peptides may be stored at –80 °C for several months.

3.5 TMT Labeling

1. Make peptide resuspension solution of 70 % ethanol and 30 % Dissolution Buffer.

2. Resuspend lyophilized peptides in 100 µL 70 % ethanol/30 % Dissolution Buffer, vortex for 1 min, and centrifuge at $12,000 \times g$ for 1 min.

3. Add 40 µL anhydrous acetonitrile to each TMT aliquot, vortex for 1 min, centrifuge at $12,000 \times g$ for 1 min.

4. Add resuspended TMT label to corresponding peptide sample. Vortex each sample for 1 min and centrifuge at $12,000 \times g$ for 1 min.

5. Incubate for 1 h at room temperature.

6. Reduce the total volume of each sample to ~30 µL in vacuum centrifuge (approximately 30 min).

7. Aliquot 1 mL of 40 % acetonitrile in 0.1 % acetic acid in a fresh tube.

8. Combine all TMT samples into a single tube (*see* **Note 5**).

9. Add 40 µL TMT Wash Solution to each tube, vortex 1 min, centrifuge at $12,000 \times g$ for 1 min, add rinse to sample tube.

10. Repeat rinsing procedure (done two times total).

11. Bring sample to dryness in vacuum centrifuge (small, dark pellet should form at the bottom of the tube).

12. Dried sample can be stored at −80 °C for several months.

3.6 Precolumn Preparation (*See* Note 6)

1. Cut approximately 20 cm long fused silica capillary with fused silica cutter.

2. Make frit mix by mixing Kasil and formamide (5:1) in microcentrifuge tube, vortexing briefly and centrifuging at $12,000 \times g$ for 1 min (*see* **Note 7**).

3. Dip one end of the column into the tube until material rises into the capillary about 0.5–1 cm.

4. Bake fritted columns at 100 °C for 10 min, ensuring that fritted end is not in contact with any surfaces.

5. Using a helium pressure injection cell (a.k.a. column packing bomb) on top of a magnetic stir plate, flush the column with acetonitrile at 400 psi for 5 min (*see* **Note 8**).

6. Resuspend small amount of YMC ODS-A beads in a glass vial containing 80 % acetonitrile/20 % isopropanol and add a magnetic stir bar.

7. Place the vial into the helium pressure injection cell with the stir plate turned on.

8. Pack beads with 500 psi until column bed length reaches 10 cm from the end of the frit.

9. Wash the column with 0.1 % acetic acid at 400 psi for 10 min.

10. Dry the column with helium at 400 psi for 10 min.

11. Cut dried column 1–2 cm from end of bead bed.

12. Prepare fresh frit mix.

13. Dip the second end of the column into the frit mix until material rises into the capillary about 0.5–1 cm.

14. Cure the second frit using a heat gun.

15. Wash the column with 0.1 % acetic acid at 400 psi for 10 min.

16. Condition precolumn with 500 fmol angiotensin.

17. Remove excess angiotensin by washing with acetonitrile at 400 psi for 5 min.

18. Remove organic solvent and recondition column by washing with 0.1 % acetic acid at 400 psi for 5 min.

3.7 Immunoprecipitation

1. Wash 60 μL Protein G Plus agarose beads with 300 μL IP Buffer. For all wash steps: combine in a microcentrifuge tube, place on rotator at 4 °C for 5 min, centrifuge at 4 °C for 1 min at $4000 \times g$, remove supernatant removing as much liquid but as few beads as possible using a gel loading pipette tip.

2. Resuspend beads with 300 μL IP Buffer and add 12 μg of each antibody to the washed beads.

3. Allow the mixture to incubate on a rotor at 4 °C for 6–8 h.

4. Wash the beads with 400 μL IP Buffer.

5. Resuspend TMT pellet in 400 μL TMT IP Buffer by vortexing.

6. Check pH of sample with 2 μL on pH strip, comparing with IP Buffer. If pH is lower than 7.4, add 5 μL of Tris Buffer, vortex, and measure again. Repeat until sample pH matches IP Buffer pH.

7. Add TMT sample to washed beads and incubate on rotor at 4 °C overnight (>12 h).

8. Centrifuge sample for 1 min at 4° C at $4000 \times g$, collect supernatant in a fresh microcentrifuge tube, store at −80 °C.

9. Wash the beads once with 400 μL TMT IP Buffer.

10. Wash the beads three times with 400 μL IP Rinse Buffer.

11. After final wash, add 70 μL of IP Elution Buffer and incubate at room temperature on rotor for 30 min.

12. Load eluted sample onto an IMAC column.

3.8 IMAC Enrichment

1. Prepare an IMAC column for metal affinity enrichment of phosphopeptides.

2. Rinse the IMAC column with EDTA Solution for 10 min at a flow rate of 10 μL/min.

3. Wash the IMAC column with MilliQ water for 10 min at a flow rate of 10 μL/min.

4. Load the IMAC column with Iron Chloride at a flow rate of 10 μL/min for 10 min.

   (a) Optional: Flip the column to flow in opposite direction at 10 min.

5. Rinse the IMAC column with 0.1 % acetic acid for 10 min at a flow rate of 10 μL/min.

6. To collect the non-retained, nonphosphorylated peptides, attach a flow-through precolumn to the IMAC with a Teflon connector, test junction by flowing 0.1 % acetic acid at 800 psi.

7. Determine pressure needed to generate a flow rate of 1 μL/min through the IMAC and precolumn in series.

8. Replace 0.1 % acetic acid with eluate from immunoprecipitation, load sample at 1 μL/min (*see* **Note 9**).

9. Remove precolumn containing flow-through peptides.

10. Rinse the IMAC column with Organic Rinse for 5 min at 10 μL/min (*see* **Note 10**).

11. Rinse with 0.1 % acetic acid for 5 min at 10 μL/min.

12. Place a fresh precolumn on the IMAC column with a Teflon connector, test junction by flowing 0.1 % acetic acid at 800 psi.

13. Determine pressure needed to generate a flow rate of 2 μL/min through the IMAC and precolumn in series.

14. Replace 0.1 % acetic acid with IMAC Elution Buffer, flow 40 μL Elution Buffer over IMAC and precolumn in series (*see* **Note 11**).

15. Rinse precolumn with HPLC Solvent A for 10 min prior to LC-MS/MS analysis.

3.9 LC-MS/MS Analysis

1. Analyze peptides eluted from IMAC by LC-MS/MS using reverse-phase chromatography performed in line with a Q Exactive mass spectrometer.

2. Elute peptides using a 120-min gradient (0–100 % HPLC Solvent A to Solvent B).

3. Acquire data using the mass spectrometer in targeted acquisition mode.

   (a) Acquire SIM scans at 70k resolution for each pair of endogenous peptide and its heavy isotope standard, with isolation window set to include both ions.

   (b) Acquire MS/MS scans for both endogenous peptide and heavy isotope standard. Typical settings include an MS1 isolation width of 2 m/z, MS2 fragmentation collision energy of 35.0, MS2 maximum ion injection time of 2 s, and an AGC target of 3e6 (this large AGC target is chosen to maximize dynamic range).

4. Fragmentation of the synthetic peptide produces a standard curve, with calibration points covering the concentration range of peptides that were originally added to each biological sample. This step also provides a control for the linear dynamic range and noise floor of the TMT marker ions.

5. Total endogenous peptide is calculated by comparing the signal intensity of endogenous peptide with the standard peptide in the SIM scan.

6. Endogenous peptide concentrations in each sample can be calculated by apportioning the total amount of endogenous peptide between input conditions using the fractional reporter ion intensities generated by MS2 fragmentation of the endogenous peptide precursor.

3.10 Synthetic Peptide Design and Testing

1. Peptides should be synthesized containing identical sequences to endogenous target peptides, according to the specificity of the selected proteolytic enzyme: e.g., for trypsin, peptides would span from the residue immediately prior to the N-terminal K or R residue to the C-terminal K or R residue.

2. Peptides must contain at least one (but can have many) heavy isotope encoded amino acid residues. Note that larger peptides might require two heavy isotope encoded amino acid residues to ensure adequate separation between the endogenous and synthetic peptides during precursor isolation.

3. Synthetic peptides should be quantified by amino acid analysis to obtain accurate concentrations.

4. Multiple standard peptides may be pooled to create a single peptide cocktail that can be added to lysates.

5. Peptides containing frequently occurring missed cleavage sites may also be synthesized (these may also be pooled to create a second mixture, but should be kept separate from standard tryptic peptide cocktail to be added after the digestion step).

6. Analyze standard peptide mixture using reverse-phase chromatography performed in line with a Q Exactive mass spectrometer, eluting peptides with a 120-min gradient (0–100 % HPLC Solvent A to Solvent B).

7. Determine target peptide elution windows from extracted ion chromatogram (XIC) using calculated peptide precursor m/z ratios.

8. Create Inclusion List for targeted MS analysis. This should include:

(a) Peptide precursor m/z ratios for synthetic standard peptides and endogenous peptides at multiple potential charge states (e.g., +2, +3)

(b) Elution start and end times, as determined from **step 7**.

4   Notes

1. Cell culture can be performed with a variety of cell lines, with a minimum necessary protein content of 400 µg per sample.

2. (Optional) Add 1 mM activated sodium orthovanadate to prevent phosphatase activity.

3. (Optional) In addition to protein standard, sample lysates may be compared to control lysates of known cell counts to obtain measurement of cells/µg. This can be later used to convert peptide measurement to copies/cell.

4. Example TMT 10plex scheme: 3 pmol, 1 pmol, 300 fmol, 100 fmol, 30 fmol, 10 fmol, 3 fmol, 1 fmol, 0.3 fmol, 0.1 fmol.

5. Use one pipette tip for all combining and washing steps to minimize sample loss.

6. IMAC enrichment is also compatible with commercial precolumn setups.

7. Frit mix will polymerize with time, so this step should be performed quickly. Overly polymerized mixture will not rise into the column.

8. For columns that do not flow immediately, use silicon cutter to cut a small piece of the fritted end of the column.

9. Eluate need not be removed from microcentrifuge tube with beads. Frits prevent agarose beads from flowing through columns.

10. (Optional) Flip the column to flow in opposite direction at 5 min to decrease nonspecific binding.

11. For autosampler setups, sample may be collected in a fresh autosampler vial placed inverted on top of the IMAC column.

References

1. Morris M, Chi A, Melas I et al (2014) Phosphoproteomics in drug discovery. Drug Discov Today 19:425–432

2. Jones R, Gordus A, Krall J et al (2005) A quantitative protein interaction network for the ErbB receptor using protein microarrays. Nature 439:168–174

3. Liang S, Xu Z, Xu X et al (2012) Quantitative proteomics for cancer biomarker discovery. Comb Chem High Throughput Screen 15:221–231

4. Wasinger V, Zeng M, Yau Y (2013) Current status and advances in quantitative proteomic mass spectrometry. Int J Proteomics 2013:1–12

5. Wolf-Yadlin A, Kumar N, Zhang Y et al (2006) Effects of HER2 overexpression on cell signaling networks governing proliferation and migration. Mol Syst Biol 2:1–15

6. Gerber S, Rush J, Stemman O et al (2003) Absolute quantification of proteins and phosphoproteins from cell lysates by tandem MS. Proc Natl Acad Sci U S A 100:6940–6945

7. Curran T, Zhang Y, Ma D et al (2015) MARQUIS: a multiplex method for absolute quantification of peptides and posttranslational modifications. Nat Commun 6:1–11

8. Yarden Y (2001) The EGFR family and its ligands in human cancer: signalling mechanisms and therapeutic opportunities. Eur J Cancer 37:S3–S8

9. Gan H, Cvrljevic A, Johns T (2013) The epidermal growth factor receptor variant III (EGFRvIII): where wild things are altered. FEBS J 280:5350–5370

10. Schulze W, Deng L, Mann M (2005) Phosphotyrosine interactome of the ErbB-receptor kinase family. Mol Syst Biol 1:42–54

11. Moser K, White F (2006) Phosphoproteomic analysis of rat liver by high capacity IMAC and LC-MS/MS. J Proteome Res 5:98–104

Chapter 18

Proteomic Analysis of Protein Turnover by Metabolic Whole Rodent Pulse-Chase Isotopic Labeling and Shotgun Mass Spectrometry Analysis

Jeffrey N. Savas, Sung Kyu Park, and John R. Yates III

Abstract

The analysis of protein half-life and degradation dynamics has proven critically important to our understanding of a broad and diverse set of biological conditions ranging from cancer to neurodegeneration. Historically these protein turnover measures have been performed in cells by monitoring protein levels after "pulse" labeling of newly synthesized proteins and subsequent chase periods. Comparing the level of labeled protein remaining as a function of time to the initial level reveals the protein's half-life. In this method we provide a detailed description of the workflow required for the determination of protein turnover rates on a whole proteome scale in vivo.

Our approach starts with the metabolic labeling of whole rodents by restricting all the nitrogen in their diet to exclusively nitrogen-15 in the form of spirulina algae. After near complete organismal labeling with nitrogen-15, the rodents are then switched to a normal nitrogen-14 rich diet for time periods of days to years. Tissues are harvested, the extracts are fractionated, and the proteins are digested to peptides. Peptides are separated by multidimensional liquid chromatography and analyzed by high resolution orbitrap mass spectrometry (MS). The nitrogen-15 containing proteins are then identified and measured by the bioinformatic proteome analysis tools Sequest, DTASelect2, and Census. In this way, our metabolic pulse-chase approach reveals in vivo protein decay rates proteome-wide.

Key words Proteomics, Mass spectrometry, Protein half-life, Protein decay dynamics, Stable isotope labeling of mammals, Nitrogen-15, SILAC, SILAM, Extremely long-lived proteins

1 Introduction

To determine the rate of protein decay, new and old versions of each protein must be discernable and ideally both be measurable. Typically cells are initially "pulsed" with a traceable molecular label (such as methionine enriched with sulfur-35 atoms) which are incorporated into newly synthesized proteins and subsequently "chased" with a normal containing methionine (sulfur-32 fraction of 95.02 %) [1, 2]. By comparing the initial amount of a specific labeled protein to that remaining as a function of time, a measure

Salvatore Sechi (ed.), *Quantitative Proteomics by Mass Spectrometry*, Methods in Molecular Biology, vol. 1410,
DOI 10.1007/978-1-4939-3524-6_18, © Springer Science+Business Media New York 2016

of protein half-life can be obtained [3]. These analyses can also provide key information on the possibility that different pools of the same protein exist and have dissimilar decay kinetics. Recently, proteomic technologies have been applied to gain insight into the analysis of protein turnover dynamics on a proteome-wide scale. By combining the yeast whole genome tap-tag gene library, translation inhibition with cycloheximide, and epitope tag western blot analysis, it was determined that on average the lifetime of a yeast protein is about 43 min [4]. In cultured HeLa and C2C12 cells, "pulse-only" stable isotope labeling of cells in culture (SILAC) for several durations with time course mass spectrometry (MS) analysis showed average protein half-life in mouse and human cells <2 days [5]. In mice, by using MS analysis to measure the rate by which isotopes are metabolically incorporated into proteins and modeling it has been suggested that on average proteins in brain tissue have a lifetime of 9.0 days, liver 3.0 days, and blood 3.5 days [6, 7].

We have developed a straightforward systematic approach to monitor protein decay dynamics on a global scale in the most relevant biological context, in vivo. Our approach has verified previously reported rapid degradation dynamics for nearly all proteins. Unexpectedly we also find a limited number of intracellular extremely long-lived proteins (ELLPs) which reside in the nucleus and cytoplasm of postmitotic neurons [8, 9]. Our approach also confirmed the existence ELLPs in the myelin sheath and eye lens [10–12]. The application of our approach to proteinopathy disease mouse models (such as Alzheimer's, Parkinson's, and Huntington's disease) could provide new insight into pathogenic mechanisms by identifying disease-specific long-lived proteins.

2   Materials

Buffers and solutions for MS analysis should be prepared with analytical chromatography grade solvents, and for biochemical experiments we prepared buffers with ultra-pure water (Milli-Q® Water Purification Systems, 18-megohm-cm deionized water). Solutions should be stored at room temperature unless otherwise indicated. To minimize keratin contamination gloves should be worn during the preparation of all buffers and samples.

2.1   Nitrogen-15 or Nitrogen-14 Spirulina Rodent Chow

1. Nitrogen-15 enriched spirulina algae: Nitrogen-15 enriched (>94 %) spirulina were purchased from Cambridge Isotopes [13], Cambridge, MA, USA, or can be grown and prepared in-house as previously described [14–16].

2. Rodent chow: Rodent chow has been prepared by mixing nitrogen-15 or nitrogen-14 spirulina with protein-free diet mixture powder (Harlan TD 93328) in a 1 to 3 ratio. Pellets were

prepared by adding ultra-pure water to the power mixture and working the mixture into dough shaped into cylinders. Individual ~2-cm discs we cut from the cylinders and dried at 60 °C for 2–4 h and then at 35 °C overnight on screen trays in an Excalibur food dehydrator [17]. Alternatively, nitrogen-15 spirulina containing chow can be purchased pre-prepared from CIL/Harlan Laboratories Inc. with 22 % protein/65 % carbohydrate (carbon, hydrogen, oxygen as CHO), 13 % fat composition (*see* **Note 1**).

2.2 Representative Protein Fractionation

1. Tissue homogenization buffer: 0.32 M sucrose, 4 mM Hepes (pH 7.4), 1 mM MgCl$_2$, and protease inhibitors (Sigma) (*see* **Note 2**). 1 M Hepes, add 600 mL water to a glass beaker, weigh and add 238.3 g of Hepes, add stir bar to dissolve on a stir plate. Determine pH and adjust with HCl or NaOH to pH 7.4 final. Transfer to a graduated cylinder and add water to 1 L. Store at 4 °C. 1 M MgCl$_2$, add 500 mL water to a glass beaker, weigh and add 203.3 g of MgCl$_2$ 6H$_2$O, add stir bar to dissolve on a stir plate. Transfer to a graduated cylinder and add water to 1 L. To a 250 mL glass beaker, a stir bar, add 50 mL of water, 5 mL of 1 M Hepes (pH 7.4), 0.1 mL of 1 M MgCl$_2$, and 10.9 g of Sucrose. Transfer to a 100 mL graduated cylinder and add water to 100 mL [18].

2. Sucrose gradient buffers (0.85 M/1.0 M/1.2 M/2.0 M): Weigh 28.9, 34.0, 40.9, 69.1 g and prepare 100 mL of buffer as described above except substitute the indicated amount of sucrose for each buffer.

3. 2,2,2-Trichloroacetic acid (TCA) buffer: prepare a 100 % (wt/vol) TCA solution with water.

2.3 Protein Digestion

1. Urea protein denaturation buffer: Dissolve 0.395 g of solid Ammonium bicarbonate (AMBC) in 100 mL of water to prepare 50 mM adjust to pH 7.5 as described above; aliquot and store at −20 °C. Add 0.240 g of urea to 320 µl of AMBC buffer to prepare 8 M solution (*see* **Note 3**).

2. ProteaseMAX surfactant buffer: Dissolve solid ProteaseMAX in 500 µl of AMBC to prepare 0.2 % solution or 100 µl for 1 % (*see* **Note 4**).

3. Reduction buffer: Dissolve solid Tris (2-carboxyethyl) phosphine hydrochloride (TCEP) in AMBC to prepare 0.5 M solution (*see* **Note 5**).

4. Alkylation buffer: Dissolve solid Iodoacetamide in AMBC and prepare 1 M solution.

5. Trypsin buffer: Dissolve 20 µg vial of lyophilized trypsin (Promega) in 40 µl of buffer (*see* **Note 6**).

2.4 Liquid Chromatography

1. HPLC buffer A: 95 % water, 5 % acetonitrile, and 0.1 % formic acid (vol/vol).

2. HPLC buffer B: 20 % water, 80 % acetonitrile, and 0.1 % formic acid (vol/vol).

3. HPLC buffer C: 500 mM ammonium acetate, 5 % (vol/vol) acetonitrile, and 0.1 % (vol/vol) formic acid.

4. Make Kasil frit and prepare multidimensional protein identification (MudPIT) column by bomb packing strong cation exchange (SCX)/reversed phase resins as previously described [19–21].

2.5 Mass Spectrometer

1. Tune and calibrate electrospray high resolution orbitrap mass spectrometer (Thermo Scientific™ Orbitrap Velos Pro or Orbitrap Tribrid Fusion) per the manufacturer's instructions with Pierce LTQ Velos ESI Positive Ion Calibration Solution (*see* **Note 7**).

2.6 Proteomic Analysis Software

1. IP2 (Integrated Proteomic Analysis environment is commercially available; http://integratedproteomics.com/).

2. RawExtractor (Spectra extraction tool is freely downloadable; http://fields.scripps.edu/researchtools.php).

3. Sequest/Prolucid (Protein database search algorithm is freely downloadable; http://fields.scripps.edu/researchtools.php).

4. DTASelect2 (protein dataset filtering tool is freely downloadable; http://fields.scripps.edu/researchtools.php).

5. Census (protein quantitation software is freely downloadable; http://fields.scripps.edu/researchtools.php).

3 Methods

Perform all procedures at room temperature unless noted.

3.1 Metabolic Labeling of Whole Rodents

1. Obtain two recently weaned female rats and allow acclimating in the university approved animal facility for several days (*see* **Note 8**).

2. Replace standard rodent chow with nitrogen-15 containing spirulina chow and house for >10 weeks (*see* **Note 9**).

3. Introduce male breeder rat into breeding cages and monitor female rat for weight gain indicative of successful pregnancy (*see* **Note 10**).

4. Closely monitor cages for pups and document successful breeding. Continue feeding with exclusively nitrogen-15 containing spirulina chow while pups are nursing (*see* **Note 11**).

5. Once pups are weaned, feed with exclusively nitrogen-15 containing spirulina chow for additional 3–4 weeks.

6. Start chase period by switching to regular nitrogen-14 rodent chow (*see* **Note 12**).

3.2   Tissue Harvest

1. Sacrifice time = 0 animal with CO_2 as the primary mechanism and secondarily by decapitation.

2. Harvest all tissues with standard dissection procedures and carefully label and freeze each tissue in a separate tube in liquid nitrogen and then store at –80 °C.

3. Sacrifice littermates at additional time points and repeat dissection and tissue harvesting as needed (*see* **Note 13**) (Fig. 1).

3.3   Representative Protein Fractionation from Brain Tissue

1. Homogenize rat brain in 12 mL of tissue homogenization buffer on ice and centrifuge at 4 °C, 1500 × *g* for 15 min, and the supernatant was collected (postnuclear supernatant).

2. Centrifuge supernatant at 4 °C, 18,000 × *g* for 20 min, collect the resulting supernatant (cytosol) and pellet (crude membrane).

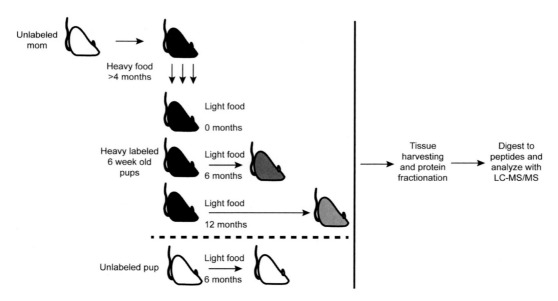

Fig. 1 Metabolic pulse chase labeling of rats workflow to measure protein turnover dynamics in vivo. Freshly weaned female rat (first generation) is obtained and the diet is switched completely to nitrogen-15 containing food for 10–16 weeks. Male rat is introduced and female rat remains on nitrogen-15 diet while pregnant and during the nursing of her pups. Pups (second generation) are sacrificed at several time points including time = 0, before switching to nitrogen-14 chow. For the identification and analysis of extremely long-lived proteins, we found 6 month and 12 month reliable chase durations. As a negative control, we analyze an unlabeled pup after feeding regular nitrogen-14 chow. After the animals are sacrificed, their tissues are dissected, proteins solubilized and then fractionated. The proteins are then digested to peptides prior to LC-MS and bioinformatic analysis

3. Resuspend pellet in homogenization buffer and load it onto a 0.85 M/1.0 M/1.2 M sucrose gradient and centrifuge at 4 °C, 78,000 ×g for 120 min, and collect the material focused at the 1.0 M/1.2 M interface (synaptosomes).

4. Add Triton X-100 to 0.5 % final concentration and extract at 4 °C, by end-over-end agitation for 20 min.

5. Centrifuge the extract at 4 °C, 32,000 ×g for 20 min, and collect the supernatant (soluble synaptosome).

6. Resuspend pellet in homogenization buffer and load onto a 1.0 M/1.5 M/2.0 M sucrose gradient and centrifuge at 4 °C, 170,000 ×g for 120 min [18].

7. Collected material at the 1.5 M/2.0 M interface (postsynaptic density, PSD).

8. Add 0.5 % Triton X-100 and detergent soluble material extracted at 4 °C, by end-over-end agitation for 10 min.

9. Centrifuge extract at 4 °C, 100,000 ×g for 20 min, and resuspend the pellet in homogenization buffer (purified PSD).

3.4 Protein Digestion and Peptide Preparation

1. To each fraction (100 µg) add TCA to 20 % (vol/vol) final concentration, vortex, incubate on ice at 4 °C for 4 h to overnight (*see* **Note 14**).

2. Centrifuge at 14,000 ×g for 45 min at 4 °C.

3. Discard supernatant and wash the pellet with 1 mL of ice-cold acetone.

4. Centrifuge the tube at 14,000 ×g for 10 min at 4 °C.

5. Remove the acetone and wash the pellet with 1 mL of ice-cold acetone (two washes in total).

6. Centrifuge the tube at 14,000 ×g for 10 min at 4 °C.

7. Remove supernatant and air-dry the pellet at room temperature.

8. Add 50 µl of urea buffer and resuspend dry protein pellet and vortex for at least 1 h.

9. Add 50 µl of 0.2 % (wt/vol) ProteaseMAX and vortex for at least 1 h.

10. Add 1 µl of TCEP buffer and vortex the mixture for at least 1 additional hour.

11. Add 2 µl of IAA buffer, mix well, and incubate in the dark for 20 min.

12. Squelch alkylation reaction by adding 5 µl of TCEP buffer.

13. Add 150 µl of AMBC and mix well (*see* **Note 15**).

14. Add 2.5 µl of 1 % (wt/vol) proteaseMAX and briefly vortex.

15. Add 2–4 μg of sequencing-grade trypsin and incubate the mixture overnight at 37 °C with shaking.

16. Recover the samples and store them at –80 °C (*see* **Note 16**).

3.5 Loading the Peptides on the Column

1. Thaw peptides and acidify to a 5 % (vol/vol) final concentration with formic acid.

2. Centrifuge the tube at $14,000 \times g$ for 15 min at room temperature and transfer supernatant to a new tube.

3. Directly Load peptide sample onto SCX/RP column with a bomb at a pressure of 500–1000 psi [22] (*see* **Note 17**).

4. Wash column with buffer A for 30 min on bomb.

5. Pull a 15-cm tip of 100-μm glass capillary and use bomb to pack RP resin.

6. Flow buffer B for 15 min to wash the analytical tip.

7. Flow buffer A for 15 min to equilibrate the analytical tip.

3.6 Liquid Chromatography/Mass Spectrometry

1. Connect the MudPIT column (frit connected to the analytical tip with an IDEX union) to the HPLC pump and start buffer A to ensure stable flow rates and pressure without leaks.

2. Generate 11-step LC and MS methods with Xcalibur software.

3. Start the 11-step LC/MS sequence with the Xcalibur software on the MS computer (*see* **Note 18**). The analysis will be performed over a 22–24 h time period per sample analysis.

3.7 Bioinformatic Data Analysis

1. Process acquired .RAW files by first extracting them to .MS1 and .MS2 format with RawExtractor software on the mass spectrometer's PC [23].

2. Upload all files (33 total, 11 .RAW, 11 .MS1, and 11 .MS2) into the IP2 software.

3. Perform Prolucid heavy and light database search with the rat (species matched) protein database and parameters such as a fixed modification of 57.02146 on cysteine, possessing at least one tryptic terminus, and with unlimited missed cleavages [24] (*see* **Note 19**) (Fig. 2).

4. Filter and control false-discovery rate for each dataset individually with DTASelect with target-decoy strategy (concatenated forward-reverse amino acid sequence protein database) to ensure a 0–1 % false discovery rate at the protein level [25].

5. To view the proteins which are identified (based on matched MS scan) only in the heavy search, run "heavy only" DTASelect analysis.

6. Perform peptide quantitation and enrichment analysis with Census software within IP2 [26–28] (*see* **Note 20**) (Fig. 3).

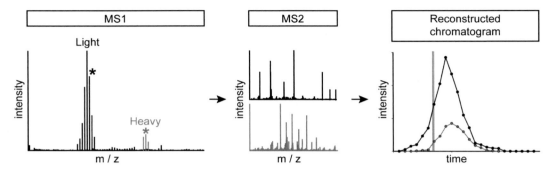

Fig. 2 Bioinformatic spectral analysis paradigm. Theoretical representation of a zoomed MS1 spectral scan, *starred peaks* are selected for MS2 and indicate identification of both the abundant nitrogen-14 light (*starred peak*) and the low abundance nitrogen-15 heavy (*starred peak*) isotopic peaks. MS1 ion abundance is analyzed as reconstructed chromatograms based on the identification of the light or heavy peak (*grey bar*). To determine the peptide abundances, the area under *each curve* is calculated and compared to determine the relative abundances of the light "new" and heavy "old" peptides

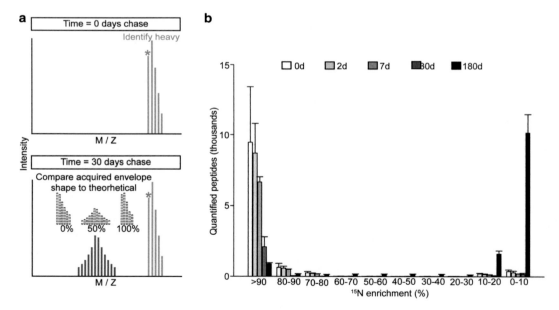

Fig. 3 Incorporation of MS1 isotopic envelope shape measurement into protein turnover analysis workflow increases confidence and shows system-wide protein degradation dynamics. (**a**) Theoretical MS1 isotopic spectral envelope after 0 or 30 day nitrogen-14 chase periods, both showing identification of the fully heavy labeled peptide species (100 % of nitrogen atoms are nitrogen-15). The corresponding "light" isotopic envelope enrichment is determined by comparing the acquired *m/z* isotopic envelope shape to a broad range of predicted enrichment peak patterns to determine the percentage of nitrogen-15 atoms. (**b**) Binned peptide nitrogen-15 enrichment distribution from synaptosome extracts after 0, 2, 7, 30, or 180 days of nitrogen-14 chase

4 Notes

1. Spirulina algae have been successfully grown on nitrogen-15 salts in research labs or can be purchased commercially [7, 29]. We have found it to be most efficient to purchase the nitrogen-15 spirulina already prepared as ready to eat chow.

2. We present here a representative protein fractionation scheme to enrich for postsynaptic density proteins. Any protein fractionation or enrichment procedure (that is compatible with MS analysis) could be utilized for the investigation of protein turnover dynamics depending on the protein's specific localization characteristics.

3. We find that 50 mM AMBC is best aliquoted into single-use tubes and stored at –20 °C and 8 M urea should be prepared fresh for each experiment.

4. ProteaseMax can be freeze thawed a few times without any significant decrease in efficacy.

5. TCEP should be aliquoted into single-use tubes at 20 μl per tube.

6. IAA should be aliquoted into single-use tubes at 10 μl per tube.

7. We believe that for success the MS instrument used for these experiments must be clean, high resolution, and fast scanning. It is our experience that older instruments such as Orbitrap XL do not have the necessary analytical power required for these experiments. The MS should be maintained, cleaned, tuned, and calibrated regularly and as described by the manufacturer.

8. Acquire a recently weaned animal in accordance with the university policies and IACUC approval. All animal use must be performed in compliance with the relevant regulations and governmental guidelines. Make sure all the lab members who will be handling animals are capable and proficient with all animal procedures prior to starting this work.

9. We suggest providing the nitrogen-15 rodent chow ad libitum. It has been our experience that mice will eat 2–3 g and rats will eat 5–6 g of spirulina per day. These are rough guidelines and the animals will eat less or more depending on their age and if they are pregnant.

10. As a cost-saving measure to reduce the amount of nitrogen-15 chow necessary for these experiments, we have found that introducing the male rodent only at night into the female's cage during labeling to be sufficient for sucessful breeding. Each morning we remove the male animal and re-introduce at the end of the day.

11. Identifying a litter of pups on the day of birth is critical for the time = 0 time point; thus we suggest checking for pups every day

once a pregnancy is detected. We have found that on occasion it is difficult to identify pregnant rodents if the litter size is very small; however standard practices (such as checking for a plug) can provide some guidance. When the litter size is large (>4), it is easy to identify the pregnancy, at which time the male rat should not be introduced any more.

12. For the chase period we have found that using "normal chow" (chow containing a nitrogen-14 fraction of 99.636 %) to be sufficient for these experiments. The alternative of using special food specifically composed with enriched nitrogen-14 would be a more perfect yet more expensive approach.

13. We have used several chase period time points to protein decay/turnover and to identify extremely long-lived proteins. It has been suggested that a log scale should be used since it will provide a broad range of analytical coverage [30].

14. We recommend determining the protein concentration and aliquoting 100 μg for each MS analysis prior to precipitating the proteins or digesting to peptides.

15. It is critical that the urea concentration be ≤2 M so that trypsin activity will not be inhibited.

16. We find that once the proteins are digested to peptides they can be stored at –80 °C for up to 3 months. Note, peptides should be frozen before the addition of the FA. Addition of FA prior to freezing will result in degradation and significantly compromised protein identifications.

17. We find that direct loading of samples onto LC columns is the most sensitive approach since peptide loss is certainly minimized. Details on bomb loading have been previously described [20, 21, 31]. However if the proteins of interest are sufficiently enriched by fractionation autosampler loading should be sufficient for the analysis of low abundance proteins.

18. MudPIT analysis has been previously described [32–34]. Briefly in **step 1** the peptides are eluted from the RP trap to the SCX section with increasing percentages of buffer B. Each of **steps 2–11** starts with an increasingly large salt pulse (10, 20, 30, 40, 50, 60, 70, 80, 90, 100 %) of 5 min followed by a shallow linear gradient of increasing buffer B. **Steps 2–11** provide orthogonal peptide separations which facilitates very deep MS based analysis of complex peptide mixtures. **Step 1** is typically 45 min and **steps 2–11** are 2 h each. The exact settings on the MS will vary but we recommend a full-MS from 500 to 1800 m/z and a minimum intensity threshold of 500 for MS/MS. We reject unassigned and +1 charged precursor ions and use a rolling exclusion list of 20 ions. For these experiments, we recommend using 15–20 MS/MS per MS precursor scan.

19. The protein database is critically important since in order to identify a protein with shotgun proteomics the protein sequence must be present in the protein database. We recommend using Uniprot protein databases.

20. For the nitrogen-15 stable isotope enrichment calculation, we used the Census program to perform 15N enrichment ratio calculation. Census uses the amino acid elemental composition to calculate corresponding isotopic distributions of nitrogen-15 enriched peptides. As nitrogen-15 labeling shifts the mass of peptide based on the number of nitrogen atoms present, Census uses all possible theoretical isotope distributions and maps to experimental ones to find the best match by using Linear regression. Census performs the atomic percent enrichment calculation for each peptide independently, as this can vary depending on a protein's turnover rate. A detailed description of the Census enrichment calculation analysis has been previously described [14].

Acknowledgements

Funding for JRY has been provided by National Institutes of Health grants P41 GM103533, R01 MH067880, R01 MH100175, UCLA/NHLBI Proteomics Centers (HHSN268201000035C). JNS is supported by the Pathway to Independence Award National Institutes of Health (K99DC013805). We acknowledge Martin Hetzer and Brandon Toyama for their involvement in initializing the project and also Varda Levram-Ellisman and Roger Tsien for their input.

References

1. Shi G, Nakahira K, Hammond S, Rhodes KJ, Schechter LE, Trimmer JS (1996) Beta subunits promote K+ channel surface expression through effects early in biosynthesis. Neuron 16:843–852

2. Coplen TB (2011) Guidelines and recommended terms for expression of stable-isotope-ratio and gas-ratio measurement results. Rapid Commun Mass Spectrom 25:2538–2560

3. Sastre M, Turner RS, Levy E (1998) X11 interaction with beta-amyloid precursor protein modulates its cellular stabilization and reduces amyloid beta-protein secretion. J Biol Chem 273:22351–22357

4. Belle A, Tanay A, Bitincka L, Shamir R, O'Shea EK (2006) Quantification of protein half-lives in the budding yeast proteome. Proc Natl Acad Sci U S A 103:13004–13009

5. Cambridge SB, Gnad F, Nguyen C, Bermejo JL, Kruger M, Mann M (2011) Systems-wide proteomic analysis in mammalian cells reveals conserved, functional protein turnover. J Proteome Res 10:5275–5284

6. Guan S, Price JC, Ghaemmaghami S, Prusiner SB, Burlingame AL (2012) Compartment modeling for mammalian protein turnover studies by stable isotope metabolic labeling. Anal Chem 84:4014–4021

7. Price JC, Guan S, Burlingame A, Prusiner SB, Ghaemmaghami S (2010) Analysis of proteome dynamics in the mouse brain. Proc Natl Acad Sci U S A 107:14508–14513

8. Savas JN, Toyama BH, Xu T, Yates JR 3rd, Hetzer MW (2012) Extremely long-lived nuclear pore proteins in the rat brain. Science 335:942

9. Toyama BH, Savas JN, Park SK, Harris MS, Ingolia NT, Yates JR 3rd, Hetzer MW (2013) Identification of long-lived proteins reveals exceptional stability of essential cellular structures. Cell 154:971–982

10. Fischer CA, Morell P (1974) Turnover of proteins in myelin and myelin-like material of mouse brain. Brain Res 74:51–65

11. Shapira R, Wilhelmi MR, Kibler RF (1981) Turnover of myelin proteins of rat brain, determined in fractions separated by sedimentation in a continuous sucrose gradient. J Neurochem 36:1427–1432

12. Lynnerup N, Kjeldsen H, Heegaard S, Jacobsen C, Heinemeier J (2008) Radiocarbon dating of the human eye lens crystallines reveal proteins without carbon turnover throughout life. PLoS One 3, e1529

13. Ko J, Fuccillo MV, Malenka RC, Sudhof TC (2009) LRRTM2 functions as a neurexin ligand in promoting excitatory synapse formation. Neuron 64:791–798

14. MacCoss MJ, Wu CC, Matthews DE, Yates JR 3rd (2005) Measurement of the isotope enrichment of stable isotope-labeled proteins using high-resolution mass spectra of peptides. Anal Chem 77:7646–7653

15. McClatchy DB, Dong MQ, Wu CC, Venable JD, Yates JR 3rd (2007) 15N metabolic labeling of mammalian tissue with slow protein turnover. J Proteome Res 6:2005–2010

16. Wu CC, MacCoss MJ, Howell KE, Matthews DE, Yates JR 3rd (2004) Metabolic labeling of mammalian organisms with stable isotopes for quantitative proteomic analysis. Anal Chem 76:4951–4959

17. McClatchy DB, Yates JR 3rd (2014) Stable isotope labeling in mammals (SILAM). Methods Mol Biol 1156:133–146

18. Carlin RK, Grab DJ, Cohen RS, Siekevitz P (1980) Isolation and characterization of postsynaptic densities from various brain regions: enrichment of different types of postsynaptic densities. J Cell Biol 86:831–845

19. Niessen S, McLeod I, Yates JR 3rd (2006) HPLC separation of digested proteins and preparation for matrix-assisted laser desorption/ionization analysis. CSH Protocols 2006

20. Savas JN, De Wit J, Comoletti D, Zemla R, Ghosh A, Yates JR 3rd (2014) Ecto-Fc MS identifies ligand-receptor interactions through extracellular domain Fc fusion protein baits and shotgun proteomic analysis. Nat Protoc 9:2061–2074

21. Fonslow BR, Stein BD, Webb KJ, Xu T, Choi J, Park SK, Yates JR 3rd (2013) Digestion and depletion of abundant proteins improves proteomic coverage. Nat Methods 10:54–56

22. Magdeldin S, Moresco JJ, Yamamoto T, Yates JR 3rd (2014) Off-line multidimensional liquid chromatography and auto sampling result in sample loss in LC/LC-MS/MS. J Proteome Res 13(8):3826–3836

23. McDonald WH, Tabb DL, Sadygov RG, MacCoss MJ, Venable J, Graumann J, Johnson JR, Cociorva D, Yates JR 3rd (2004) MS1, MS2, and SQT-three unified, compact, and easily parsed file formats for the storage of shotgun proteomic spectra and identifications. Rapid Commun Mass Spectrom 18:2162–2168

24. Eng JK, McCormack AL, Yates JR (1994) An approach to correlate tandem mass spectral data of peptides with amino acid sequences in a protein database. J Am Soc Mass Spectrom 5:976–989

25. Tabb DL, McDonald WH, Yates JR 3rd (2002) DTASelect and Contrast: tools for assembling and comparing protein identifications from shotgun proteomics. J Proteome Res 1:21–26

26. Park SK, Venable JD, Xu T, Yates JR 3rd (2008) A quantitative analysis software tool for mass spectrometry-based proteomics. Nat Methods 5:319–322

27. Venable JD, Wohlschlegel J, McClatchy DB, Park SK, Yates JR 3rd (2007) Relative quantification of stable isotope labeled peptides using a linear ion trap-Orbitrap hybrid mass spectrometer. Anal Chem 79:3056–3064

28. MacCoss MJ, Wu CC, Liu H, Sadygov R, Yates JR 3rd (2003) A correlation algorithm for the automated quantitative analysis of shotgun proteomics data. Anal Chem 75:6912–6921

29. McClatchy DB, Liao L, Park SK, Venable JD, Yates JR (2007) Quantification of the synaptosomal proteome of the rat cerebellum during post-natal development. Genome Res 17:1378–1388

30. Tsien RY (2013) Very long-term memories may be stored in the pattern of holes in the perineuronal net. Proc Natl Acad Sci U S A 110:12456–12461

31. MacCoss MJ, Wu CC, Yates JR 3rd (2002) Probability-based validation of protein identifications using a modified SEQUEST algorithm. Anal Chem 74:5593–5599

32. Washburn MP, Wolters D, Yates JR 3rd (2001) Large-scale analysis of the yeast proteome by multidimensional protein identification technology. Nat Biotechnol 19:242–247

33. Link AJ, Eng J, Schieltz DM, Carmack E, Mize GJ, Morris DR, Garvik BM, Yates JR 3rd (1999) Direct analysis of protein complexes using mass spectrometry. Nat Biotechnol 17:676–682

34. Liao L, McClatchy DB, Yates JR (2009) Shotgun proteomics in neuroscience. Neuron 63:12–26

INDEX

A

Absolute quantification 170, 195, 219, 253, 281–291
Antibody
    anti-peptide 135–162
    monoclonal 24, 140, 145
    polyclonal 24, 140, 145, 158
Apolipoprotein 106
    apolipoprotein L1 (APO L1) 117, 251, 253,
    255, 258, 261
Area under the curve (AUC) 66, 76, 257, 267

B

Biomarker 2, 3, 15, 16, 122, 129, 135, 195, 197
Brain 241, 294, 297–298

C

Cell culture 25–29, 34, 43, 47, 48, 94, 124–126,
    158, 260, 283, 285, 291
Chemoenzymatic labeling 92, 93
Clinical Proteomic Tumor Analysis Consortium
    (CPTAC) 224–235
    assay portal 224–235
Cryogenic cell disruption 48–49

D

Data independent acquisition (DIA) 265–278
Density gradient ultracentrifugation 105–119
Differential mass spectrometry (dMS), ... 123

E

Electron transfer dissociation
    (ETD) 92, 93, 95, 98–100
ELISA 157, 158, 249–252
Epidermal growth factor receptor (EGFR) 283

G

Glutathione S-Transferase (GST) fusion
    proteins 210–212

H

High density lipoprotein (HDL) 105, 106

I

Immuno-affinity
    enrichment 24–26, 135–162, 282
    immuno precipitation
        (IP) 30, 136, 278, 284, 288, 289
    purification 27–28, 31–32, 41, 48, 60
Isotope labeling Stable isotope labeling
    isotopic interference
        correction 183, 186, 189, 197, 202, 204
    isotopic N,N-dimethyl leucine
        (iDiLeu) 195

L

Label-free quantitation 65–76, 122
Lectin affinity 92
Liquid chromatography (LC)
    basic-reverse phase HPLC
        (bRP-HPLC) 270, 277
    high performance liquid chromatography
        (HPLC) 9, 27, 28, 30, 31, 35, 54,
        66, 67, 79, 140, 141, 144, 174, 189, 239, 243, 253,
        255, 262, 285, 289, 290, 296, 299
    IMAC 285, 288–289, 291
    multidimensional LC 3
    nano-HPLC 32
    reversed-phase liquid chromatography
        (RPLC) 3, 6, 9, 16, 17, 122,
        139, 157, 238
    strong anion exchange (SAX) 270, 277
    strong-cation-exchange
        (SCX) 35, 79, 83, 84, 95, 98, 101,
        174, 181, 189, 199–201, 203, 238, 270, 277, 296,
        299, 302
    two-dimensional liquid chromatography
        (2D-LC) 1–19
Lysine, ... 23, 55, 99, 139, 170, 183, 204
    lysine succinylation 23–36

M

Mass difference labeling 196, 197
Mouse 24, 33, 99, 197, 200, 294
    whole mouse labeling 33

Multiple reaction monitoring (MRM)
  fractionMRM 136
  MRM-MS 1–19, 139, 149, 154–155,
    157, 161, 162, 228
Multiplexed immunoaffinity 135–162
Multiplexed immuno-MRM-MS 135–162

O

O-GlcNAcome 91–101
Orbitrap mass spectrometer 32, 285, 296

P

Parallel reaction monitoring (PRM) 2, 111, 136, 228
Peripheral blood mononuclear cells (PBMCs) 123
Phosphorylated peptides 34, 281–291
Plasma 66, 260
  depletion 66, 67, 136
  fractionation 11
  HDL isolation 106–107, 109
  preparation 7–9
  protein quantitation 1–19
Post-translational modification (PTM) 24
  lysine succinylation 24
    histone deacetylases 24
    sirtuin (sirt5) 24
  O-GlcNAcylation 91
  serine/threonine phosphorylation 91
  tyrosine phosphorylation 283
Protein biomarkers 2
Protein complex 39–62, 227
Protein decay 293, 294, 302
Protein digestion 81–83, 108, 173, 179,
    200, 241, 261, 295, 298–299
Protein extraction, ...42, 78, 80, 96–97, 99,
    239, 241–242, 245
Protein fractionation 124–126, 295, 297–298, 301
Protein half-life 294
Protein networks 2
Proteinopathy 294
Protein pathways 226, 227
Proteoform 121–124, 127–130, 250, 252, 253
Proteogenomic 77–88
Proteotypic peptide 118, 209, 250
  proteotypic tryptic peptide 136
Pseudogenes 78, 80, 86, 88

R

RNA 80
  non-coding RNAs (ncRNAs) 78, 80, 86, 88

S

Selected reaction monitoring (SRM) 106, 136, 208–210,
    217–219, 249–263
Serum 2, 43, 60, 67–69, 73–75, 94, 99, 117, 260, 261
Shotgun mass spectrometry 106, 293–303
Significance Analysis of INTeractome
    (SAINT) 40, 42, 47, 55–59, 61, 62
Solid phase extraction (SPE) 8, 126, 174, 179,
    181, 239, 242, 245, 270
Spirulina 293–303
Stable isotope labeling 174, 175, 180, 237–246
  in vitro synthesized proteins as analytical
    standards 207–220
  isobaric tags
    DiLeu isobaric labeling 174, 180
    isobaric labeling 237–246
    synthesis 175
  Isotopic Differentiation of Interactions as Random or
    Targeted (I-DIRT) 42, 43, 45,
    47–48, 55–60
  iTRAQ 35, 92, 94, 238–240, 242–244, 246, 275
  metabolic labeling 42, 43, 47–48, 92, 296–297
  mTRAQ 196
  pulse-chase isotopic labeling 293–303
  stable isotope labeled standard (SIS) 2, 3, 5, 11,
    14, 15, 17, 18, 106, 113, 195, 260
  stable isotope labeling of amino acid
    (SILAC) 24–30, 33–35, 40, 92
    94–96, 99, 169, 275, 294
  TMT 35, 92, 94, 122, 170, 284, 287
SWATH 265–278

T

Top-down proteomics 121–132
Top-down quantitation 122, 124
Trypsin, ... 66, 67, 69, 71, 74, 270, 271,
    284, 286, 290, 295, 299, 302
Tryptic digest, ... 72, 254, 267, 277

U

Urine 260–261, 263

Printed in the United States
By Bookmasters